Polymer Phasing

Takeji Hashimoto

Polymer Phasing

Self-Organization of Order/Disorder under Flow Fields

Springer

Takeji Hashimoto
Kyoto University
Kyoto, Japan

ISBN 978-981-97-9647-2 ISBN 978-981-97-9648-9 (eBook)
https://doi.org/10.1007/978-981-97-9648-9

This Springer imprint is published by the registered company Springer Nature Singapore Pte Ltd.
The registered company address is: 152 Beach Road, #21-01/04 Gateway East, Singapore 189721, Singapore

To Setsuko

Preface

This book deals with self-organization of ordered structures and/or disordered structures in nonequilibrium systems open to externally imposed energy flow. Various energy flows can be applied to the systems, but we consider here shear flow and extensional flow. The systems to be considered here are semidilute polymer solutions of either crystallizable or noncrystallizable polymers or semidilute solutions of noncrystallizable polymer mixtures in a neutral solvent.

We aim to investigate "polymer phasing", i.e., phase formation of the polymer systems under externally imposed flow fields. Polymer phasing generally involves formation of a single phase, two phases or two-phase coexistence phase. It also involves formation of amorphous phase, crystalline phase, or coexistence of amorphous and crystalline phase in the case of crystallizable polymers.

The polymer phasing at the thermal equilibrium free from the external energy flow (the equilibrium polymer systems) is well-known. It involves thermally induced liquid-liquid (L-L) phase transition from single-phase to two-phase (phase separation) or from two-phase to single-phase in the case of two-component polymer mixtures. It involves also the thermally induced liquid-solid (L-S) phase transition from amorphous liquids to crystalline solids (crystallization) or solid-liquid (S-L) phase transition from crystalline solids to amorphous liquids (melting) in the case of, at least, one component being crystallizable polymers.

The polymer phasing at the nonequilibrium under the external flow fields (nonequilibrium polymer systems) is concerned with the "flow-induced polymer phasing" which is controlled by the dynamic balance between the deformation of flow-induced structures driven by the dissipation of the mechanical energy stored in the systems by the imposed flow and the growth and restoration of the deformed flow-induced structures driven by the thermodynamic driving force inherent to the system. Here, we should note the two opposing physical factors as follows. On the one hand, the dissipation of the mechanical energy stored in the systems involves the "deformation and burst" of the dissipative structures self-organized by the imposed flow at the any given flow rates. On the other hand, the thermodynamic driving force inherent to the systems involves the "growth and restoration" of the deformed dissipative structures self-organized by the imposed flow to the systems. Thus, the flow-induced polymer phasing depends on the flow rates, because the dynamic balance is also depends on the flow-rates through the flow-rate-dependent mechanical energy stored and dissipated and the flow-rate dependent thermodynamic driving force. The L-L and L-S or S-L phase transition depend also on the flow-rate.

The flow-induced phase transitions control the self-organization of order/disorder through the cascade evolution of a series of the dissipative structures under the increased or decreased flow rate or with the time spent at a fixed flow rate after the step-up and step-down flow rate. The dissipative structures under the flow are universally controlled by the work done and dissipated by the mechanical energy imposed to the systems by the various flow fields. The research along this line may provide a fundamental physical basis regarding stimuli-responsible polymer systems with switchable structures/properties with respect to mechanical stimuli.

We especially focus on investigation of the mesoscopic-scale dissipative structures spanning from a few nm to several tens of micrometers using the rheo-optical method presented here which enables real-time, in situ and simultaneous observations of small-angle light scattering, neutron scattering, birefringence, transmitted light intensity, and shear and normal stress. The mesoscopic dissipative structures focused by this Book have been relatively unexplored in comparison with the macroscopic dissipative structures explored in the the field of mathematical physics regarding the "open-nonequilibrium systems" i.e., the systems open for external energy flows.

Polymer systems with a large space-time scale provide good model systems for fundamental studies of the self-organization in the open-nonequilibrium systems because of such a unique point that large structures grow or relax very slowly, which brings about the following characteristics. Polymers are extremely sensitive to the external flow fields: even small flow rates self-organize large dissipative structures; The large space-time scales facilitate real-time and in situ investigations of the self-organization in open nonequilibrium systems with exceptionally high precision and depth that may be difficult to be attained by any other systems having small space-time scales.

The principle of the self-organization in the nonequilibrium systems elucidates the ordering of the shish-kebab and extended-chain crystal textures in the semidilute solutions of crystallizable polymers through the cascade evolution of dissipative structures, which will be specifically treated in Chap. 6.

The author is especially indebted to Prof. A. Onuki for his collaborations in some parts of this work and for his enlightening comments throughout this work. He gratefully acknowledges Prof. G. Fuller and Dr. T. Takahashi, Drs. I. Morfin, F. Boué, and P. Lindner, Dr. D. J. Pine, Profs. T. Inoue and H. Watanabe for their collaborations in some parts of this work and many useful comments. He is indebted to Dr. C. C. Han, Profs. J. S. Higgins, H. H. Winter, and J. Mewis for many valuable comments on this work.

The author takes pleasure in acknowledging the collaboration of many of his colleagues. He is especially indebted to Profs. H. Hasegawa and S. Suehiro, Drs. K. Saijyo, T. Takebe, T. Kume, H. Murase, Y. Ohta, M. Takenaka, S. Koizumi, E. Moses, S. Saito, H. Takeno, T. Izumitani, K. Matsuzaka, K. Asakawa, M. Itakura, and M.

Endo, S. Okamoto, and Y. Zhao as the lists of references and citations in the text testify.

The author is very much indebted to Dr. Shinichi Koizumi, Senior Editor, Books Chemistry and Materials Science, Springer for his support, encouragement, and the editorial assistance. Last, but not least, I want to thank my wife Setsuko Hashimoto for her cooperation to this Book. She was pleased to conduct typing the whole part of the manuscript in the Book, preparing the whole part of the artworks in the Figures employed, and typing the proof corrections.

Kyoto, Japan Takeji Hashimoto

Contents

Abbreviations

2D FFT image	2D Fast Fourier Transform images
2D SALS pattern	Two-Dimensional SALS pattern observed in the detector plane
2D SANS pattern	Two-dimensional SANS pattern observed in the detector plane
3D	Three-Dimensional
ASF	As-Spun fiber
BD	Bridging domains
BOSC	Bundles of stretched chains
Butterfly SALS pattern	SALS pattern exhibiting "butterfly scattering"
CFs	Concentration fluctuations
CHC	Cahn-Hilliard-Cook
C-P transition	Cluster-to-percolation transition
CPT	Cluster-to-percolation (CP) transition
CSCT	Coil-to-Stretch-Chain transition
d	Space dimensionality
D_F	Diameter of the single spun fiber
D_K	Diameter of kebab
DOP	Dioctyl phthalate
DR	Draw ratio
D_S	Diameter of shish
ECC	Extended Chain crystal
FD	Flow direction, Ox direction
FFT	Fast Fourier transformation
GL	Ginzburg–Landau
HFMO theory	Helfand, Fredrickson, Milner and Onuki theory
HMW polymer	High-Molecular-Weight polymer
LAOS	Large amplitude oscillatory shear flow
LCST	Lower critical solution temperature
L-L phase transition	Liquid-liquid phase transition
LMW polymer	Low-Molecular-Weight polymer

L-S phase transition	Liquid-Solid phase transition
LS	Laser light scattering
MF	Mean field
ND	Neutral direction, Oz direction
NG	Nucleation and growth
OM	Transmission light microscope
OZ plot	Ornstein-Zernike plot
OZ scattering	Ornstein-Zernike scattering
PB	Polybutadiene
PCOM	Phase-Contrast optical microscopy
PCT	Percolation-to-cluster (PC) transition
PLWCF	Plane wave concentration fluctuations
POM image	Polarizing optical microscope image
PS	Polystyrene
PS/DOP	PS in DOP solvent
PS/PB (50:50)/DOP	50/50 wt. % mixture of polystyrene and polybutadiene (PB) in DOP solvent
PS/PB/DOP	Mixture of PS and PB in DOP
PS548/DOP	PS with molecular weight 5.48×10^6 in DOP
PVME	Poly (vinyl methyl ether)
S, L and G phase	Solid (crystal), liquid (disorder) and gaseous (void) phase
SALS butterfly pattern	SALS pattern exhibiting "butterfly scattering"
SALS-pattern intensity	Intensity distribution of SALS in the detector plane
SALS	Small-Angle light scattering
SANS: butterfly pattern	SANS pattern exhibiting "butterfly scattering"
SANS	Small-Angle neutron scattering
SD	Spinodal decomposition
Shear-OM	Optical microscopy observed under shear flow
Shear-SALS	Small-Angle light scattering observed under shear flow
SK	Shish-Kebab structure
SLCFL	Surface of lamellae containing chain-folded loops
SLD	Scattering length density
SQL	Squared Lorentzian function
TDCHC equation	Time-Dependent CHC equation
TDGL equation	Time-Dependent GL equation
TDR	Total draw ratio
TEM	Transmission electron microscopy
TRSALS	Time-Resolved Small-Angle light scattering
UCST	Upper-Critical-Solution-Temperature
UHMW a-PS	Ultrahigh-Molecular-Weight atactic PS
UHMW polymer	Ultrahigh-Molecular-Weight polymer
UHMWPS/DOP	Ultrahigh-Molecular-Weightt PS in DOP

Chapter 1
Introduction

Abstract In this chapter, we discuss general backgrounds of our research theme: the self-organization of polymer solutions and solutions of polymer mixtures under external flow fields, specifically under shear flow fields. The systems belong to the so-called open nonequilibrium systems that are open to various imposed external energy flows. Structures formed in such systems are called "dissipative structures", since they are self-organized in the dynamic balance between burst (and/or deformation) driven by dissipation of the energy stored in the systems by the imposed mechanical force and growth (and/or restoration) of the burst structures driven by the thermodynamic force inherent to the systems. We focus on to investigate mesoscopic-scale dissipative structures spanning from a few nm to several tens μm as a function of flow rate. We investigate also the flow-induced phase transitions from two-phase to single-phase and vice versa. The investigation utilizes the rheo-optical method which enables to observe *real-time* and *in situ* small-angle light scattering and small-angle neutron scattering in reciprocal space as well as birefringence, transmitted light intensity, shear and normal stress in real space. The chapter discusses the basic physics underlying the flow-induced phase transition. Polymer systems provide a good model system for fundamental studies of self-organization in open nonequilibrium systems because of such a unique point that large structures grow or relax very slowly in the polymer systems, which brings about the following characteristics.

(1) Polymers are extremely sensitive to external flow fields: Even under small flow rates large dissipative structures are self-organized in polymer systems.
(2) The large space–time scales facilitate real-time and in situ investigations of the self-organization in open nonequilibrium systems with exceptionally high precision and depth that may be difficult to be attained by any other systems having small space–time scales.

T. Hashimoto, *Polymer Phasing*, https://doi.org/10.1007/978-981-97-9648-9_1

1.1 General Background

In this chapter, we discuss self-organization of polymer solutions and solutions of polymer mixtures under external flow fields, specifically under shear flow fields, as observed by *real-time* and in situ small-angle light scattering, small-angle neutron scattering, and small-angle X-ray scattering in reciprocal space as well as optical microscopy, birefringence, transmitted light intensity, shear and normal stress in real space. Although we shall discuss shear-induced structures of solutions of polymer mixtures in some parts of this chapter, our discussion can be in principle extended to the flow-induced structures of binary polymer mixtures in bulk. This is because the solvent used in the solutions is a neutral solvent which effectively weakens segmental interactions of the constituent polymers as will be detailed in Sect. 4.2.

Our research theme to be dealt with here belongs to the general theme on pattern formation in nature. The pattern formation in various systems in nature occurs under the influence of external fields imposed to the systems and is inherently related to the self-organization of the patterns in the so-called **open nonequilibrium systems**, i.e., systems that are imposed or open to various external flows of energy flows. We define here **self-organization** as a general term which includes self-organizing processes, mechanisms, and structures of molecules.

Figure 1.1 presents a thermally equilibrium system (part a) and nonequilibrium system (part b). In the equilibrium system, a flow of energy (or molecules) into the

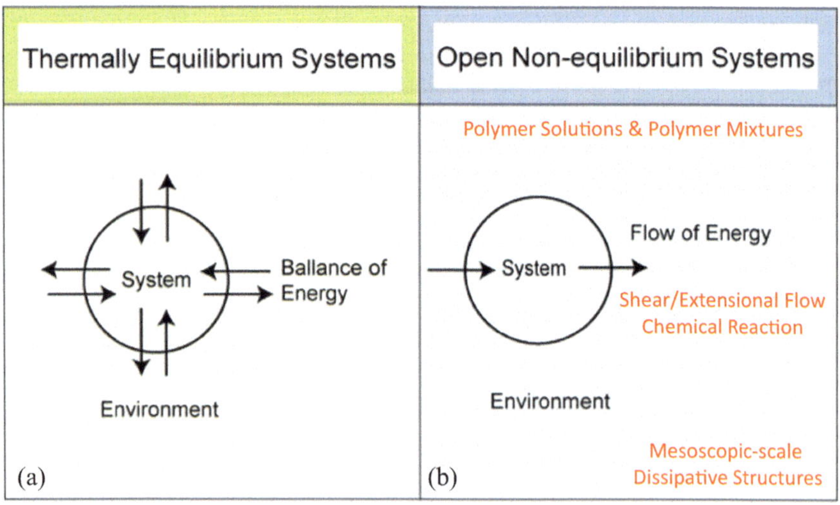

Fig. 1.1 Part **a**: Thermally equilibrium systems with no macroscopic flow of energy across the system. Part **b**: nonequilibrium systems open for energy flow such as shear flow, extensional flow, and chemical reactions. In the open nonequilibrium systems, we focus on the mesoscopic-scale steady-state or transient dissipative structures evolved in noncrystallizable or crystallizable polymer solutions and noncrystallizable polymer mixtures under the flow

system is microscopically balanced with a flow of energy (or molecules) out of the system everywhere in the boundary between the system and its surrounding. Thus, there is no macroscopic flow of the energy (or molecules) across the system: the system is close to the macroscopic energy flow.

In the nonequilibrium systems, there is the macroscopic flow across the system: the system is open for the energy flow. The applied external energy flow from the environment to the system brings the system in nonequilibrium states. Since the system is open for the energy flow, the system is called "open" nonequilibrium system. Structures formed in the open nonequilibrium system are called "dissipative structures", since the structures are self-organized through various processes by which the energy stored in the system due to the externally applied field is dissipated. The dynamic stability of the open nonequilibrium system is maintained by the process, in which the energy imposed to the system from the environment is partially dissipated to change the system states or structures and the rest of the imposed energy is released into the environment.

In this book, we describe flow-induced phase transitions from two-phase to single-phase structures and from the single-phase to two-phase structures and the dissipative structures evolved in the processes of the flow-induced phase transitions. We specifically focus on to elucidate the dissipative structures in the mesoscopic-scale spanning from a few nm to several tens μm.

We can think of various types of the flow of the energy to be applied to the system such as shear flow, extensional flow, temperature gradient [1–3], and chemical reactions [4–6]. This book is concerned mainly with shear flow and partially with extensional flow. Regarding the systems to be focused, we are concerned with noncrystallizable and crystallizable polymer solutions and polymer mixtures in neutral solvents.

Polymer systems provide good model system for fundamental studies of self-organization in open nonequilibrium systems because of the following unique point. **Large structures grow or relax very slowly**.

This unique point may give rise to the following characteristics.

(1) The pattern formation (self-organization) in polymers is extremely sensitive to external flow fields. Even under small flow rates, large dissipative structures are self-organized.
(2) The large space–time scales involved in the pattern formation facilitate real-time and in situ investigations of the self-organization in open nonequilibrium systems with exceptionally high precision and depth that may be difficult to be attained by any other systems having small space–time scales.

As one of the open nonequilibrium phenomena in polymers, we discuss crystallization into shish-kebab and extended-chain crystal texture under applied external flow fields. The key issue here is that the ordering into these textures demands the reduction of enormously large conformational extropy of high polymers. This demand is never fulfilled in the equilibrium system. How does the nature overcome this difficulty? What is the principle of self-organization in the nonequilibrium

Fig. 1.2 Statistical mechanical studies of open nonequilibrium systems

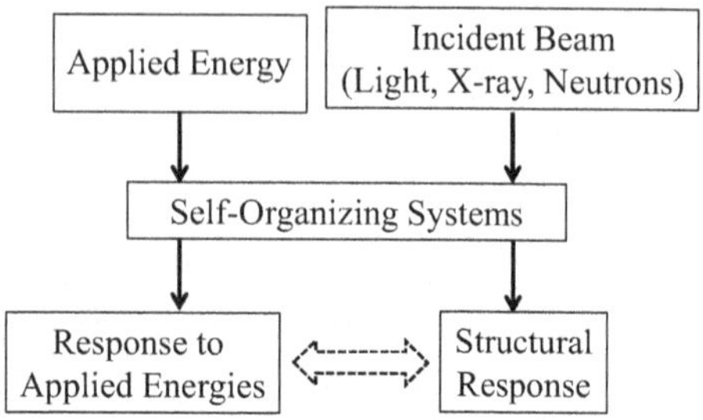

systems which leads to the ordered textures as the minimum entropy products from dissipative structures through fluctuations?

We may consider our systems as a model of **soft matters** (or **complex liquids**) that comprise matters such as polymers, liquid crystals, membranes, gels, micelles, colloids, supercooled liquids, and multicomponent liquids. They are assemblies of molecules and/or supramolecules which commonly have characteristic structures and properties in between solids and liquids, depending on the space–time scale of our observations.

We are particularly interested in various **dissipative structures**, which are developed accompanied by the dissipation of the external energy imposed to systems. How can we observe the dissipative structures? Fig. 1.2 shows a general principle of our method. We impose the applied energy (e.g., strain and strain rate) to our systems and observe the output response (e.g., stress) to the applied energy. Simultaneously with the imposed external energy, we send incident beams (e.g., light, X-ray, neutrons) to our systems to monitor structural responses within the systems to the applied external energy over a wide length scale ranging from ~ nm to several tens μm. We aim to correlate the physical properties (i.e., the response to the applied energies) and structural responses elucidated by *in situ*, real-time, and simultaneous scattering and optical microscopy experiments with the guide of statistical mechanical theories, as shown by the broken arrow in the figure. As for the applied fields, we shall discuss steady, oscillatory, and step-up or step-down shear flow.

1.2 Principles of Rheo-Optics

We apply the so-called **rheo-optical methods** to explore the dissipative structures. The principle of rheo-optics is illustrated in Fig. 1.3 [7]. Rheo-optical methods involve simultaneous measurements of both rheological and optical properties of systems to clarify relationships between rheological properties and responses of fundamental structural units of the systems over a range of length scale as wide as possible (e.g., several 0.1 nm to several 10 μm).

Upon applying mechanical excitations to our systems, we detect and analyze mechanical responses. **Rheology** clarifies the law of causality between mechanical

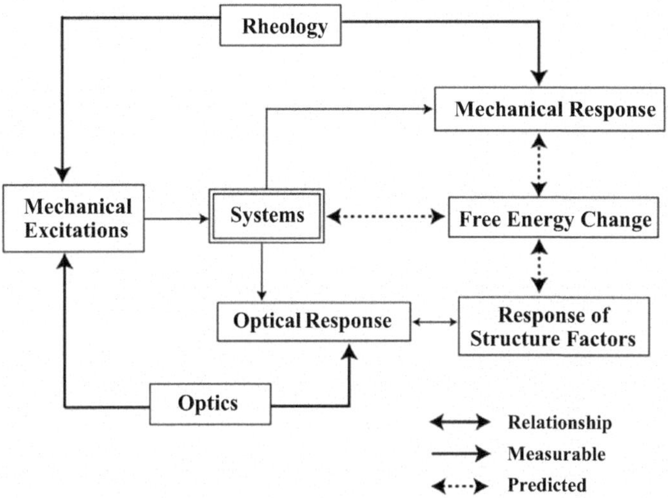

Fig. 1.3 Principle of rheo-optics: a simultaneous measurement of rheological and optical properties of systems under flow of energy for elucidating relationships between mechanical responses and responses of various structural elements existing in systems over wide space–time scales. Based on Hashimoto, T. In *Soft Matter Characterization*; Borsali, R., Pecora, R., Eds., Springer. New York, 2008; Chapter 8, pp378-462 [7]

stimuli and responses. Simultaneously, we detect and analyze optical responses or responses of structure factors based on principles established in **optics**. The analysis of optical responses will help the development of nonequilibrium statistical mechanical theories to predict the free energy associated with the applied mechanical excitations and the relevant dissipative structures evolved. The prediction is then fed back to predict further specific mechanical responses associated with the specific optical responses analyzed by the rheo-optical methods. Thus, the rheo-optics are anticipated to give, directly, the molecular or structural origin of rheological properties beyond phenomenological or mathematical understanding of the properties. As for optical responses, one can think of optical images such as those obtained by transmission and reflection optical microscopy, fluorescence and Raman microscopy, phase-contrast microscopy, and polarizing light microscopy. One can think of various scattering methods such as light, X-ray, and neutron scattering as well as spectroscopy such as Raman and infrared spectroscopy. Birefringence and form optical dichroism also provide useful methods.

1.3 Shear-Rheo-Optics

Let us discuss a shear-rheo-optical method, as one example of the rheo-optical methods. Figure 1.4 represents a schematic illustration of the rheo-optical apparatus under shear deformation and/or flow [7, 8]. In this apparatus, a rheometer to measure the stress–strain behavior is linked with an apparatus to measure a transmission light microscope (OM) and birefringence and an apparatus to measure a laser light scattering (LS) and transmittance of the incident light beam. Samples are placed in a shear cell composed of two parallel plates or a set of cone and plate which are transparent for the incident beam, transmitted beam, and the scattered beam. In this figure, shear strain is applied by rotating the lower part of the cell (either a plate or cone); shear force and normal force from the sample are measured by a stress transducer fixed to the upper part of the cell. The shear (or flow) direction, strain (or velocity) gradient direction, and neutral (or vorticity) direction are conventionally taken, respectively, along O*x*-, O*y*-, and O*z*-axes in the Cartesian coordinate O*xyz* shown in the figure. In this setup, the OM images and small-angle light scattering (SALS) patterns are observed *in situ* and *at real time* in the plane parallel to O*xz* plane as a function of time t and shear rate $\dot{\gamma}$ after imposing a given $\dot{\gamma}$ as well as a function of variables inherent to systems under our consideration, e.g., temperature, etc. They are observed also as a function of angular frequency ω, amplitude of strain γ_0, strain phase ϕ_s and time after imposing oscillatory shear. Wu et al. reported a

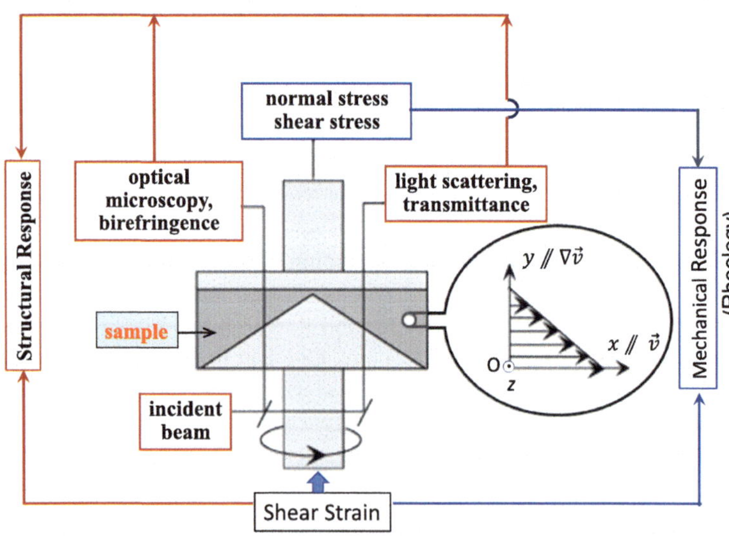

Fig. 1.4 Schematic diagram for a shear-rheo-optical apparatus; light scattering, transmitted light beam, optical microscopic images, and birefringence are simultaneously measured with shear stress and normal stress for samples under shear flow to explore the relationship between structural and mechanical responses in mesoscopic scale at real time and in situ. Based on Hashimoto [2, 3] permission with Oxford University Press

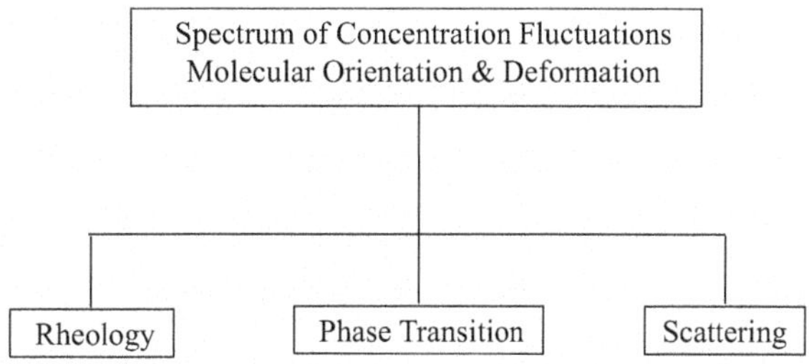

Fig. 1.5 Physical background in the rheo-optical studies of mixtures under the flow [10]

shear-LS apparatus which allows measurement of the scattering in the Oxy plane in which the velocity gradient exists [9].

It is worth noting that the rheo-optics on multicomponent systems involve three major fields in physical science: rheology, phase transition, and scattering/spectroscopy, as illustrated in Fig. 1.5 [10]. Fundamental quantities underlying systems of interest are (i) a scattering structure factor or spectrum of concentration fluctuations in the case of multicomponent systems or density fluctuations in the case of single component systems and (ii) orientation and deformation of molecules and structural units forming hierarchical structures in general. The scattering and spectroscopy directly explore these quantities under shear. Statistical mechanical studies concerning simultaneously captured rheological and optical (scattering and spectroscopic) responses of such hierarchically structured systems as described above are very important as one of challenging research topics in physical science of the open nonequilibrium systems.

1.4 Basic Physics Underlying Flow-Induced Phase Transition

Why does the flow affect the phase transition? Let us consider a binary mixture of polymer liquids with the space–time thermal concentration fluctuations of constituent polymer segments and the deformation of chain molecules as order parameters. The thermal concentration fluctuations are described by a superposition of Fourier modes of the fluctuations: the Fourier mode defined as **q**-Fourier mode is given by a sinusoidal concentration variation in space **r** with the amplitude $\eta(\mathbf{q})$, wave vector **q**, wavelength $\Lambda(\mathbf{q}) \equiv (2\pi/|\mathbf{q}|)$, and relaxation rate $\Gamma(\mathbf{q})$. Let us consider the **q**-Fourier mode at a given direction of **q** with wave number $q = 2\pi/\Lambda(q)$ and relaxation rate $\Gamma(q)$. $\Gamma(q)$ generally increases with q as schematically shown in Fig. 1.6. Figure 1.6 presents the effects of the shear flow with given shear rate $\dot{\gamma}$:

(i) In region (1) where $\dot{\gamma} > \Gamma(q)$, the deformation time by the shear flow, $t_{\text{shear}} \equiv 1/\dot{\gamma}$ is shorter than the relaxation time of the concentration fluctuations $\tau_{\text{q}} \equiv 1/\Gamma(q)$; the concentration fluctuations last during the deformation time. Thus, the flow deforms the concentration fluctuations with $q < q_{\text{shear}}$ where q_{shear} is

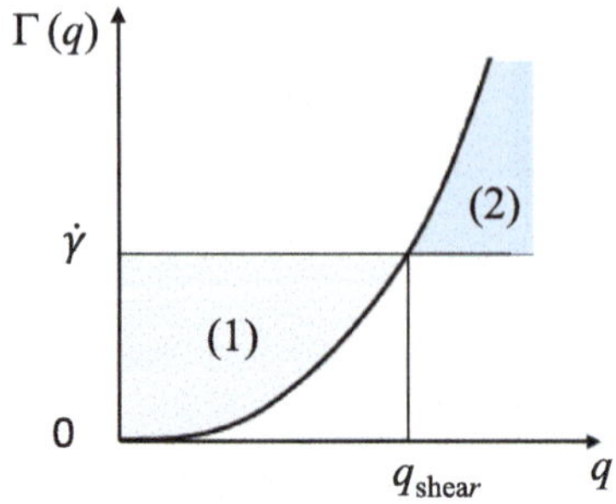

Fig. 1.6 Schematic illustration of a relationship between the relaxation rate of the thermal concentration fluctuations of q-Fourier modes $\Gamma(q)$ and a given shear rate imposed to the system $\dot{\gamma}$. The shear effect on the thermal concentration fluctuations is weak in the region (2) where $\Gamma(q) > \dot{\gamma}$ at $q > q_{\text{shear}}$ but strong in the region (1) where $\Gamma(q) < \dot{\gamma}$ at $q < q_{\text{shear}}$

defined by the shear rate which satisfies.

$$\dot{\gamma} = \Gamma(q) \text{ at } q \equiv q_{\text{shear}}. \tag{1.4.1}$$

Thus, the shear affects the Fourier modes of the fluctuations with the small q values, smaller than the q_{shear} or with the large length scales, larger than the $1/q_{\text{shear}} \equiv l_{\text{shear}}$;

In the region (2) where $\dot{\gamma} < \Gamma(q)$ (as shown in Fig.1.6), the relaxation time τ_{q} for the q-Fourier mode is shorter than the shear deformation time t_{shear}.In other words, the high q modes rapidly relax; hence they cannot last during the shear deformation time t_{shear}, so that the high q modes higher than the q_{shear} are unaffected by the flow. The shear flow represented in Fig. 1.6 partially affects the concentration fluctuations of the system, i.e., only the fluctuations with the small wave number $q < q_{\text{shear}}$ or the large length scale $l > l_{\text{shear}}$.

For the mixture under the shear flow, there are three characteristic rates, $\dot{\gamma}$, Γ_{conc}, and Γ_{MD}, characterizing states of this nonequilibrium system: the applied shear rate $\dot{\gamma}$, the characteristic relaxation rate Γ_{conc} for the concentration fluctuations with the thermal correlation length ξ, and the molecular deformation (MD) rate Γ_{MD} of the chain molecules with the characteristic end-to-end distance R_{D}, as schematically illustrated in part (a), (b), and (c) in Fig. 1.7, respectively. Here, it should be noted that:

$$\Gamma_{\text{conc}} \equiv \Gamma(q) \text{ at } q = q_{\text{c}} \sim 1/\xi, \tag{1.4.2}$$

$$\xi > R_{\text{D}}, \text{ and } \Gamma_{\text{conc}} < \Gamma_{\text{MD}}. \tag{1.4.3}$$

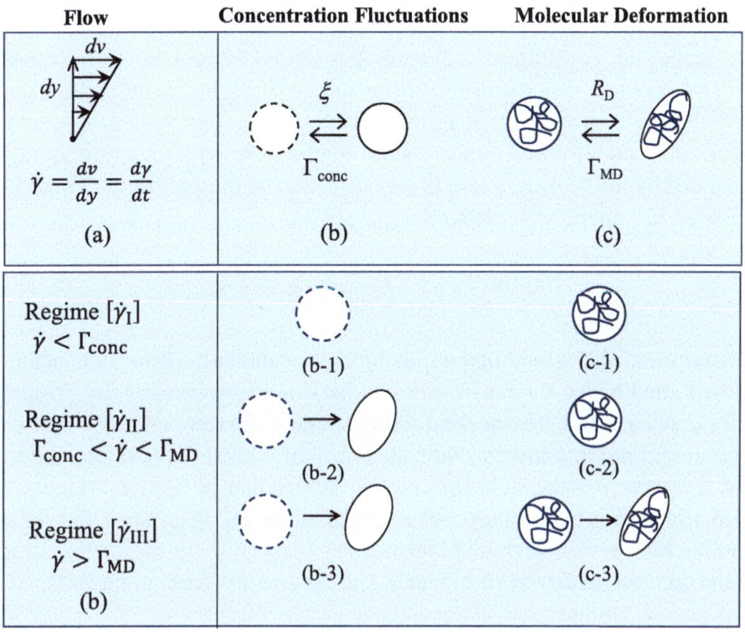

Fig. 1.7 Shear flow with the flow direction along Ox axis and the velocity gradient along Oy axis (part **a**), relaxation rate Γ_{conc} for the thermal concentration fluctuation with the thermal correlation length ξ (part **b**), and relaxation rate Γ_{MD} for the molecular deformation with the characteristic length R_{D} (part **c**). The shear Regime $[\dot{\gamma}_{\mathrm{I}}]$, Regime $[\dot{\gamma}_{\mathrm{II}}]$, and Regime $[\dot{\gamma}_{\mathrm{III}}]$ and the corresponding state of the concentration fluctuations (part (b-1), part (b-2), and part (b-3)) and the molecular deformation (part (c-1), part (c-2), and part (c-3)), respectively. Regime $[\dot{\gamma}_{\mathrm{I}}]$: weak shear with respect to the concentration fluctuations (b-1) and the molecular deformation (c-1), Regime $[\dot{\gamma}_{\mathrm{II}}]$: strong shear with respect to the concentration fluctuations (b-2) but weak shear with respect to the molecular deformation (c-2), and Regime $[\dot{\gamma}_{\mathrm{III}}]$: strong shear with respect to both the concentration fluctuations (b-3) and the molecular deformation (c-3)

Γ_{conc}, q_{c}, and ξ characterize the quantities which represents the concentration fluctuations of the system as a whole rather than a part of the concentration fluctuations of the system.

Part (b) schematically presents thermal excitation (from the left broken circle to the right solid circle) and relaxation (from the right solid circle to the left broken circle) of the concentration fluctuation with the characteristic length ξ in dynamic equilibrium and the characteristic relaxation rate Γ_{conc}. Part (c) schematically presents thermal excitation (from the left to right) and relaxation (from the right to left) of the MD of the chain molecules with the characteristic length R_{D} and characteristic relaxation rate Γ_{MD} in dynamic equilibrium. The system under the shear flow can be classified into the following three shear regimes with respect to the relative size of the characteristic rates (part b).

$$\textbf{Regime}\left[\dot{\gamma}_{\mathrm{I}}\right] : \dot{\gamma} < \Gamma_{\mathrm{conc}}$$

In this case, the concentration fluctuations of the system as a whole and MD thermally relax before they are deformed under the shear. In this case, the shear does not affect the concentration fluctuations and MD; hence, the state of the mixture is characterized by undeformed concentration fluctuations with the correlation length ξ as illustrated by the broken sphere in part (b-1) and by the undeformed random-coil chain of spherical shape as illustrated in part (c-1).

$$\textbf{Regime}\left[\dot{\gamma}_{\text{II}}\right] : \Gamma_{\text{conc}} < \dot{\gamma} < \Gamma_{\text{MD}}$$

The concentration fluctuations cannot thermally relax in the time scale of the given applied field, and hence they are deformed by shear as schematically presented in part (b-2); consequently, the shear affects the state of mixture and may hence affect the liquid–liquid phase transition. In fact, a drop of critical temperature under shear flow was reported for some mixtures as will be detailed in Sect. 4.4. However, the MD thermally relaxes before they are deformed under the shear. Thus, the shear does not affect the MD, as demonstrated in part (c-2). In this regime, dissipative structures are formed and the phase transition occurs in the systems without the MD.

$$\textbf{Regime}\left[\dot{\gamma}_{\text{III}}\right] : \dot{\gamma} > \Gamma_{\text{MD}}$$

The concentration fluctuations and MD cannot thermally relax in the time scale of the given applied field, and thereby they are deformed under the shear, as schematically indicated in parts (b-3) and (c-3), respectively. Thus, the system state changes with respect to the concentration fluctuations and MD.

Figure 1.8 summarizes relations between the two characteristic wave numbers $1/\xi$ and $1/R_{\text{D}}$ on the q scale (the wave number of the dominant Fourier modes of the concentration fluctuations or magnitude of the scattering vector) (part b) and the corresponding characteristic relaxation rates Γ_{conc} and Γ_{MD} on the $\Gamma(q)$ scale (part c) together with the $\dot{\gamma}$ scale (part d). The criterion for the weak shear for the concentration fluctuations of the system is attained in Regime $\left[\dot{\gamma}_{\text{I}}\right]$ at $\dot{\gamma} = \dot{\gamma}_{\text{I}} < \Gamma_{\text{conc}}$ (part e), where only the Fourier modes with small wave number q smaller than $q_{\text{c}} \sim 1/\xi$, are affected (part b-1 and c-1 in Fig. 1.7). The criterion for the strong shear with the concentration fluctuations but the weak shear with MD is attained in Regime $\left[\dot{\gamma}_{\text{II}}\right]$ at $\Gamma_{\text{conc}} < \dot{\gamma} = \dot{\gamma}_{\text{II}} < \Gamma_{\text{MD}}$, where the shear strongly affects the concentration fluctuations but hardly the MD (part b-2 and part c-2 in Fig. 1.7 and part e in Fig. 1.8). This criterion may bring about the flow-induced two-phase to single-phase liquid–liquid (L-L) phase transitions for dynamically symmetric systems and single-phase to two-phase L-L phase transitions for the dynamically asymmetric systems upon increased shear rate. The criterion of the strong shear with respect to both the concentration fluctuation and the MD is attained in Regime $\left[\dot{\gamma}_{\text{III}}\right]$ at $\dot{\gamma} = \dot{\gamma}_{\text{III}} > \Gamma_{\text{MD}}$ (part b-3 and c-3 in Fig. 1.7). This brings about cascade phase transitions: first the L-L phase transition followed by a liquid–solid phase transition (L-S) (crystallization)

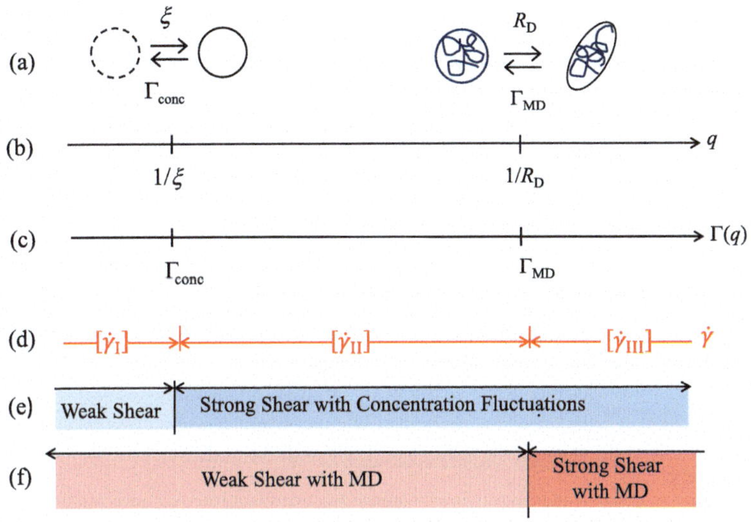

Fig. 1.8 Part (a) illustrates the characteristic length ξ and the characteristic rate Γ_{conc} for the dominant Fourier modes of the concentration fluctuations and the characteristic length R_D and the characteristic rate Γ_{MD} for the molecular deformation. Part (b) presents the characteristic wave number $1/\xi$ and $1/R_D$ for the concentration fluctuations and molecular deformation, respectively, on the q-scale. Part (c) presents the characteristic rate Γ_{conc} and Γ_{MD} for the dominant Fourier modes of the concentration fluctuations and the molecular deformation, respectively, on the characteristic-rate scale $\Gamma(q)$. Part (d) defines shear-rate Regimes $[\dot{\gamma}_I]$, $[\dot{\gamma}_{II}]$, and $[\dot{\gamma}_{III}]$ classified by the set of characteristic parameters $(1/\xi, \Gamma_{conc})$ and $(1/R_D, \Gamma_{MD})$. Part (e) specifies the weak and strong shear regime with respect to the concentration fluctuations. Part (f) specified the weak shear and strong shear regime with respect to the molecular deformation MD

upon increased shear rate for dynamically asymmetric mixtures having at least one crystallizable component.

When polymers are one of the components of a binary mixture, the polymer molecules can be temporarily deformed under shear, regardless of the case of $\dot{\gamma} \leq \Gamma_{MD}$ or $\dot{\gamma} \geq \Gamma_{MD}$. The deformation energy comes into play in the constitutive equation regarding the time evolution of the concentration-fluctuation field $\delta\phi(\mathbf{r}, t)$ and velocity field $\mathbf{v}(\mathbf{r}, t)$ where $\delta\phi(\mathbf{r}, t)$ is local concentration fluctuations of one of the components in the binary mixture, and $\mathbf{v}(\mathbf{r}, t)$ is the average velocity of the system. The time evolutions of $\delta\phi$ and $\mathbf{v}$ are given in general by a simultaneous nonlinear equation in the context of the HFMO (Helfand and Fredrickson [11], Milner [12], and Onuki [13]) theory,

$$\frac{\partial \delta\phi}{\partial t} = -\nabla \cdot (\delta\phi \mathbf{v}) + \Lambda \nabla^2(\mu_0 + \mu_{el}) + \theta_t = f(\delta\phi, \mathbf{v}), \tag{1.4.4}$$

$$\rho \frac{\partial \mathbf{v}}{\partial t} = -\nabla p - \phi_0 \nabla \frac{\delta F}{\delta \phi} + \nabla \cdot \overset{\leftrightarrow}{\sigma} + \eta_s \nabla^2 \mathbf{v} + \zeta = g(\delta\phi, \mathbf{v}), \tag{1.4.5}$$

where Λ is the Onsager kinetic coefficient, $\mu_0 = (\delta F/\delta\phi)_0$ is the chemical potential for the concentration fluctuations, $(-\nabla\mu_0)$ is the thermodynamic driving force to create the concentration fluctuations), μ_{el} is the elastic part of the chemical potential associated with the deformation of the constituent polymer(s), θ_t and ζ are contributions of the random thermal force to $\delta\phi$ and $\mathbf{v}$, respectively, ρ is density, p is pressure, ϕ_0 is the average concentration of one polymer component in binary mixtures, $(\delta F/\delta\phi)_0$ is the variational derivative of the free energy functional $F\{\delta\phi\}$ of the mixture at ϕ_0, $\overleftrightarrow{\sigma}$ is the stress tensor, and η_s is the viscosity. In the above equation, the elastic effect due to the deformation of polymers affects $\overleftrightarrow{\sigma}$ and μ_{el} which depend on the $\dot{\gamma}$. In the dynamic symmetric systems, the terms $\overleftrightarrow{\sigma}$ and the terms μ_{el} are absent in Eqs. 1.4.4 and 1.4.5.

References

1. Kumaki, J., Hashimoto, T., and Granick, S. (1996) *Phys. Rev. Lett.*,**77** (10), 1990–1993.
2. Bodycomb, J., Funaki, Y., Kimishima, K., and Hashimoto, T. (1999) *Macromolecules*, **32**, 2075–2077.
3. Mita, K., Tanaka, H., Saijo, K., Takenaka, M., and Hashimoto, T. (2007) *Macromolecules*, **40**, 5923-5933.
4. Hashimoto, T., Tanaka, H., Koizumi, S., Naka, K., and Chujo, Y. (2007) *J. Appl. Cryst.*, **40**, s73–s77.
5. Tanaka, H., Koizumi, S., Hashimoto, T., Itoh, H., Satoh, M., Naka, K., and Chujo, Y. (2007) *Macromolecules*, **40**, 12, 4327–4337.
6. Hashimoto, T., and Koizumi, S. (2012) In *Polymer Science: A. Comprehensive Reference*, Matyjaszewski, K. and Möller, M. (ed), Elsevier Vol. 2, Hashimoto, T. Spiess, H.W., and Takenaka, M. (vol ed), Chapter 2.13, pp. 381–398.
7. Hashimoto, T. (2008) In *Soft Matter Characterization*, Borsali, R. and Pecora, R. (ed), Springer. New York; Chapter 8, pp. 378–462.
8. Hashimoto, T. (2005) *Bull. Chem. Soc.*, Jpn, **78**, 1–39.
9. Wu, X-L., Pine, D.J., and Dixon, P.K. (1991) *Phys. Rev. Lett.* **66**, 2408.
10. Hashimoto, T., Takebe, T., and Fujioka, K. (1990) in *Dynamics and Patterns in Complex Fluids*, Onuki, A. and Kawasaki, K. (ed), Springer, Berlin, pp. 86–99.
11. Helfand E. and Fredickson, G. H. (1989) *Phys. Rev. Lett.*, **62**, 2468.
12. Milner, S. T. (1993) *Phys. Rev. E*, **48**, 3674.
13. Onuki, A, (1997) *J. Phys., Condens. Matter.,* **9**, 6119.

Chapter 2
Flow-Induced Phase Transitions

Abstract Shear flow changes the equilibrium state of systems. In one case, the increased shear rate brings about shear-induced two-phase to single-phase transition, whereas in the other case, it brings about shear-induced single-phase to two-phase transition. The former and latter occur in dynamically symmetric systems and dynamically asymmetric systems, respectively. In dynamically symmetric mixtures, component molecules A and B have equal mobility, whereas in dynamically asymmetric mixtures, component molecules have different mobilities as in the case of mixtures of polymers having different self-diffusivities. The thermally built-up local concentration fluctuations generally build up local stress and its stress relaxation with time, which affects the free energy of the system. In the dynamically symmetric systems, this stress is equally divided into the two components, and the stress relaxation rate is faster than the growth rate of the concentration fluctuations. Thereby, the stress built-up and its relaxation give no influence on the structural growth. On the other hand, in the dynamically asymmetric systems, the stress born by the fast component relaxes faster than the stress born by its counterpart, so that the local stress is primarily born by the smaller mobility component. This local stress and the stress relaxation affect the free energy functional and hence the cooperative diffusivity of the slower mobility component of the systems. This stress-diffusion coupling inherent to the asymmetric systems gives rise to intriguing effects on the dynamics of the composition fluctuations and pattern formation. The chapter describes some general backgrounds on self-organization of dynamically symmetric and asymmetric systems under shear flow.

2.1 Two Opposing Phase Transitions: Flow-Induced Mixing and Demixing

The effect of shear flow on phase transition was for the first time reported by Silberberg and Kuhn [1, 2] more than a half-century ago. They reported shear-induced mixing of the phase-separated polymer solution comprised of a polymer mixture in

T. Hashimoto, *Polymer Phasing*, https://doi.org/10.1007/978-981-97-9648-9_2

a solvent. Since then, there has been a considerable number of reports as detailed later in Sect. 4.5.

A series of our studies of the shear flow effects on various systems revealed that the shear flow changes the equilibrium state of systems. In one case, the increased shear rate brings about phase-separated binary mixtures into single-phase homogeneous mixtures (**shear-induced homogenization or shear-induced two-phase to single-phase transition**), but in the other case, the increased shear rate brings about homogeneous single-phase solution into phase-separated solution (**shear-induced phase separation or shear-induced single-phase to two-phase transition**), as schematically presented in Fig. 2.1a, b, respectively. These two effects are quite opposite.

Figure 2.2 demonstrates the opposite effects on the steady-state SALS patterns under the flow, which were observed in the Oxz plane: the flow is in the vertical direction (Ox-axis) and both the velocity-gradient direction and incident-beam propagation direction are along the Oy-axis [3] (see Fig. 1.4). The left half (parts a and b) shows a disappearance of the strong SALS pattern with increased $\dot{\gamma}$ from 40 to $250\,s^{-1}$ at 68 °C (8 °C below the cloud point at $\dot{\gamma} = 0$), whereas the right half (parts c and d) shows an appearance of strong SALS pattern with increased $\dot{\gamma}$ from 0.4 to $6.3\,s^{-1}$ at 21 °C.

The system in Fig. 2.2a, b is composed of the 50/50 wt. % mixture of polystyrene (PS) having weight-average molecular weight $M_w = 2.1 \times 10^5$ and polybutadiene (PB) having $M_w = 3.1 \times 10^5$ which are dissolved in an approximately neutral solvent of dioctyl phthalate (DOP) for both polymers with a total polymer concentration of 3.3 wt. % where the two polymers are weakly overlapping both in the two-phase solution and in the single-phase solution. The system is hereafter designated PS/PB (50:50)/ DOP 3.3 wt. %. The system in Fig. 2.2c, d is composed of a very high molecular weight PS with $M_w \cong 1.0 \times 10^7$ dissolved in DOP with polymer concentration of 2.0 wt. % (designated PS1000/DOP 2.0 wt. % or UHMWPS/DOP 2.0 wt. % as in Fig. 2.2c, d) where the PS chains are highly entangled with $c/c^* = 20$ where c^* is the overlap concentration of the polymer [4].

In parts (b) and (c), the bright halo around the dark circle at the center of the patterns, which corresponds to an incident-beam stopper, is a tail part of the incident beam leaked from the beam stopper and does not represent the true small-angle

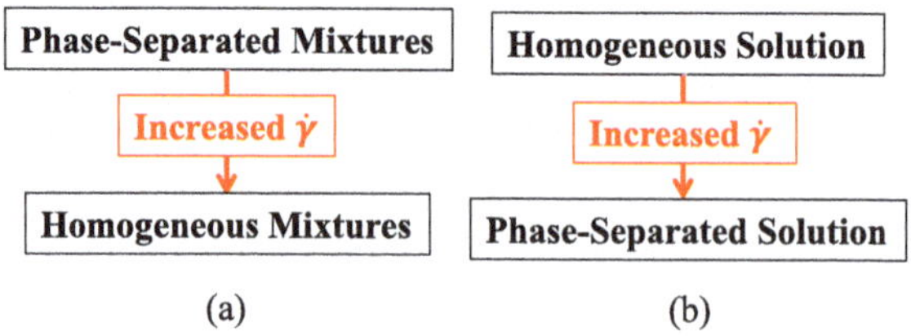

Fig. 2.1 Effects of increased shear rate $\dot{\gamma}$ on shear-induced homogenization or two-phase to single-phase transition (part **a**) and shear-induced phase separation or single-phase to two-phase transition (part **b**)

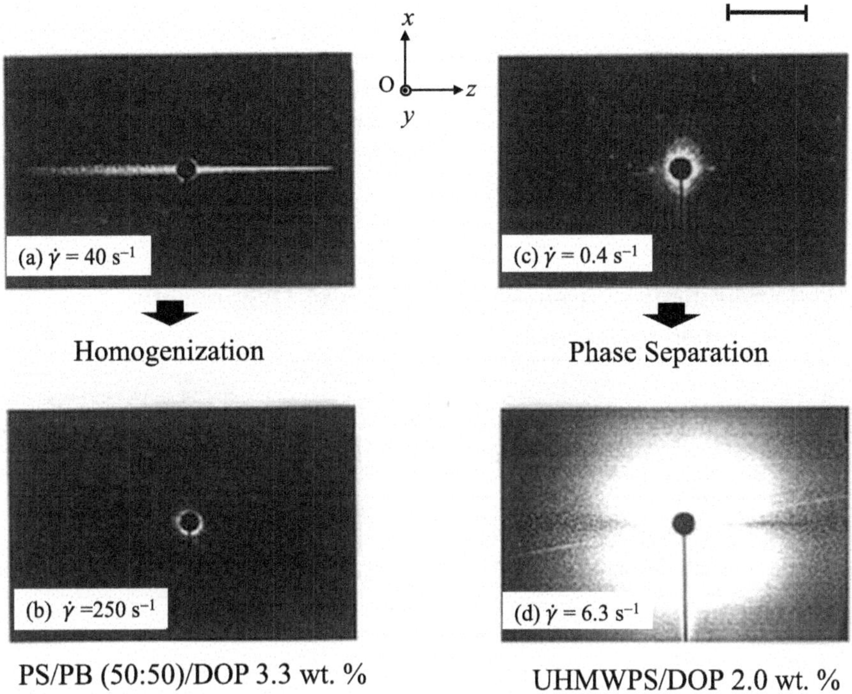

Fig. 2.2 Changes in steady-state scattering patterns with increased shear rate. The change from pattern **a** to **b** indicates shear-induced single-phase formation, whereas that from pattern **c** to **d** indicates shear-induced phase separation or concentration fluctuations. The flow direction (x-axis) is along the vertical direction, and the bar in the top-right corner corresponds to the scattering angle $\theta = 7°$ in the solutions. Based on Hashimoto, T. In *Soft Matter Characterization*; Borsali, R. and Pecora, R., (ed)., Springer. New York, 2008; Chapter 8, pp. 378–462 [3]

scattering. A quantitative detection of the weak scattering intensity in the dark region of the pattern shows that the system under the shear flow is in a single-phase solution as will be detailed later. The strong scattering in part (a) represents highly elongated domains designated as **strings** oriented along the flow direction, as will be also detailed later in Sect. 4.3.2, which are domains rich in one of the polymer solutions in a matrix of a solution rich in the other polymer. The disappearance of the domains is then indicative of the solution being brought into a single-phase state [3, 5].

On the other hand, in the PS/DOP system, a strong scattered intensity of ~ 100 times as high as the scattered intensity from the single-phase solution appears along the flow direction, when the imposed shear rate exceeds a critical shear rate. The strong scattering pattern, the so-called **butterfly scattering pattern,** reveals that the shear induces strong concentration fluctuations or even phase separation for the single-phase solution in a quiescent state.

What are the physical factors which give rise to the quite contrasting effects of shear-induced single-phase formation or shear-induced phase separation? From a series of our studies, we propose that **shear-induced single-phase formation** occurs in **dynamically symmetric systems** with sufficiently weak repulsive segmental interaction between the two constituent polymers or sufficiently weak interfacial tension between the two phases, whereas the **shear-induced phase separation** occurs in **dynamically asymmetric systems** [3, 5]. We shall discuss each of the two cases in the following sections in this chapter.

2.2 Dynamically Symmetric Systems and Dynamically Asymmetric Systems

In **dynamically symmetric mixtures**, component molecules A and B have equal mobility or self-diffusion coefficients. Many small-molecule systems belong to this family as illustrated in part (a-2) in the right half of Fig. 2.3a [3, 5]. The nonequilibrium dynamics and processes for this family have been relatively well explored. Polymer pairs having about equal degree of polymerization (DP) and monomeric frictional coefficients also belong to this family as illustrated in part (a-1) in the left half of Fig. 2.3a.

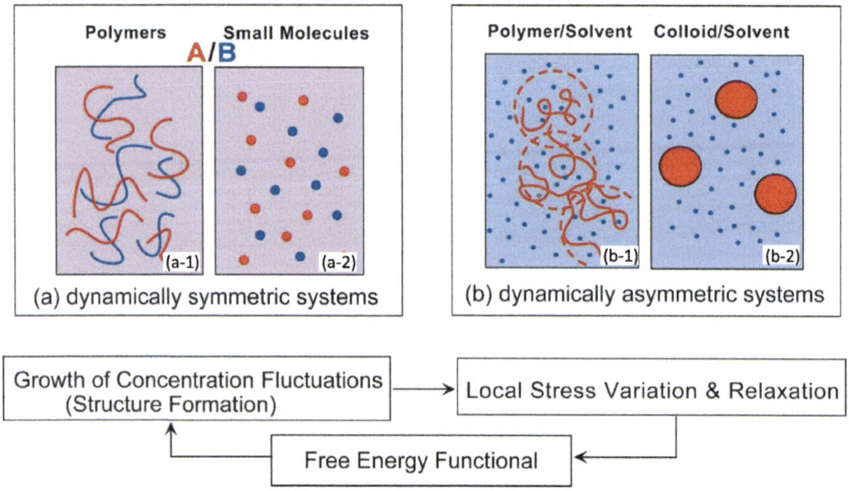

Fig. 2.3 Dynamics of concentration fluctuations of systems with dynamic symmetry and asymmetry. Part (**a**): dynamically symmetric binary mixtures of polymer A/polymer B (a-1) and small molecule A/small molecule B (a-2). Part (**b**): dynamically asymmetric polymer solution (polymer/solvent) (b-1) and colloidal dispersion (colloid/solvent) (b-2). Below part (a) and (b), a general principle common to the two systems is presented: "Growth of Concentration Fluctuations (Structure Formation)" gives rise to build up "Local Stress Variation & Relaxation", thereby affecting "Free Energy Functional", which in turn reflects back to further "Growth of Concentration Fluctuations (Structure Formation)". Based on [5], permission with Oxford University Press

In **dynamically asymmetric mixtures**, component molecules have different mobilities (self-diffusivities) as in the case of mixtures of polymers having different molecular weights, an extreme case of which is observed for a high molecular weight polymer in a solvent as schematically shown in part (b-1) in the left part of Fig. 2.3b, or a dispersion of colloidal particles in a solvent, as schematically shown in part (b-2) in the right part of Fig. 2.3b. Although the colloidal particles themselves are not molecules, they can be treated as big spherical molecules. The dynamically asymmetric systems have been relatively less explored than their symmetric counterparts, despite the fact that such systems are more frequently found and more general in nature.

The structure formation or structure growth driven by thermodynamic instability (spinodal decomposition) of binary mixtures composed of A and B involves the growth of spatial composition fluctuations $\phi_K(r, t)$ (K = A or B) with time as shown in Fig. 2.4a, where

$$\phi_{\mathrm{A}}(r, t) + \phi_{\mathrm{B}}(r, t) = \phi_{\mathrm{A0}} + \phi_{\mathrm{B0}} \equiv \phi_0, \tag{2.2.1}$$

and ϕ_{K0} and ϕ_0 are the average concentration of Kth polymer and the net polymer concentration in the solution, respectively. The built-up local concentration fluctuations (Fig. 2.4a) naturally build up local stress and its stress relaxation with time as

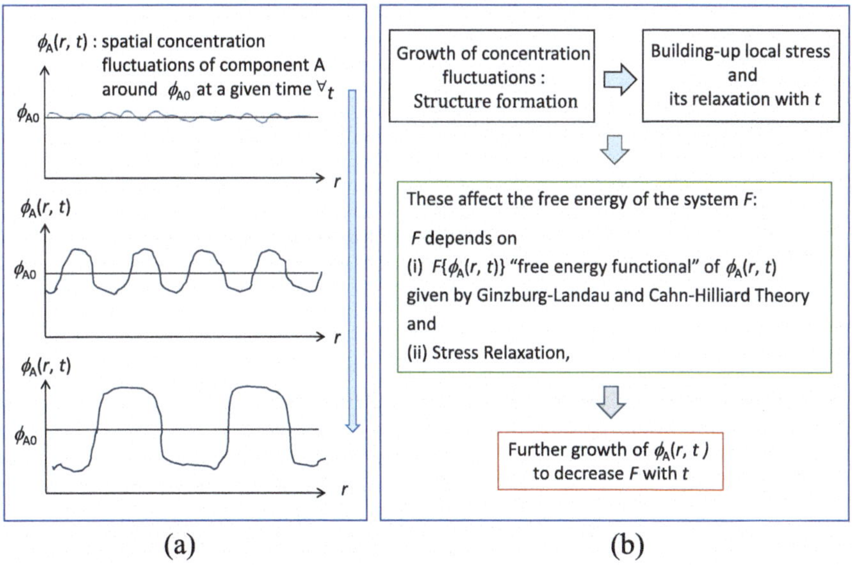

Fig. 2.4 The growth of the concentration fluctuations of component A, $\phi_{\mathrm{A}}(r, t)$, around the concentration, ϕ_{A0}, with time. Part (a) shows three graphs depicting the increasing fluctuations over distance r with time, as indicated by the arrow from the top to bottom. Part (b) describes effects of the increasing fluctuations with time on local stress, free energy functional, and further growth of the concentration fluctuations

noted in the top of Fig. 2.4b. The built-up local stress and its relaxation affect the free energy of the system as shown by the scheme in the middle part of Fig. 2.4b: the free energy F is the functional of $\phi_K(r, t)$ depending on the spatial composition fluctuations in the mixture as a whole [6, 7], and the stress relaxation. The free energy functional F controls further growth of the fluctuations or the structures with t in such a way that F decreases with$t$ as presented by the scheme shown in the bottom part of Fig. 2.4b.

In the dynamically symmetric systems, this stress is equally born by the two components, and the stress relaxes at equal rates during structure growth. Furthermore, the stress relaxation rate is faster than the growth rate of the concentration fluctuations (CFs), as schematically illustrated in Fig. 2.5a. Thereby, the stress built-up and its relaxation are predicted to give no influence on the structural growth.

On the other hand, in the dynamically asymmetric systems, the stress σ_B born by the fast component having a larger mobility relaxes faster than the stress σ_A born by its counterpart, i.e., the slow component having a small mobility, so that the local stress is primarily born by the smaller mobility component as schematically illustrated in Fig. 2.5b, giving rise to the so-called stress imbalance [8]. This "imbalance-induced" local stress and the stress relaxation affect the free energy functional and hence the cooperative diffusivity of the slower mobility component of the systems, which should be incorporated into the dynamics of the composition fluctuations and the growth of pattern in the systems, as summarized in Fig. 2.5b.

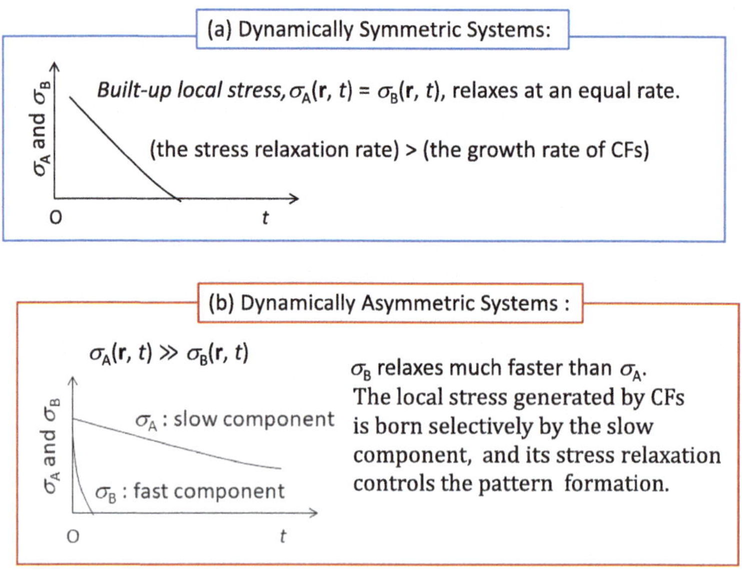

Fig. 2.5 Summary on stress built-up by local concentration fluctuations and its relaxation with time for (part **a**) dynamically symmetric and (part **b**) asymmetric two-component systems A/B

This **stress-diffusion** coupling inherent in the asymmetric systems is anticipated to give intriguing effects on the dynamics of the composition fluctuations and pattern formation, theories of which have been pioneered by Onuki [8] for various systems. It is believed that various pattern formations in nature, including various materials and various systems of living bodies, may involve the stress-diffusion coupling. The unique pattern formation due to this coupling effect is expected to be one of the most important problems to be unveiled, both experimentally and theoretically. Below, we shall present some examples of the unique pattern formation in the dynamically asymmetric systems.

2.3 Some Anticipated Effects of Dynamical Asymmetry on Self-Organization in the Quiescent State

Let us first consider the growth of the concentration fluctuation during the spinodal decomposition in an entangled polymer solution. In Fig. 2.6a a homogeneous entangled polymer solution is simplified by straight lines aligned in a square lattice filled with solvent molecules in which lattice points correspond to the entangled coupling points of the chains. Moreover, parts of chains between the entanglement coupling points are simplified as the straight lines. The trajectory of each chain fluctuates around each straight line. Since the polymer chains are the slow component, and the solvent molecules are the fast component, the concentration fluctuations, driven by the thermodynamic instability of the homogeneous solutions, will be built up by the fast interdiffusion of solvent molecules rather than the slow interdiffusion of polymer chains. Figure 2.6a focuses given lattices, marked by broken red circles, to be enriched by solvent through interdiffusion of solvent molecules from the lattices neighbored to the given lattices. Such interdiffusion as described above develops the solvent-rich lattices indicated by solid circles marked by S and polymer-rich lattices shaded by red as shown in Fig. 2.6b. The built-up concentration fluctuations via the solvent interdiffusion may involve a free energy barrier due to the elastic deformation of the network chains. However, this barrier is outweighed by the thermodynamic driving force for the growth of the concentration fluctuations, and moreover, the transient elastic deformation of chains is relaxed through molecular rearrangements of chains during the growth of the concentration fluctuations.

These solvent-rich regions indicated by the solid circles marked by S are created by squeezing solvent molecules from the neighbored lattice cells as marked by red-shaded cells in Fig. 2.6b. The squeezing involves interdiffusion of solvent molecules, as illustrated by arrows, from the red cells into the swollen cells marked by S where the original cells are indicated by broken red circles in parts (a) and (b). The solvent-squeezing process is predicted to create the interconnected red regions, i.e., the network phase rich in polymer, in the matrix phase rich in solvent. The network phase is percolated in 3D space and co-continuous with the matrix phase.

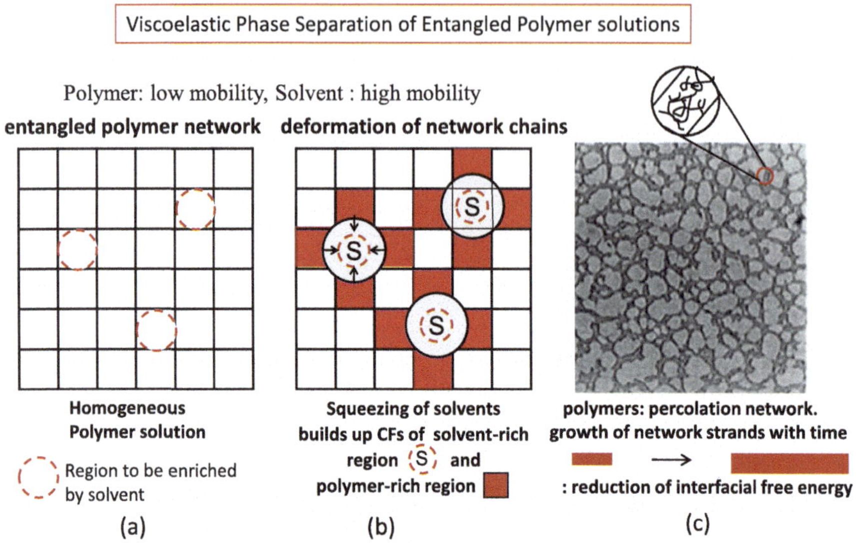

Fig. 2.6 Viscoelastic phase separation of entangled polymer solutions: a typical dynamically asymmetric two-component system. A model of viscoelastic phase separation via spinodal decomposition based on the solvent squeeze (parts **a** and **b**) and an experimental example (part **c**)

Figure 2.6c presents a snapshot of optical microscope (OM) image taken during a coarsening process of the phase separation driven by SD for 7 wt.% poly (vinyl methyl ether) in water at 310.3 K [9]. The image presents the percolated network domain (the dark domain rich in polymer) in the matrix rich in solvent (bright phase). The network strands between junctions of the network domains as marked by the red circle in Fig. 2.6c are composed of entangled chains as sketched in the inset of the figure. The network strands grow with respect to the longitudinal length and the lateral width with time during the coarsening process as schematically illustrated by the change in the red bar from the left one to the right one, as indicated underneath the OM image in Fig. 2.6c, to reduce interfacial free energy or interfacial tension via reduction of the interfacial areas between the two phases.

Figure 2.7 presents the void formation of charged colloidal particles dispersed in water as observed by laser-scanning confocal microscopy under the reflection mode (refl.-LSCM) [5, 10, 11]. The states of dispersion are indicated by the three-dimensional (3D) phase diagram, as shown in the center part of Fig. 2.8, in the parameter space of salt concentration, charge density of the particles, and particle concentration, where S, L, and G designate the solid (Crystal), liquid (disorder), and gaseous (void) phase, respectively, and (S + L) and (L + G) designate the coexistence of S and L phase and that of L and G phase, respectively. The representative sliced image taken with refl.-LSCM for the L, S, (S + L), and (L + G) phase are shown in the top-left, top-right, bottom-left, and bottom-right corner of the figure, respectively.

The void formation was discovered under a special condition of a very small or almost zero salt concentration, a high charge density (e.g., 0.48 $\mu C/cm^2$) per

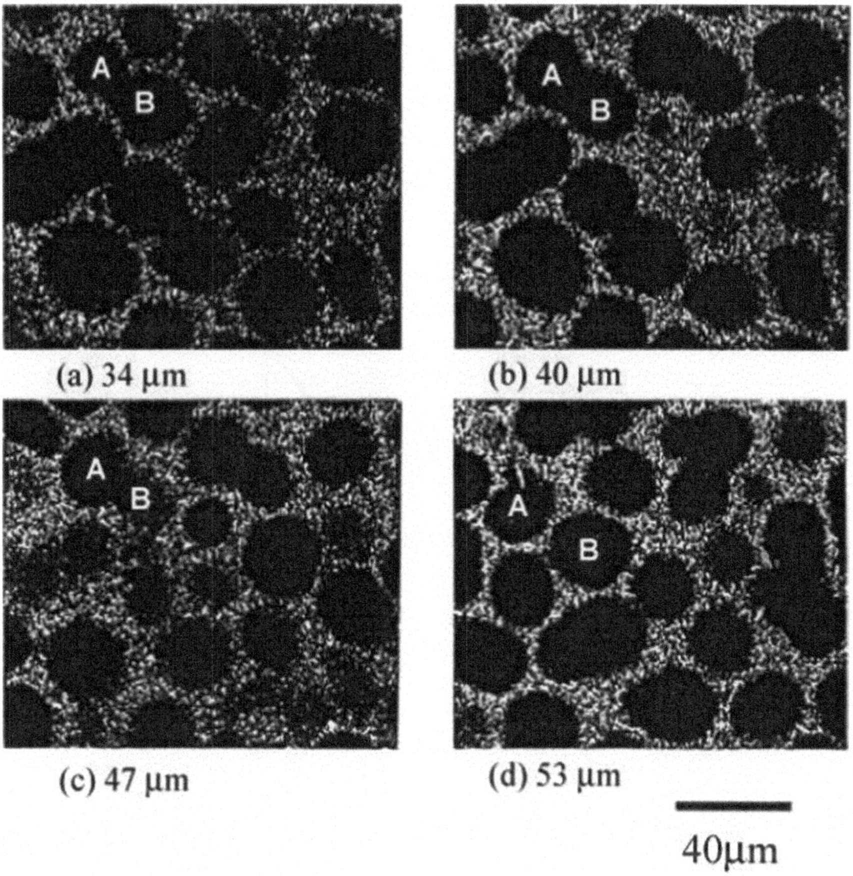

Fig. 2.7 Unique pattern formation in the dilute suspension of charged colloidal particles in density-matched H_2O/D_2O medium as observed by a series of laser-scanning confocal microscope (LSCM) images on the same area of the sample but sliced at different depths along incident beam (the depth direction). Based on [5], permission with Oxford University Press

particle or an effective charge number of $1.4 \times 10^3/(\text{particle})$, and a small particle concentration (0.10 vol. %) as shown in the bottom right corner of the diagram (shaded with green color marked by L + G). The density of the particles was carefully matched to water to avoid gravitational effects on the pattern formation. In Figs. 2.7 and 2.8, the particles having an average diameter of 0.12 μm appear as bright dots in the sliced images with depth resolution of 0.5 μm, because of the contrast given by scattering from the particles in liquid-like state. The series of patterns in Fig. 2.7 was obtained at two months after homogenization of the dispersion via vigorous shaking of the cell. Figure 2.9a, b represents the sliced images of the dispersion after the homogenization taken by the refl.-LSCM, whereas Fig. 2.9c, d represents the sliced images of the dispersion obtained at two months after the homogenization [10–12].

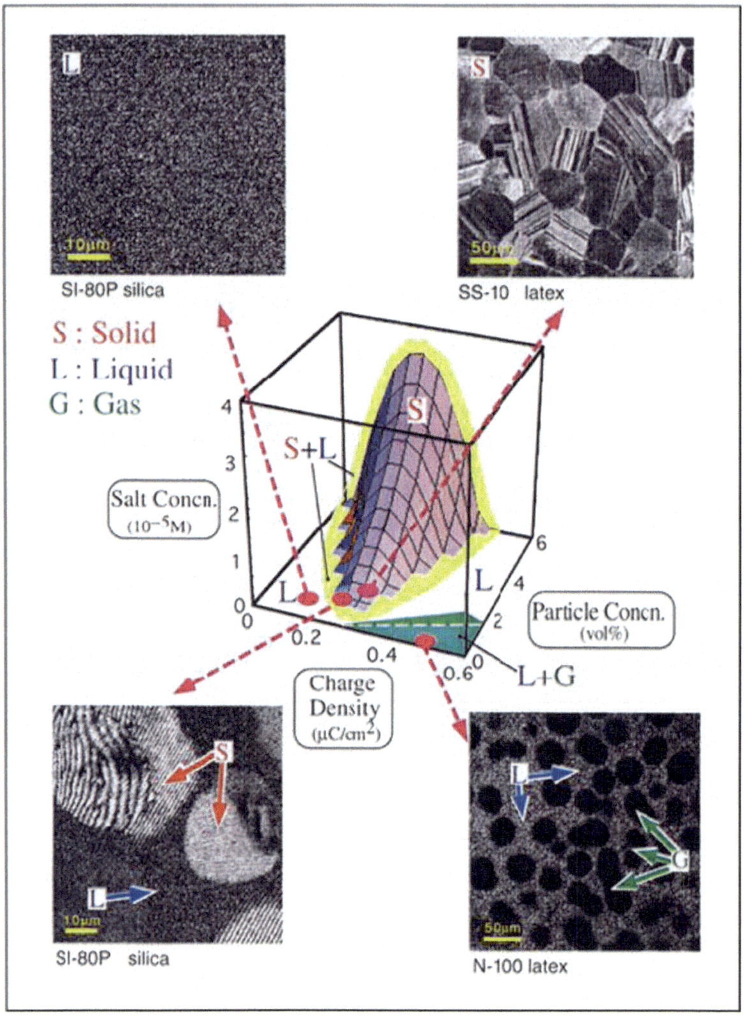

Fig. 2.8 3D phase diagram of charged colloidal particles dispersed in water in the parameter space of charge density, particle concentration, and salt concentration at the center of the figure and typical LSCM images in liquid state (L, top left), solid state (S, top right), S/L coexisting state (S + L, bottom left) and spinodal-decomposed state (S + G, bottom right)

The sliced images presented in Fig. 2.7 were taken in the same area of the sample but by changing the slicing position along the incident beam axis, i.e., the depth direction of the dispersion [5, 10]. Statistically identical patterns are observed everywhere in the sample cell, indicating that the pattern was not formed by the gravitational effect but rather by long-range electrostatic interactions. The patterns are composed of regions rich in the particles, where the particles are in the liquid state, and voids where almost no particles are observed under the given period to obtain the sliced

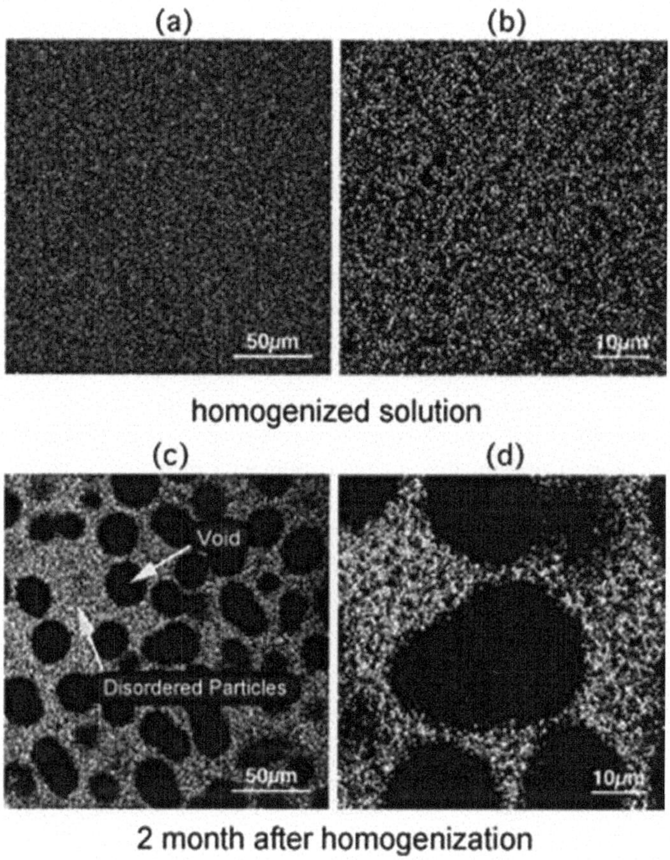

Fig. 2.9 LSCM images of homogenized charged colloidal particles dispersed in water (part (**a**) and its magnified image shown in part (**b**)). A phase-separated dispersion of the colloidal particles at 2 months after the homogenization (part c and its magnified image shown in part d). The thermal equilibrium state of the dispersion is the phase-separated state of (L + G) in the phase diagram shown in Fig. 2.8 where the L and G phase macroscopically phase-separate. Based on [10], reprinted by AIP Publishing with permission

image. A careful tracing of the series of the sliced images along the depth direction reveals that the voids are not isolated but are interconnected along the depth and hence co-continuous with the network-type particle-rich regions. For example, two voids A and B in the slice shown in part (d) appear to be separated, but they are interconnected in the focal planes located at 6 μm (slice c), 13 μm (slice b), and 19 μm (slice a) above the slice (d).

It is quite striking to note that such a very dilute dispersion of strongly charged colloidal particles forms a 3D bicontinuous network structure rich in particles in the matrix of the void phase under the field of long-range electrostatic interactions, regardless of the origin of the interactions, either DLVO-type repulsive interactions

[13, 14] or net attractive interactions [10]. If the repulsive interactions play a role, our intuition may predict a homogeneous liquid-like spatial distribution of the particles, as shown in Figs. 2.9a, b, and 2.10a, or clusters of the particles isolated in water, as shown by Fig. 2.10c. Even in the case when the net attractive interactions exist, formation of the isolated clusters as in Fig. 2.10c seems to be more reasonable than the 3D percolated co-continuous network formation.

Figure 2.10 schematically presents a model for the pattern formation in the colloidal dispersion [3, 5]. When the systems having a uniform spatial distribution of particles under the long-range electrostatic interaction (part *a*) start to build up concentration fluctuations of the particles driven by the thermodynamic instability (spinodal decomposition, SD) brought about in the system, the particles tend to first form the volume-filling assemblies of the particles impinged at their grain boundaries: within each grain, the particles form the percolated, periodic network structures due to the SD (part *b*) rather than clusters of the particles discretely dispersed in the matrix phase (part *c*), because they are interacting with each other in the electrostatic

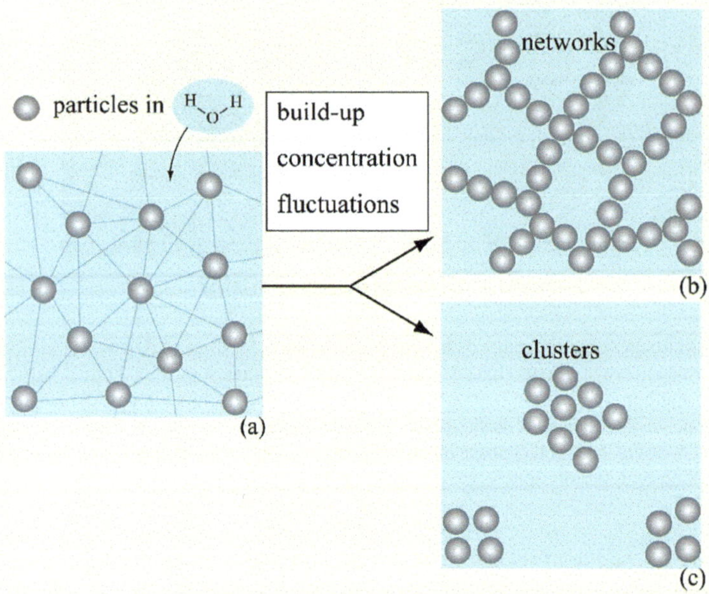

Fig. 2.10 Schematic illustrations of possible changes in spatial distributions of charged colloidal particles in water, which belongs to a typically dynamic-asymmetric system, when the system is brought into a thermodynamic instability condition to build up concentration fluctuations of the particles: **a** uniform distribution of the particles under the long-range electrostatic interactions as schematically illustrated by a network composed of the thin lines and **b** phase-separated particles exhibiting the network-type domain structure where the network phase and water phase are co-continuous in the sample space and **c** cluster-type inhomogeneous distribution of the particles. The medium of water phase is composed of the mixture of H_2O/D_2O whose density is matched with that of the colloidal particles. Based on [5], permission with Oxford University Press

field where the stress built-up by the concentration fluctuations is born only by the particles. Part *b* presents only a part of the structure within a grain, whereas part c presents a few assemblies of the clusters of the particles. In the stress relaxation process, solvent-rich regions may be created via a local solvent-squeezing process, i.e., a process involving squeezing of solvents from the regions rich in the particles, i.e., the regions where the local concentration of the particles is higher than the average concentration, into the regions poor in the particles, resulting in the network structure formation, as in the case discussed in Fig. 2.6. The network-domain structure coarsens with time. In the long time limit (hydrodynamic limit), the network shown in part (b) will be eventually transformed into clusters of the particles as shown in part (c) through percolation-to-cluster transition. We may treat this phenomenon as the viscoelastic phase separation via SD coupled by the viscoelastic stress relaxation of the colloidal particles.

A similar network formation was reported in phase separation processes of polymer solutions. For example, poly (vinyl methyl ether) (PVME) in water, a highly dynamic-asymmetric system, showed a percolated network-domain structure rich in polymers in the matrix of water, as shown in Fig. 2.11 [9], because of stress-diffusion coupling and viscoelastic phase separation as discussed earlier in conjunction with Fig. 2.6. The percolated network-domain structures grow as presented from part (a)–(c) via the break-up processes of a part of the network strands between the neighboring junction points and the degeneration processes of the broken network strands to the neighboring networks. The break-up and degeneration processes cause a coarsening of the network-domain structure to reduce the total interface area and hence the interfacial free energy of the percolated network-domain structures. The break-up processes involve the built-up of local stress, and the degeneration processes involve its stress relaxation.

In the growth process, the entangled polymer chains in the strands are pulled against one another, which involves **rubber-like elastic force** and thinner parts of the strands will be broken because of higher stress concentration. Consequently, the network strands between the network junctions do not have uniform diameters, lengths, and smooth interfaces against the water phase, as highlighted in parts (c) and (d), and the mesh size of the network domain has a broad distribution, as highlighted in parts (a) and (b). Furthermore, growth of network domains does not display self-similarity. These characteristics are believed to be a consequence of the strands being comprised of entangled polymer solution where the entangled polymer chains are in the field of **rubber-like elastic forces** during this pattern formation. This elastic force field extends over a large length scale through the domain-network structures.

In the long-time limit of the coarsening process, the macroscopic percolation of the network domain is unable to be maintained, because (1) its phase volume (0.10 vol %) is very small as expected from the phase diagram of the solution presented in Fig. 2.9, and because (2), as a consequence of the reason (1) described above, the number density of the junction points is very small, and it further decreases with time to cause the percolation-to-cluster transition. The percolated network domain will be eventually fragmented as evidenced by fragmented smaller network domains as shown in Fig. 2.11d: the fragmented network domains will be collapsed into

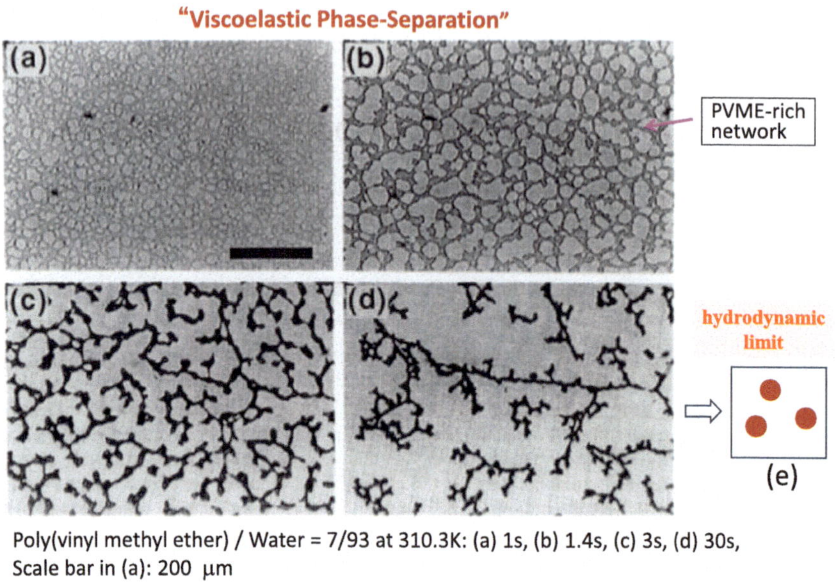

Fig. 2.11 Viscoelastic phase separation of a dynamically asymmetric two-component system composed of poly (vinyl methyl ether) and water with 7 and 93 wt. %, respectively, at 1 s (**a**), 1.4 s (**b**), 3 s (**c**), and 30 s (**d**) after onset of the phase separation as observed by transmission optical microscope. Based on [9], reprinted with permission, American Chemical Society

clusters of droplets rich in the polymer discretely dispersed in the matrix rich in the solvent in the hydrodynamic limit, as schematically illustrated by part (e) in Fig. 2.11. The reported experiment here did not yet reach hydrodynamic limit as predicted by part (e).

Note that the pattern presented in Fig. 2.11a for the PVME/water system has the characteristic length (the spacing of the network domain) of about 30 μm, already very large close to a macroscopic length relative to the radius of gyration R_g of the individual polymer chains, reflecting a very late stage of phase separation. What happens in much earlier stages of the phase separation? What can we learn from a very early stage of phase separation in such systems? How different is the behavior from that for a dynamically symmetric system? These fundamental issues have not yet been fully explored and are not well understood. We conducted a work along this line [15, 16].

Figure 2.12 represents time-resolved small-angle light scattering (TRSALS) profiles after the onset of the phase separation via spinodal decomposition for (a) a semidilute solution of very high molecular weight PS in DOP at the quench depth $\Delta T = 3.0\,°\text{C}$ below the cloud point temperature $T_{cl} = 13.8\,°\text{C}$ [15], a typical system having the dynamical asymmetry, and (b) a binary polymer mixture of PB and SBR, poly(styrene-*r*-butadiene), a typically dynamically symmetric system [17] phase-separated at $T = 40.0\,°\text{C}$. $I(q, t)$ is the light scattering intensity distribution with

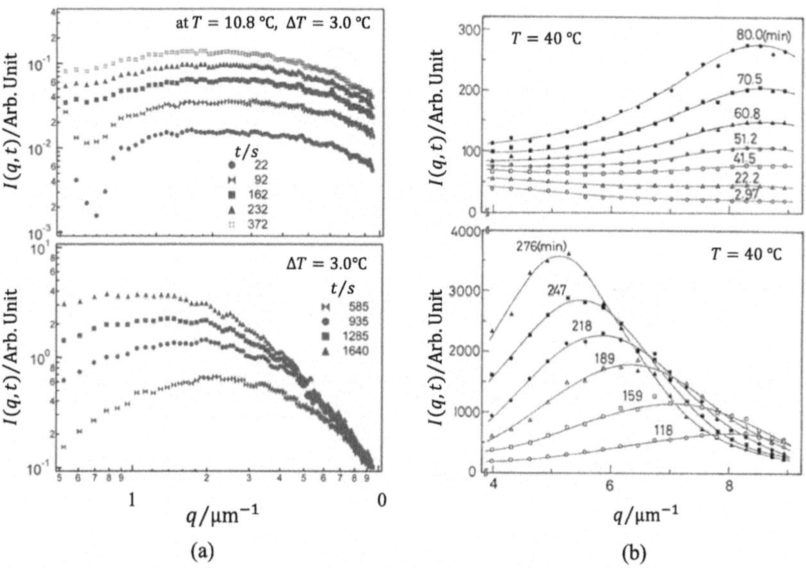

Fig. 2.12 Typical time-evolution of light scattering profiles for (**a**) the dynamically asymmetric system of PS548/DOP 6.0 wt. % at quench depth $\Delta T = 3.0\,°C$ from the cloud point temperature $T_{cl} = 13.8\,°C$ and (**b**) the dynamically symmetric mixture of SBR1/PB19 with the composition of 30 vol % / 70 vol %, respectively, phase-separated at $T = 40.0\,°C$ right after the homogenization via so-called Baker's Transformation described in ref. [17, 18]. The top and bottom parts in (**a**) and (**b**) correspond to the early-stage SD and the late-stage SD, respectively. Based on [5], permission with Oxford University Press

q at time t after the phase separation started, where q is the magnitude of the scattering vector $\mathbf{q}$, defined by $q = (4\pi/\lambda)\sin(\theta/2)$ with λ and θ being the wavelength of incident light and the scattering angle (both in the sample), respectively. PS has M_w of 5.48×10^6, and the PS concentration is 6 wt. %, equal to 6.7 times overlap concentration [5]. This system is hereafter designated as PS548/DOP 6.0 wt. %.

Figure 2.12a presents the result for the solution at quench depth of $\Delta T = 3.0\,°C$ below $T_{cl} = 13.8\,°C$, displayed in the double logarithmic scale with respect to $I(q, t)$ and q [15]. In the early stage shortly after the quench (see the top part of part a), a very broad scattering maximum appears; the maximum intensity increases with time, while the peak position with respect to q (defined q_m) remains almost unchanged with time, indicative of an early-stage spinodal decomposition (SD) [18–20] where the composition fluctuations with a large length scale and over a wide length scale, ranging $2\pi/q = 1$ to 6 μm, are developed in this system: the characteristic length defined as $2\pi/q_m$ is $\sim$ 3 μm. As time elapses, the peak position q_m rapidly shifts toward small q and the peak intensity further increases as clearly shown in the bottom part of part (a). It is quite striking to note that the low q modes increase their intensity with time, whereas the high q modes hardly increase or decrease their intensity with time as revealed in the bottom part of part (a). We note here that this time-evolution behavior is very different from that observed for dynamically symmetric mixtures of simple liquids or polymers, as will be presented immediately below.

Figure 2.12b presents typical results for nearly dynamically symmetric polymer mixture which are displayed in the linear scale with respect to $I(q, t)$ and q [17]. In the early-stage SD (top part of b), a particular Fourier mode of the composition fluctuations is selected to grow faster than the other modes, so that the scattering peak at $q_{\rm m}$ becomes sharper with time without changing $q_{\rm m}$, in contrast to the asymmetric system where the peak tends to become broader with time. In the late stage (bottom part of Fig. 2.12b), the mode–mode coupling occurs, so that the intensity of high q-modes decay and that of the low q-modes grow; consequently, the peak position $q_{\rm m}$ shifts toward low q with increased time t. In contrast to this trend, for the asymmetric mixture only the low q modes grow, and as a consequence the peak becomes broad and shifts toward small q. Apparently the profiles at different times in the bottom part of Fig. 2.12a cannot be scaled with the time-dependent $q_{\rm m}$ so that the domain-growth is not self-similar, i.e., the scaled scattering intensity profiles $q_{\rm m}^3 I(q/q_{\rm m}, t)$ cannot be universal with t, albeit the time evolution of the profiles in the bottom part of Fig. 2.12b can be clearly scaled with the time-dependent $q_{\rm m}$, i.e., the scaled profiles $q_{\rm m}^3 I(q/q_{\rm m}, t)$ are universal with t [19–21]. What do these unique characteristics found for the dynamic asymmetric system mean? To understand the fundamental differences in the two systems, we analyze below the growth rate of various Fourier modes in the early-stage SD as a function of q.

Suppose the system composed of the entangled polymer network swollen by solvent molecules has a uniform concentration of polymers with the average mesh size of the entangled networks ξ_0 at the length scale $l \gg \xi_0$, as schematically shown in Fig. 2.13a, where polymers are schematically illustrated by lines, and solvent molecules are represented by small open circles or coarse-grained by the blue medium [5, 15]. Let us consider first the case where the concentration fluctuations with a characteristic length r are built up in the system by thermal excitation as shown in Fig. 2.13b, so that the system has regions rich in polymers and regions poor in polymers, represented respectively by the darker and brighter blue regions.

The concentration fluctuations will build up stress, and this stress is born only by the polymers, since the stress on solvent molecules relax much faster than polymers as shown in Fig. 2.5b. The built-up spatial variations of local stress will contribute to the free energy functional and hence to the cooperative diffusion of polymers. This in turn will affect the concentration fluctuations of polymers (Fig. 2.3) and hence the pattern formation of the system, suppressing the dynamics of the fluctuations, because the building up of the concentration fluctuations involves an extra cost of stretching free energy. This intuition about the suppression will be proven to be reasonable later in conjunction with Eqs. 2.4.12 and 2.4.21.

Let us next consider the case where the concentration fluctuations of the large-length scale r, as shown in Fig. 2.13c, are thermally excited so that we have the entangled networks with mesh size ξ smaller than ξ_0 in polymer-rich regions and the entangled networks with mesh size ξ larger than ξ_0 in polymer-poor regions. Such fluctuations as described above will be developed through very large length-scale molecular rearrangements and will inevitably occur very slowly. Under these conditions, the stress developed in the systems will be completely relaxed during the process where such concentration fluctuations are built up via the long-range

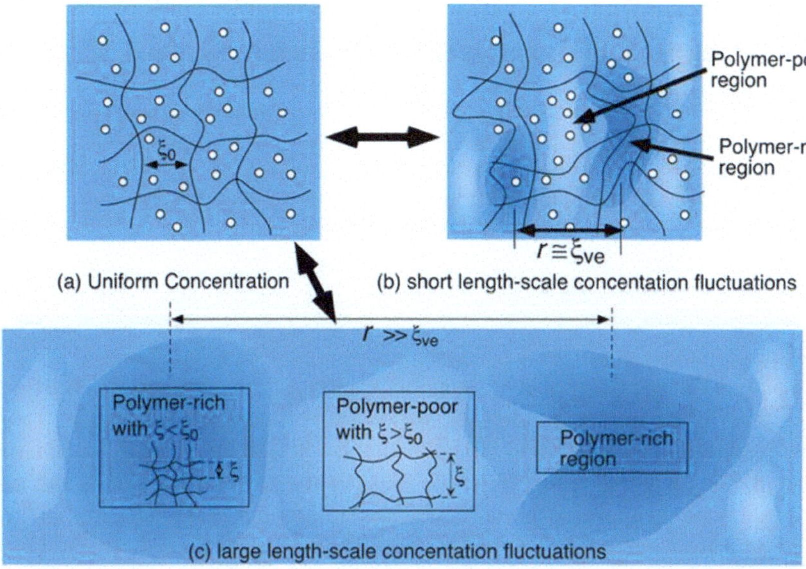

Fig. 2.13 Schematic representation of the concept of the stress-diffusion coupling and the viscoelastic effects in polymer solution. The lines and circles in parts (**a**) and (**b**) respectively represent polymer chains and solvent molecules. Part (**a**) represents homogeneous semidilute solutions, whereas parts (**b**) and (**c**) represent solutions with spatial concentration fluctuations at a short length scale and at a large length scale, respectively. ξ_0 designates an average mesh size of entangled polymer networks for a statistically homogeneous semidilute solution. In parts (**b**) and (**c**), the dark blue regions designate polymer-rich regions with the average mesh size $\xi < \xi_0$, while the light blue regions designate polymer-poor regions with $\xi > \xi_0$. Parts (**b**) and (**c**) differ in the characteristic length r for the concentration fluctuations. In part (**c**) solvent molecules are not shown but coarse-grained in the blue matrix. Based on [5], permission with Oxford University Press

rearrangements of entangled polymers, i.e., the process involved here is slow. Thus, the stress-diffusion coupling is completely screened out in this case, which leads to the concept of a "**screening length**" for the stress-diffusion coupling effects. This screening length was named **viscoelastic length** ξ_{ve} [8, 15]: the stress-diffusion coupling is screened out if $r > \xi_{ve}$, but if $r < \xi_{ve}$ the coupling is effective. This length should depend on dynamical properties of component elements, as shown quantitatively by Eqs. 2.4.2, 2.4.3, and 2.4.22 later.

2.4 Time-Evolution Equation of Concentration Fluctuations in Dynamically Asymmetric Systems at Equilibrium

Doi and Onuki [22, 23] formulated a general time-evolution equation for concentration fluctuations in dynamically asymmetric solutions in the quiescent state:

$$\frac{\partial}{\partial t}\delta\phi(\mathbf{r},t) = \Lambda\nabla\cdot\left[\nabla\frac{\delta F}{\delta\phi} - \alpha_{\mathrm{a}}\nabla\cdot\overleftrightarrow{\sigma}(\mathbf{r},t)\right] + \zeta(\mathbf{r},t) + (\text{hydrodynamic term}), \tag{2.4.1}$$

where $\delta\phi(\mathbf{r},t) \equiv \phi(\mathbf{r},t) - \phi_0$ represents fluctuations of local concentration $\phi(\mathbf{r},t)$ of a polymer in a polymer solution at position $\mathbf{r}$ and time t from its average value ϕ_0. Note that the above equation is applicable also to the dynamically symmetric binary mixtures where $\alpha_{\mathrm{a}} = 0$. Here, we consider the polymer solution at thermal equilibrium without applied external flow. The solutions under applied external flow will be described in Sect. 5.7.1. In Eq. 2.4.1, $F = F(\mathbf{r},t) = F\{\delta\phi(\mathbf{r},t)\}$ is the free energy functional of $\delta\phi(\mathbf{r},t)$, and $\delta F/\delta\phi$ is the variational derivative of F with respect to $\delta\phi(\mathbf{r},t)$. Λ is the Onsager transport coefficient in the small-q limit or in the large-r (length-scale) limit. α_{a} is the so-called **dynamical asymmetry parameter** defined by,

$$\alpha_{\mathrm{a}} \equiv |D_{\mathrm{A}}N_{\mathrm{A}} - D_{\mathrm{B}}N_{\mathrm{B}}|/(D_{\mathrm{A}}N_{\mathrm{A}}\phi_{\mathrm{B}} + D_{\mathrm{B}}N_{\mathrm{B}}\phi_{\mathrm{A}}), \tag{2.4.2}$$

for binary mixtures, where D_K, N_K, and ϕ_K (K = A or B) are respectively self-diffusivity, degree of polymerization (DP), and volume fraction of the Kth component in mixtures composed of A and B. In Eq. 2.4.2,

$$\alpha_{\mathrm{a}} = \phi_0^{-1} \tag{2.4.3}$$

for polymer solutions, $\overleftrightarrow{\sigma}(\mathbf{r},t)$ is the local stress tensor, and $\zeta(\mathbf{r},t)$ is the random thermal noise term expressed by the following fluctuation–dissipation theorem:

$$\langle\zeta(\mathbf{r},t)\zeta(\mathbf{r}',t')\rangle = 2k_{\mathrm{B}}T\Lambda\nabla^2\delta(\mathbf{r}-\mathbf{r}')\delta(t-t'), \tag{2.4.4}$$

where k_{B} and T are the Boltzmann constant and absolute temperature, restrictively, $\langle x\rangle$ denotes the thermal average of quantity x, and δ is Dirac's delta function. Cahn–Hilliard–Cook (CHC) [7, 24] and Ginzburg–Landau (GL) [6] originally introduced the first and third terms on the right-hand side (rhs) of Eq. 2.4.1, the so-called time-dependent CHC equation (TDCHC) or GL equation (TDGL), while Kawasaki-Ohta introduced the hydrodynamic term [25]. Doi-Onuki further generalized the equation by incorporating the stress term associated with the dynamical asymmetry and the stress-diffusion coupling (the second term on the rhs of Eq. 2.4.1) [22].

If the systems to be considered are dynamically symmetric,

$$D_A N_A = D_B N_B, \text{ and hence } \alpha_a = 0, \tag{2.4.5}$$

so that the second term associated with the concentration-fluctuations-induced stress-tensor on the rhs of Eq. 2.4.1 vanishes. Consequently, Eq. 2.4.1 reduces to the well-known **TDCHC equation** [7, 24]**or TDGL equation** [6].

The evolution of the concentration fluctuations of the polymer in the polymer solution per unit volume at a local position **r** and time t is brought about by the diffusional flux $\mathbf{J}_{\text{chem}}(\mathbf{r}, t)$ driven by the chemical potential $\mu(\mathbf{r}, t)$, the diffusional flux $\mathbf{J}_{\text{stress}}(\mathbf{r}, t)$ driven by the stress, the random thermal noise, and hydrodynamic interaction,

$$\frac{\partial}{\partial t}\delta\phi(\mathbf{r}, t) = -\text{div}\mathbf{J}_{\text{chem}}(\mathbf{r}, t) - \text{div}\mathbf{J}_{\text{stress}}(\mathbf{r}, t) + \zeta(\mathbf{r}, t) + (\text{hydrodynamic term}) \tag{2.4.6}$$

$\mathbf{J}_{\text{chem}}$ is given by the linear response theory of Onsager (Onsager's law),

$$\mathbf{J}_{\text{chem}}(\mathbf{r}, t) = \Lambda\mathbf{X}_{\text{chem}}(\mathbf{r}, t), \tag{2.4.7}$$

where $\mathbf{X}_{\text{chem}}(\mathbf{r}, t)$ is the chemical force associated with the chemical potential $\mu(r, t)$,

$$\mathbf{X}_{\text{chem}}(\mathbf{r}, t) = -\nabla\mu(\mathbf{r}, t). \tag{2.4.8}$$

$\mu(\mathbf{r}, t)$ is given by the variational derivative of the free energy functional with respect to the concentration fluctuations $\delta\phi(r, t)$,

$$\mu(\mathbf{r}, t) = \frac{\delta F\{\phi(r, t)\}}{\delta\phi}. \tag{2.4.9}$$

Equations 2.4.6–2.4.9 lead to the first term in the rhs of Eq. 2.4.1.

$\mathbf{J}_{\text{stress}}$ is given by

$$\mathbf{J}_{\text{stress}} = \Lambda\mathbf{X}_{\text{stress}}(\mathbf{r}, t), \tag{2.4.10}$$

where $\mathbf{X}_{\text{stress}}$ is the force arising from the relaxation of the imbalanced stress $\overleftrightarrow{\sigma}(r, t)$ given by

$$\mathbf{X}_{\text{stress}}(\mathbf{r}, t) = \alpha_a\nabla\cdot\overleftrightarrow{\sigma}(\mathbf{r}, t). \tag{2.4.11}$$

Equations 2.4.6, 2.4.10, and 2.4.11 lead to the second term in the rhs of Eq. 2.4.1.

If $\delta\phi(r, t)$ is small and the hydrodynamic term can be neglected, Eq. 2.4.1 can be linearized, and the linearized time-evolution equation for the composition fluctuations in q-space is given by,

$$\frac{\partial}{\partial t}\delta\phi(q,t) = \Lambda(q)q^2\left[\left(r_0 - C_g q^2\right)\delta\phi(q,t) - \frac{4}{3}\alpha_a^2\int_0^t dt' G\left(t-t'\right)\frac{\partial}{\partial t'}\delta\phi\left(q,t'\right)\right] + \zeta(q,t). \quad (2.4.12)$$

Here, the first term on the rhs of Eq. 2.4.12 is obtained based on the GL or CHC expansion of the free energy functional. r_0 is a parameter related to the **thermodynamic driving force for the growth of the concentration fluctuations** defined by

$$r_0 \equiv -(\partial^2 f/\partial\phi^2)_0 = 2(k_B T/v_0)(\chi - \chi_s) \quad (2.4.13)$$

with f being the free energy density of the solution. In the mean field (MF) approximation r_0 is given by the second line of the above equation where χ is the segmental interaction parameter between the polymer segment and solvent molecule, χ_s is the χ parameter at spinodal point, v_0 is the volume of the polymer segment and solvent molecule (here the volume being assumed to be identical for brevity), and $k_B T$ is the thermal energy. The driving forcer_0 is positive in the thermodynamically unstable solution where $(\partial^2 f/\partial\phi^2)_0 < 0$ and $(\chi - \chi_s) > 0$, which therefore tends to grow the fluctuations with t, whereas r_0 is negative for a single-phase state, which signifies the decay of thermally activated fluctuations with t. C_g is a positive constant related to the gradient free energy associated with the nonlocality of interactions given in the context of MF by,

$$C_g = \frac{k_B T}{v_0}\frac{a^2}{18\phi_0} \quad \text{for polymer solutions,} \quad (2.4.14)$$

$$C_g = \frac{k_B T}{v_0}\frac{a^2}{18\phi_0(1-\phi_0)} \quad \text{for polymer mixtures,} \quad (2.4.15)$$

and a is the segmental length [4, 26].

In Eq. 2.4.12, $G(t)$ is a relaxation function for the shear modulus given by

$$G(t) = \sum_{i=1}^{n} G_i \exp\left(-\frac{t}{\tau_i}\right), \quad (2.4.16)$$

where G_i and τ_i are the strength and relaxation time for the ith stress relaxation process, and $\zeta(q,t)$ satisfies the fluctuation dissipation theorem in q-space,

$$\left\langle \zeta(q,t)\zeta\left(q',t'\right)\right\rangle = 2k_B T\Lambda(q)q^2\delta\left(t-t'\right)(2\pi)^3\delta\left(q+q'\right). \quad (2.4.17)$$

The second term in the rhs of Eq. 2.4.12 represents the contribution of the stress built-up and its relaxation driven by the increase of the concentration fluctuations $\partial\delta\phi\left(q,t'\right)/\partial t'$ at t' to the time evolution of the concentration fluctuations at

t, $\partial\delta\phi(q,t)/\partial t$. Since the integral value in the second term of the rhs of Eq. 2.4.12 is always positive, the stress term negatively contributes to $\partial\delta\phi(q,t)/\partial t$ and thereby tends to suppress the growth rate of concentration fluctuations in two-phase solutions and the relaxation rate of the concentration fluctuations in single-phase solutions, as we naturally anticipate.

Thus, the dynamics of the concentration fluctuations are affected by the stress relaxation process and hence by the mechanical properties of the system. In the case where the rate of the longest stress relaxation, $1/\tau_1$, is faster than the characteristic rate of the concentration fluctuations, Γ_{conc}, which is usually the case near the spinodal temperature due to the critical slowing down of the dynamics of concentration fluctuations, this integral is simplified such that the term $\partial\delta\phi(q,t')/\partial t'$ can be assumed to be independent of t', having a constant value at t, $\partial\delta\phi(q,t)/\partial t$, so that it can be put outside the integral.Therefore, the 2nd term of the rhs of Eq. 2.4.12, designated as ("Stress Term") in the equation below, can be rewritten as follows,

$$\begin{aligned}(\text{Stress Term}) &\cong -\frac{4}{3\phi_0^2}\frac{\partial\delta\phi(q,t)}{\partial t}\sum_{i=1}^{n}G_i\int_0^t dt'\exp\left[-\frac{(t-t')}{\tau_i}\right]\\ &\cong -\frac{4}{3\phi_0^2}\frac{\partial\delta\phi(q,t)}{\partial t}\sum_{i=1}^{n}G_i\tau_i.\end{aligned} \tag{2.4.18}$$

Here, we note that

$$\int_0^t dt'\exp\left[-\frac{(t-t')}{\tau_i}\right] = \int_0^t dy\exp(-y/\tau_i) \cong \tau_i, \tag{2.4.19}$$

since $t \gg \tau_i$.

In this situation and at $q \ll 1/R_g$ where $\Lambda(q)$ in Eq. 2.4.12 becomes independent of q, equal to $\Lambda(0)$, i.e., a limiting value $\Lambda(q)$ at $q = 0$, we obtain,

$$\frac{\partial}{\partial t}\delta\phi(q,t) = R(q)\delta\phi(q,t) + \zeta(q,t). \tag{2.4.20}$$

The q-dependent growth rate $R(q)$ for the concentration fluctuations in the early-stage SD is given by [15]

$$R(q) = \frac{\Lambda(0)}{1+\xi_{\mathrm{ve}}^2 q^2}q^2\left(r_0 - C_g q^2\right), \tag{2.4.21}$$

where ξ_{ve} defined as follows,

$$\xi_{\mathrm{ve}} \equiv \left[\frac{4}{3}\alpha_a^2\Lambda(0)\eta_0\right]^{1/2}. \tag{2.4.22}$$

ξ_{ve} has the dimension of length and hence is called **viscoelastic length**. η_0 is the zero-shear viscosity given by

$$\eta_0 \equiv \sum_{i=1}^{n} G_i \tau_i. \quad (2.4.23)$$

$\Lambda(0)$ in Eq. 2.4.21 is expressed by

$$\Lambda(0) = \phi_A \phi_B (D_A N_A \phi_B + D_B N_B \phi_A) v_0 / k_B T, \quad (2.4.24)$$

where v_0 is defined by

$$v_0 \equiv (\phi_A / v_A + \phi_B / v_B)^{-1} \quad (2.4.25)$$

with $v_K (K = \text{A or B})$ being the molar volume of Kth monomeric unit.

The **viscoelastic effect** gives a **renormalization effect** on the Onsager kinetic coefficient $\Lambda(0)$ in the small q range via the term of $(1 + \xi_{ve}^2 q^2)^{-1}$, giving rise to a q-dependent effective suppression of $\Lambda(0)$. This suppression depends on a product of the viscoelastic length ξ_{ve} and q; the larger the value q is, the larger is the suppression of the renormalized Onsager coefficient. This explains the **anomalous time-evolution behavior** of $I(q, t)$ in the early stage of SD as already discussed in the upper part of Fig. 2.12a in comparison with the normal behavior shown in the upper part of Fig. 2.12b. ξ_{ve} for the system in Fig. 2.12a was found to be very large [15], as large as $\xi_{ve}/R_g = 15.6$ where R_g is the radius of gyration of PS548 in DOP. ξ_{ve} is a unique and important characteristic length of the system which depends on the dynamical properties: most importantly on the dynamical asymmetry parameter α_a (see Eqs. 2.4.3 and 2.4.22) and also on the zero-shear viscosity and the small q-limit of the Onsager coefficient $\Lambda(0)$.

It is important to note that the effects of dynamical asymmetry disappear in the case when $q \ll 1/\xi_{ve}$, or $r \gg \xi_{ev}$ or $t \gg \tau_{ve} \equiv \xi_{ve}^2 / D_{app}$ (τ_{ve} and D_{app} being the **viscoelastic time**, i.e., the relaxation time for the viscoelastic effects, and the cooperative diffusion coefficient, respectively). In other words, the dynamically asymmetric effects are insignificant in the hydrodynamic regime of a large space–time scale of observation.

Figure 2.14 presents the growth rate $R(q)$ versus q (part a) and $R(q)/q^2$ versus q^2 (part b) based on Eq. 2.4.21 [3, 15]. The experimental growth rate $R(q)$ was estimated by the best fit of the experimental results on $I(q, t)$ versus t presented in the upper part of Fig. 2.12a with theoretical expression given by

$$I(q, t) = I(q, \infty) + \left[I(q, 0) - I(q, \infty)\right] \exp\left[2R(q)t\right], \quad (2.4.26)$$

where $I(q, \infty)$ and $I(q, 0)$ are the virtual scattered intensity due to the random thermal noise and $I(q, t)$ at $t = 0$, respectively.

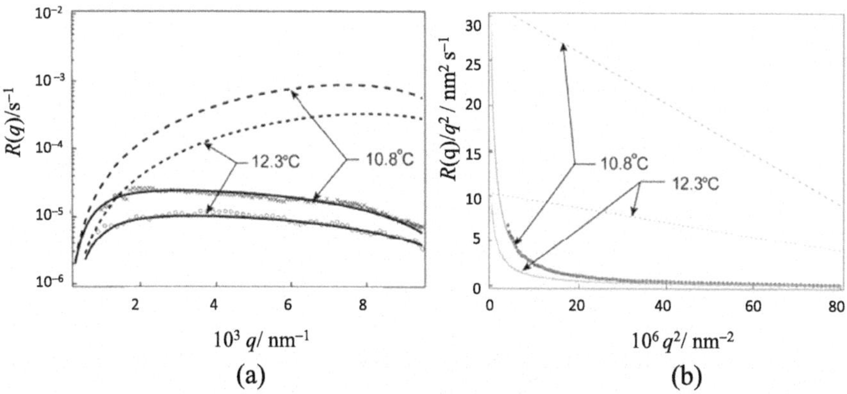

Fig. 2.14 Part (**a**) presents growth rate $R(q)$ versus q measured for the PS 548/DOP 6.0 wt. % with cloud point $T_{cl} = 13.8$ °C which are phase-separated at 12.3 °C (open circles) and at 10.8 °C (cross symbols). The solid lines are the best-fitted theoretical curves for the data at 10.8 °C and 12.3 °C by using Eq. 2.4.21 with $\xi_{ve} = 1.0$ μm and 0.9 μm, respectively, whereas the broken and dotted curves are fictitious ones with $\xi_{ve} = 0$ at 10.8 and 12.3 °C, respectively. Part (**b**) presents the Cahn's plot of $R(q)/q^2$ versus q^2. The symbols (open circles and crosses) as well as solid, broken, and dotted lines have the same meaning as in part a. Strong deviations of both experimental results shown by the symbols and the best-fitted theoretical results shown by solid lines from the fictitious ones of $\xi_{ve} = 0$ (broken and dotted straight lines), as appeared in part (**a**) and in the downward curves in part (**b**), are a consequence of a strong suppression of $R(q)$ at $q\xi_{ve} > 1$. Based on Hashimoto, T. 2008, In *Soft Matter Characterization*; Borsali, R., Pecora, R., Eds., Springer. New York, Chapter 8, pp. 378-462 [3]

The growth rate $R(q)$ was evaluated as a function of q by the nonlinear best fit of $\ln I(q, t)$ versus t based on the experimental data shown in Fig. 2.12a with $I(q, t)$ given by Eq. 2.4.26. $R(q)$ was evaluated at the phase separation temperature $T = 10.8$ °C (or at a quench depth $\Delta T = T_{cl} - T = 3.0$ °C, $T_{cl} = 13.8$ °C being the cloud point temperature), and the result was plotted in Fig. 2.14a (cross symbols) [15]. The figure also contains $R(q)$ versus q at $T = 12.3$ °C or ($\Delta T = 1.5$ °C) (open circles). The solid lines in Fig. 2.14a present the theoretical predictions best-fitted with the experimental results using Eq. 2.4.21 with $\xi_{ve} = 1.0$ μm and 0.9 μm for the PS/DOP system at 10.8 °C (the solid line fitted to cross symbols) and at 12.3 °C (the solid line fitted to open symbols), respectively. The best fittings were conducted by using ξ_{ve} and Λ_0 as adjustable parameters: ξ_{ve} affects the q dependence of $R(q)$ and Λ_0 affects the absolute value of $R(q)$. The figure also presents the fictitious cases of $\xi_{ve} = 0$, which ignored the viscoelastic suppression for $R(q)$, by the broken line at 10.8 °C and the dotted line at 12.3 °C. It is striking to note that a large suppression of $R(q)$ at $q > \xi_{ve}^{-1} \cong 1 \times 10^{-3} nm^{-1}$ compared with the case of $\xi_{ve} = 0$: the suppression tends to vanish as $q \to 0$, or approaching to the case of $\xi_{ve} = 0$. ξ_{ve} is as large as 1.0 μm and about 15 times larger than R_g, $\xi_{ve}/R_g \cong 15$. Owing to the large suppression of $R(q)$ at $q\xi_{ve} > 1$, the Cahn plot in Fig. 2.14b exhibits a strong downward deviation from the linear relationship between $R(q)/q^2$ versus q^2 expected for the fictions cases of $\xi_{\mathrm{ve}} = 0$ as shown by broken and dotted straight lines. This deviation is another indication of the strong suppression of $R(q)$ at $q\xi_{\mathrm{ve}} > 1$.

2.5 General Background on Dynamics of Concentration Flucutations under Shear Flow: Dynamically Symmetric Systems vs Dynamically Asymmetric Systems

There are two important parameters that characterize both dynamically symmetric and asymmetric systems: (i) the shear rate $\dot{\gamma}$ that one can externally vary and impose on the systems and (ii) the characteristic rate $R(q)$ inherent to the systems. $R(q)$ generally depends on the wave number q of the Fourier modes of concentration fluctuations or composition fluctuations and hence depends on the magnitude of the scattering vector q as predicted by Eq. 2.4.21. In the single-phase state of solutions and mixtures, $(\partial^2 f/\partial\phi^2)_0 > 0$, and thereby r_0 defined in Eq. 2.4.13 is negative: C_{g} is always positive (Eqs. 2.4.14 or 2.4.15). Consequently, $R(q)$ becomes negative. Hence, the thermally activated fluctuations decay with time. The decay rate $\Gamma(q)$ defined by

$$\Gamma(q) \equiv -R(q) \tag{2.5.1}$$

is given from Eq. (2.4.21) as follows:

$$\Gamma(q) = \Lambda(0)q^2\left(-r_0 + C_{\mathrm{g}}q^2\right)/\left(1 + q^2\xi_{\mathrm{ve}}^2\right) \tag{2.5.2}$$

In the case when $\xi_{\mathrm{ve}} = 0$, $\Gamma(q)$ is quadratic at $q \ll q_c$, but quartic at $q \gg q_c$ with respect to q, as indicated by the blue solid line in Fig. 2.15, where the crossover wave number q_c is given by,

$$q_{\mathrm{c}} = \left(-r_0/C_{\mathrm{g}}\right)^{1/2} = 1/\xi_{\mathrm{T}}, \tag{2.5.3}$$

and ξ_{T} is the thermal correlation length of the concentration fluctuations of the system. Using this q_{c}, the reduced relaxation rate $\tilde{\Gamma}(q)$ of the concentration fluctuations defined by the first line of Eq. 2.5.4 below is given as follows:

$$\tilde{\Gamma}(q) \equiv \frac{\Gamma(q)}{\Lambda(0)(-r_0)} = \frac{1}{1+\xi_{\mathrm{ve}}^2 q^2}\left(q^2 + \frac{q^4}{q_{\mathrm{c}}^2}\right). \tag{2.5.4}$$

Figure 2.15 presents $\tilde{\Gamma}(q)$ as a function of q.in the double logarithmic scale.

In the case of $\xi_{\mathrm{ve}} \neq 0$, $\Gamma(q)$ is quadratic with respect to q at $q\xi_{\mathrm{ve}} \ll 1$ (which naturally satisfies $q \ll q_{\mathrm{c}}$ as $\xi_{\mathrm{ve}} \gg \xi_{\mathrm{T}}$ or $q_{\mathrm{ve}} \ll q_{\mathrm{c}}$) as in the case of $\xi_{\mathrm{ve}} =$

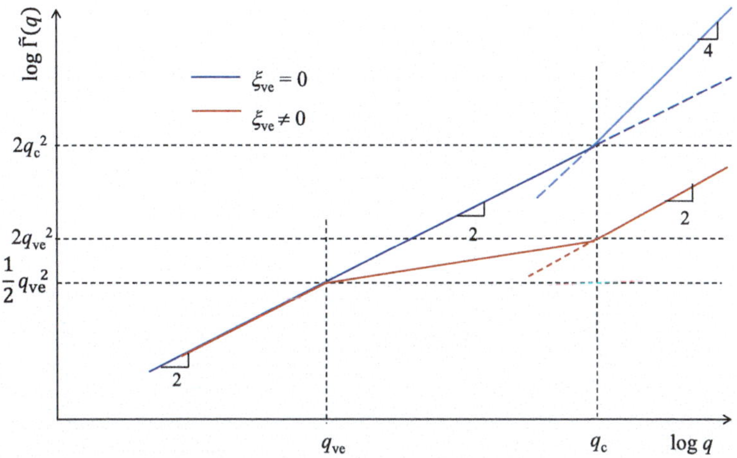

Fig. 2.15 Reduced relaxation rate $\tilde{\Gamma}(q)$ versus q, in the double logarithmic scale, for the concentration fluctuations of two-component systems with the dynamic symmetry ($\xi_{ve} = 0$ shown by the blue line) and the dynamic asymmetry ($\xi_{ve} \neq 0$ shown by the red line)

0. Furthermore, $\Gamma(q)$ is quadratic also with respect to q even at $q \gg q_c$ rather than being quartic as in the case of $\xi_{ve} = 0$, as indicated by the solid red line in Fig. 2.15. It is revealed that the viscoelastic effect ξ_{ve} gives rise to the increasingly large suppression of $\Gamma(q)$ with increased q at $q > q_{ve}$, where q_{ve} is defined by

$$q_{ve} \equiv 1/\xi_{ve}. \tag{2.5.5}$$

The larger the q value is, the faster is the decay rate $\Gamma(q)$ in all cases as depicted by the blue and red solid curves in Fig. 2.15.

For the shear flow with a given shear rate $\dot{\gamma}$, we define the q Fourier modes of the concentration fluctuations satisfying $\Gamma(q) = \dot{\gamma}$ as q_{shear}, then q-modes with q larger than q_{shear} satisfy

$$\Gamma(q) > \dot{\gamma} \text{ at } q > q_{shear}, \tag{2.5.6}$$

as shown in Fig. 1.6, so that these modes of fluctuations relax before the shear deforms the fluctuations, and hence they are not affected by the shear. However, the slow modes with $q < q_{shear}$ satisfy

$$\Gamma(q) < \dot{\gamma} \text{ at } q < q_{shear}, \tag{2.5.7}$$

so that these slow modes of fluctuations do not relax but rather stay under shear flow. Hence, they are deformed and affected by the shear.

In the case where the relaxation rate $\Gamma(q)$, given by Eq. (2.5.2), at the characteristic wave number q_c, $\Gamma(q = q_c)$, satisfies

$$\Gamma(q = q_c) < \dot{\gamma} \tag{2.5.8}$$

or $q_{\text{shear}} > q_c$, most of the concentration fluctuations in the system is affected by the shear flow, and we therefore define the shear as "**strong shear**" for this system. This q_c corresponds to the inverse of the thermal correlation length ξ_T (Eq. 2.5.3). In the case where

$$\Gamma(q = q_c) > \dot{\gamma} \tag{2.5.9}$$

or $q_{\text{shear}} < q_c$, the systems as a whole is not affected by the shear; only the Fourier modes of the small wave numbers q smaller than q_{shear} are affected by the shear. In this case, the shear is defined as "**weak shear**" for this system.

Unique points in polymers are anticipated to arise from $\Gamma(q)$ being very small compared with $\Gamma(q)$ for small molecules. Hence, even **very small shear rates affect polymer systems** compared to the case of small-molecule systems. In the dynamically asymmetric systems, the viscoelastic effect further suppresses $\Gamma(q)$ as elucidated in Fig. 2.15. Hence, the systems become even more sensitive to the small shear rates.

References

1. Silberberg, A. and Kuhn, W. (1952) *Nature*, **170**, 450.
2. Silberberg, A. and Kuhn, W. (1954) *J. Polym. Sci.*, **13**, 21.
3. Hashimoto, T. (2008) In *Soft Matter Characterization*, Borsali, R. and Pecora, R. (ed), Springer, New York, Chapter 8, pp. 378–462.
4. de Gennes, P.G. (1979) *Scaling Concept in Polymer Physics*, Cornell University, Ithaca.
5. Hashimoto, T. (2005) *Bull. Chem. Soc., Jpn.* **78**, 1.
6. Landau, L.D. and Lifshitz, E.M. (1964) *Statistical Physics*, Pergamon, New York.
7. Cahn, J.W. and Hilliard, J.E. (1958) *J. Chem. Phys.* **29**, 258.
8. Onuki, A. (2002) *Phase Transition Dynamics*, Cambridge Univ. Press., Cambridge, UK.
9. Tanaka, H. (1992) *Macromolecule*, **25**, 6377.
10. Yoshida, H., Ise, N., and Hashimoto, T. (1995) *J. Chem. Phys.*, **103**, 10146–10151.
11. Yamanaka, J., Yoshida, H., Koga, T., Ise, N., and Hashimoto, T. (1998) *Phys. Rev. Lett*, **80**, 5806–5089.
12. Yamanaka, J., Yoshida, H., Koga, T., Ise, N., and Hashimoto, T. (1999) *Langmuir*, **15**, 4198.
13. Derjaguin, B.V. and Landau, L. (1941) *Acta Physiochim.* **14**, 633.
14. Verwey, E.J.W. and Overkeek, J.Th.G. (1948) *Theory of the Stability of Lyophobic Colloids*, Elsevier, Amsterdam.
15. Toyoda, N., Takenaka, M., Saito, S., and Hashimoto (2001) T. *Polymer*, **42**, 9193.
16. Takenaka, M., Takeno, H., Hasegawa, H., Saito S., Hashimoto, T., and Nagao M. (2002) *Phys. Rev. E*, **65**, 021806–1.
17. Izumitani T. and Hashimoto, T. (1985) *J. Chem. Phys.*, **83**, 3694.
18. Cahn, J. W. (1965) *J. Chem. Phys.*, **42**, 93.
19. Hashimoto, T. (1993) In *Materials Science and Technology*, Cahn, R.W., Kramer, E.J., and Haasen, P. (ed), *Vol. 12, Structure and Properties of Polymers*, Thomas, E.L. (vol ed), VCH, Weinheim, Chapt. 6, p. 251–300.
20. Hashimoto, T. (1988) *Phase Transitions*, **12**, 47.

21. Takenaka, M., Izumitani, T., and Hashimoto, T. (1990) *J. Chem. Phys.*, **92**, 4566.
22. Doi, M. and Onuki, A. (1992) *J. Physique II*, 2, 1631.
23. Onuki, A. (1994) *J. Non-Crystalline Solids*, **172**, 1151.
24. Cook, H.E. (1970) *Acta Metall.*, **18**, 297.
25. Kawasaki, K. and Ohta, T. (1978) *Progr. Theor. Phys.*, **59**, 362.
26. Hashimoto, T. (2022) *Principles and Applications of X-ray, Light and Neutron Scattering*, Springer Nature, Singapore.

Chapter 3
Methodology

Abstract This chapter describes the methodology of shear-rheo-optics which enables *real-time* and *in situ* simultaneous measurements of shear and normal stress, optical microscopy (OM), small-angle light scattering (SALS), light transmittance, and birefringence for exploration of dissipative structures in mesoscopic length scale developed in the nonequilibrium systems. We paid a special attention to design a versatile apparatus which would permit to measure shear and normal stress ranging from very low to very large one, and SALS and OM also covering large dynamic ranges and to measure rheo-optical properties under continuous shear, oscillatory shear, and step-up and step–down shear. For this purpose, one of the most reliable rheometers commercially available is modified to install a SALS apparatus and an OM without sacrificing the high quality as a rheometer. Data acquisitions of SALS and OM can be synchronized with a given strain phase under oscillatory shear flows, which enables strain-phase dissolved measurements of SALS and OM. The chapter describes details of the rheometer, shear SALS, Shear OM, and temperature enclosure enabling a fine temperature control of specimens, both space-wise and time-wise, under cone-plate fixture, indispensable for investigation of critical phenomena under shear flow fields.

In this chapter, we describe the methodology of shear-rheo-optics which enables *real-time* and in situ exploration of dissipative structures in mesoscopic length scale developed in systems under applied shear flow together with shear stress and normal stress.

T. Hashimoto, *Polymer Phasing*, https://doi.org/10.1007/978-981-97-9648-9_3

3.1 Simultaneous Measurements of Stress, Optical Microscopy, Light Scattering, Light Transmittance, and Birefringence

3.1.1 Overall Features

We paid a special attention to design, as much as possible, a versatile apparatus which would permit *simultaneous measurements* of rheological properties, small-angle light scattering (SALS), and transmission optical microscopic (OM) images, birefringence, and transmitted light intensity. The instrument should be suitable for measuring shear and normal stress ranging from very low to very large one, and scattering intensity and optical contrast also covering large dynamic ranges. For this purpose, we modified one of the most reliable rheometers commercially available to install a light scattering apparatus and an optical microscope into the rheometer without sacrificing the high quality as a rheometer [1]. The apparatus measures in situ, *simultaneously* and *at real time,* a two-dimensional (2D) SALS pattern and its intensity distribution, an optical micrograph, birefringence, and transmitted light intensity, and rheological properties under continuous shear, oscillatory shear, and step-up and step–down shear.

Data acquisitions of SALS and OM can be synchronized with a given strain phase ϕ_s under a dynamic shear mode, which enables strain-phase dissolved SALS and OM measurements. Here ϕ_s is defined by $\phi_s \equiv \sin^{-1}(\gamma/\gamma_0)$, where γ is strain and γ_0 is strain amplitude. Charge coupled device (CCD) cameras are used as detectors for SALS and OM measurements, which enables us to detect quantitatively 2D SALS with a weak intensity level and OM images with a weak contrast level.

Figure 3.1 presents an overall look of the apparatus encompassed by white broken line (part A) in Fig. 3.1 [1, 2]. The apparatus is a modified version of the RMS-800® (commercially available in the past from Rheometric Scientific Inc., now from TA Inc.). We modified the rheometer to install a SALS apparatus encompassed by white dotted line (part B) in Fig. 3.1 and a transmission OM set encompassed by white dotted line (part C) in Fig. 3.1. Figure 3.2 presents a close-up of the main parts (B) and (C) shown in Fig. 3.1: parts (c) and (d) in Fig. 3.2 show a part of the SALS apparatus (B) in Fig. 3.1 on the right-hand side of a shear cell (part B); parts (c) and (e) to (g) in Fig. 3.2 show a part of the OM set (C) in Fig. 3.1 on the left-hand side of the shear cell. These parts (a) to (g) in Fig. 3.2 are more clearly shown in Fig. 3.3.

Figure 3.3 shows the scheme of the whole apparatus [1, 2]. The rheometer, the shear SALS, and the shear microscopy are controlled by using dedicated computers A5, B18, and C14, respectively. Additionally, another computer D1 is used for synchronization of the optical and scattering measurements with rheological measurements. The flow chart of data and that of control signals are shown by black and gray arrows, respectively.

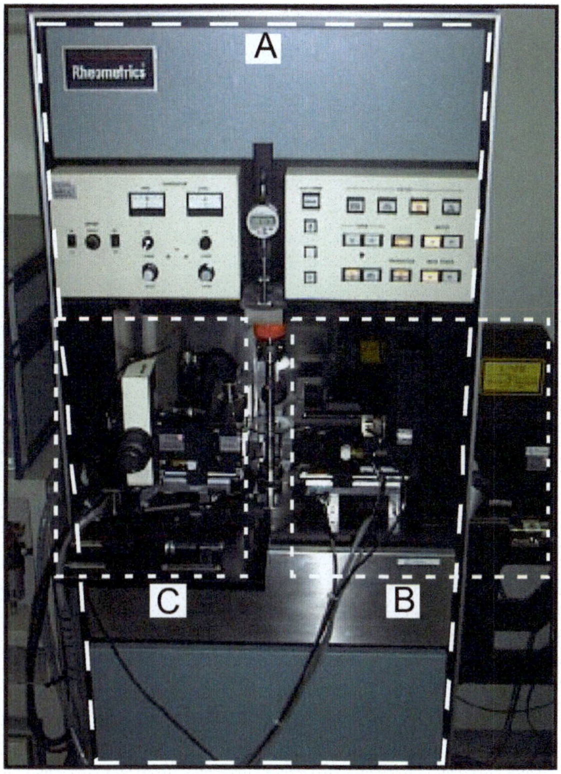

Fig. 3.1 Overall look of the shear-rheo-optical apparatus. (A) rheometer encompassed by a thick white broken line: (B) a part of light scattering apparatus, and (C) a part of optical microscope unit, both being encompassed by thin white broken lines. From Hashimoto [2]

3.1.2 Rheometer

A rheometer RMS-800 permits various kinds of rheological measurements, i.e., constant shear, oscillatory shear, constant stress, stress relaxation, and application of an arbitrary wave shape for the strain. The position of transducer (part a in Fig. 3.2, part A3 in Fig. 3.3) is modified in collaboration with Rheometric Scientific Inc. with a particular attention paid to keep its performance as a rheometer inherent to RMS 800; the transducer was set at a position higher than that for a normal RMS-800 by 110 mm to enable an expanded working space for the installation of the optical and scattering instruments. Two types of shear cell are employed: a parallel plate and a cone-and-plate (part *b* in Fig. 3.2, part A1 in Fig. 3.3) made of transparent quartz for the optical measurements. Their diameter is 80 mm, and the cone angle of the cone is 1.00°. The lower plate or cone is mounted on an actuator (A4) for an application of the strain in Fig. 3.3. The cone and plate are set in the lower and upper part of the shear cell, respectively, to reduce the effects of multiple reflections of the incident-light beam at the surfaces of the cone-plate on the 2D SALS patterns to tolerable intensity level; the shear cell is mounted on the transducer (A3), which allows measuring both torque and normal force. The capacity of the transducer is

Fig. 3.2 A close-up view of the main part of the shear-rheo-optical apparatus: **a** transducer, **b** a cone-plate fixture (or a parallel plate fixture); **c** temperature enclosure (two parts labeled *c* come close to each other and are connected into one by sliding each part on a set of parallel rails when it is actually used; **d** light scattering unit; **e** objective lens: **f** focusing microscope body; **g** CCD camera. From Hashimoto [2]

from 2×10^{-5} to 0.2 Nm for a torque measurement and from 2×10^{-2} to 20 N for a normal force measurement, respectively. Therefore, we can measure not only the viscosity of water ($\sim$ 1 mPa) but also polymer in bulk having a high shear modulus ($\sim 10^9$ Pa). However, we must be watchful of the machine compliance effect, which may be important for the shear storage modulus G′ at $G' > 10^4$ Pa at $\omega > 0.1$ rad/s with such a large diameter cone-and-plate.

3.1.3 Shear SALS

The whole optics of the shear-SALS instrument (part B in Figs. 3.1 and 3.3) together with one half of the temperature enclosure A2 for specimens in the shear cell in Fig. 3.3 is mounted on a set of parallel rails laid on the right-hand side of the rheometer to facilitate an alignment of the optics on the rails as shown in part B in Fig. 3.3: the optics can be slid in and out from the rheometer as shown in Part B in Figs. 3.1 and 3.3.

A 10 mW He–Ne laser (B1 in Fig. 3.3) is used as an incident light source (wavelength in vacuum $\lambda_0 = 632.8$ nm). A beam splitter (B2 in Fig. 3.3) is placed to monitor the intensity of the incident beam by a photodiode (B3 in Fig. 3.3). The incident laser beam can be attenuated by a set of neutral density (ND) filters (B4 in Fig. 3.3) with the transmitted intensity being adjustable in the range from 0.01% to

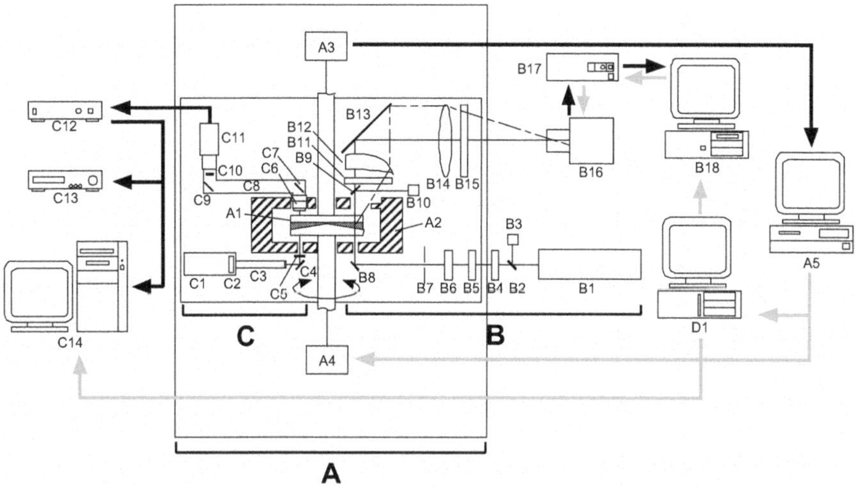

Fig. 3.3 Schematic illustration of the whole apparatus: (A) rheometer, (B) light scattering apparatus: (C) optical-microscope unit; (A1) a cone-plate fixture (or a parallel-plate fixture); (A2) temperature enclosure; (A3) transducer; (A4) actuator; (A5) personal computer for rheology measurement; (B1) He–Ne laser; (B2) half mirror; (B3) photodiode; (B4) ND filter, (B5) $1/4\lambda_0$ plate; (B6) polarizer; (B7) pinhole: (B8) mirror; (B9) incident beam trap; (B10) photodiode; (B11) analyzer; (B12) aspherical lens; (B13) mirror; (B14) biconvex lens; (B15) ND filter; (B16) cooled CCD camera; (B17) CCD camera controller; (B18) personal computer for light scattering measurement; (C1) halogen lamp; (C2) heat absorbing filter; (C3) optical fiber; (C4) mirror; (C5) polarizer; (C6) objective lens; (C7) mirror; (C8) focusing microscope body; (C9) mirror; (C10) analyzer; (C11) CCD camera; (C12) analog image processor; (C13) video tape recorder; (C14) personal computer with a video board for capturing micrographs; (D1) personal computer as a synchronizer. Black and gray arrows indicate the flow of data and that of the control signal, respectively. From Hashimoto [2]

100% of the intensity of the incident beam to avoid intense scattered beam coming into the detector. A $\lambda_0/4$ plate (B5 in Fig. 3.3) and a polarizer (B6 in Fig. 3.3) are installed to rotate a polarization direction of the polarized incident beam. After passing through a pinhole (B7 in Fig. 3.3) for eliminating stray light, the incident beam is reflected upwards by a mirror (B8 in Fig. 3.3), passes through the shear cell, and irradiates a sample between a cone-plate or a plate-plate fixture (A1 in Fig. 3.3). A small mirror (B9 in Fig. 3.3) is placed behind the fixture to block the transmitted direct beam from entering the detector (B16 in Fig. 3.3), which could produce artificial signals. Intensity of the transmitted beam through the specimen reflected by the mirror is also measured by a photodiode (B10 in Fig. 3.3), which allows in situ measurement of the intensity of the transmitted incident beam or turbidity of the sample. An analyzer (B11 in Fig. 3.3) is set to study polarized and depolarized light scattering intensity distributions.

Since the active surface area of a cooled CCD camera (B16 in Fig. 3.3) is small (13.82×13.82 mm^2), a focusing optical system is mounted in the SALS instrument. This system consists of two lenses as the condenser for the scattered light. The first aspherical lens (B12 in Fig. 3.3) together with the second biconvex lens (B14 in

Fig. 3.3), placed behind the analyzer (B11 in Fig. 3.3), collimate the scattered light onto CCD camera B16 (in Fig. 3.3). Because of the space limitation, the aspherical lens is cut near the middle, and a large flat mirror (B13 in Fig. 3.3) is mounted between two lenses at a 45 °C angle to the incident beam. A neutral density filter (B15 in Fig. 3.3) is available for attenuation of the scattered light in the range from 0.01 to 100% of the original unattenuated scattered intensity.

Focusing a scattering pattern directly on the cooled CCD camera (B16 in Fig. 3.3) having a sufficiently large dynamic range (16-bit) enables scattered intensity distributions of a 2D scattering pattern to be quantitatively recorded. The cooled CCD camera has an active surface area of $13.82 \times 13.82\ \text{mm}^2$ with 512×512 pixels ($27 \times 27 \mu\text{m}^2$/pixel). It is possible to set any rectangular areas of the CCD chip for the data acquisition, which enables a rapid data acquisition and a reduction of the data size. We usually use an area of 280×512 pixels, because the half-size aspherical lens is used. It is also possible to divide the accessible CCD chip area by binding a rectangular data area comprised of xy pixels as a unit for data acquisition to reduce data size, where x and y are any integers satisfying $1 \leq x, y \leq 512$.

The minimum exposure time of the detector is 50 ms, and the minimum acquisition time for one measurement consisting of 280×512 pixels is 2.0 s. Therefore, a time resolution between consecutive measurements is about 2.0 s, which is sufficient for kinetic studies of phase separation behavior in polymer melts and solutions. It is possible to take data with a shorter time resolution by accessing a smaller CCD chip area, for example accessing one-dimensional (1D) area parallel or perpendicular to the flow direction, and/or by binding the pixels as described above.

A CCD camera controller (B17 in Fig. 3.3) is connected to a computer (B18 in Fig. 3.3), which allows various modes of measurements for both steady and dynamic shear flow as follows: a measurement with a fixed time interval, a measurement with a triggered signal, and one with a programmed timetable.

3.1.4 Shear Microscopy

We designed a transmission optical microscope (part C in Figs. 3.1 and 3.3) specific to the rheometer. The whole optics together with another half of the temperature enclosure is mounted on a set of parallel rails, laid on the left-hand side of the rheometer, similarly to the shear-SALS instrument. The optics (c7–c11 in Fig. 3.3) and the half of the temperature enclosure (A7 in Fig. 3.3) can be slid in and out from the rheometer on the rail. The temperature enclosures linked with the optics for SALS and OM are slid in on the set of parallel rails and fixed into one when the simultaneous experiments are performed. A halogen lamp (C1 in Fig. 3.3) (12 V 100 W, illuminance of 30,000 lx) is used as a light source. The incident beam from the lamp passes through a flexible optical fiber (C3 in Fig. 3.3) and is reflected up into the sample by the mirror (C4 in Fig. 3.3). A heat-absorbing filter (C2 in Fig. 3.3) is set in the halogen lamp unit to avoid heating the sample with the intense incident light beam. Three kinds of long-focus type objective lenses (C6 in Fig. 3.3)

(10×, 20×, and 50×) are available for the observation of submicron-size structures. A polarizer (C5 in Fig. 3.3) can be set above the mirror (C4 in Fig. 3.3) and an analyzer (C10 in Fig. 3.3) in the focusing microscope body (C8 in Fig. 3.3), which makes observations under various polarization conditions possible.

Because of a limitation of the working space available to the rheometer, a focusing microscope body (C8 in Fig. 3.3) is bent toward a CCD camera (C11 in Fig. 3.3). In the focusing microscope body, the two mirrors (C7 and C9 in Fig. 3.3) are set at 45° against the objective lens to bend the light-beam path. The CCD camera (C11 in Fig. 3.3) is connected to an analogue image processing unit (C12 in Fig. 3.3). We should note that the CCD camera is not the same as the one (B16 in Fig. 3.3) used for the shear SALS. The C12 can perform the real-time-image processing for the obtained micrographs, e.g., contrast enhancement and background subtraction, etc. Therefore, we can easily take a high contrast micrograph even when structures in a sample themselves had a weak contrast due to a small difference in refractive indices between the structures of interest and the surrounding medium.

The unit C12 is connected in parallel to a video tape recorder (VTR) (C13 in Fig. 3.3) for continuous analogue recording and to a computer (C14 in Fig. 3.3) with a video board for digital image capturing with 640 × 80 pixels and 8-bit resolution. The public domain NIH image program enables us to conduct not only automatic capturing of micrographs but also digital image analyses for these micrographs. The shortest data acquisition time per frame with 640 × 80 pixels is 1.0 s, which is short enough for observations of kinetic processes in polymeric systems under the shear flow. Since the objects under a continuous shear flow quickly move away from the camera window, an exposure time of the CCD camera should be short enough to take snapshot images. Therefore, we set 0.1 ms for the shutter speed of the CCD, during the time of which an object moves at most 5 μm in the case of the sample thickness of 0.5 mm and shear rate $\dot{\gamma}$ of 100 s^{-1}. We perform image analyses and digital image processing on thus saved micrographs by using the NIH image and Adobe Photoshop 4.0 J.

3.1.5 Temperature Enclosure

We constructed a temperature enclosure (part c in Fig. 3.2. and A2 in Fig. 3.3) which has two windows W1 and W2, one in the left-hand side for the shear microscopy and the other in the right-hand side for the shear SALS, respectively, to allow the simultaneous observation of both SALS patterns and OM images as shown in Fig. 3.4. Figure 3.4 shows a cross-sectional view of the temperature enclosure cut through the center of the shafts holding the corn-and-plate [1, 2]. Its design is basically the same as that used previously for a shear-SALS apparatus [4, 5], except for the new features associated with the two windows (W1 and W2 in Fig. 3.4) and minor changes in dimensions.

The enclosure has a double layer structure in which temperature-controlled dry air circulates as shown by the arrows. The temperature-controlled dry air-flow enters

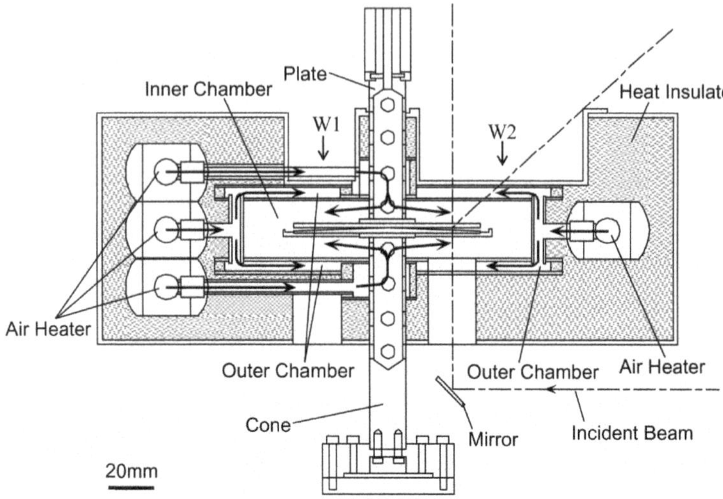

Fig. 3.4 Cross section of a temperature enclosure. Arrows indicate flow of temperature-controlled air. From Hashimoto [2]

the outer and inner chambers of the enclosure. The flow of temperature-controlled air in the outer chamber shields the inner sample chamber from an ambient temperature change outside the enclosure. It also goes through the shaft holding the cone-and-plate (or parallel plates) and then re-enters into the inner chamber through the holes made in the shaft as shown by the arrow to reduce a temperature gradient along radial direction of the cone-and-plate. The temperature control of the outer chamber by the temperature-controlled air flow, as shown by the straight arrows, enhances precise temperature control of the cone-and-plate shear cell.

A cooling system is also available for the measurements below room temperature. It cools dry air before entering the temperature enclosure. The experimental temperature range attained by the heating and cooling system is from 0 to 200 °C with an accuracy of ±0.3 °C. The temperature variation along the radial direction of the shear cell was measured to be less than ±0.3 °C.

The uniformity of temperature over the shear cell was checked by a set of thermo-sensors embedded in the various places of the "test" upper plate of the shear cell: the test upper cell is identical to the upper cell used for the shear-SALS and OM experiments except for the embedded thermo-sensors. Such temperature accuracy is essential to investigate the critical phenomena of the mixtures under the flow with respect to the SALS, which are much more sensitive to temperature than that for conventional rheological properties of systems: the SALS intensity at$q = 0$ discontinuously changes (or diverges) with temperature, whereas the rheological properties continuously change with temperature, through the critical temperature for the phase transition between the single-phase and two-phase (Figs. 4.15, 4.18 and 4.22 in Sect. 4.3.7).

3.1.6 Transmittance of Incident Light and Birefringence

The turbidity, of the systems or the transmittance of the incident beam for the systems subjected to the shear flow can be measured in situ by a photodiode (B10) in Fig. 3.3 as described already in Sect. 3.1.3. Birefringence can be also measured by detecting transmitted light intensity with the CCD camera (C11 in Fig. 3.3) under crossed polarizer (C5 in Fig. 3.3) and analyzer (C10 in Fig. 3.3), with their polarization directions oriented at $+45°$ and $-45°$ with respect to the flow direction. A compensator should be inserted in between C11 and C10 in Fig. 3.3 to convert the transmitted intensity into birefringence, although it is not included in this figure for brevity.

It should be noted that the construction of the apparatuses becomes much simpler when one wants to measure only the shear SALS and/or the shear optical microscopy [4]. A one-dimensional photo-diode array may be also used for quantitative measurements of light scattered intensity distributions from systems under shear flow [3].

3.1.7 Calibrations and Corrections

We briefly describe here various calibrations and corrections that need to be done for the shear-rheo-optical apparatus: the details are referred to the original article [1].

The rheometer was calibrated according to the specifications of the manufacturer. The shear-SALS instrument with the cooled CCD camera was calibrated in terms of both the scattering angle and the scattering intensity. The scattering angle versus the channel number of the CCD was calibrated by measuring diffraction patterns from the two kinds of diffraction gratings (500 and 1000 lines/inch). The range of the scattering angle θ detected by the CCD and the resolutions of θ, $\Delta\theta$, were determined to be $2.3° \leq \theta \leq 22.0°$ and $\Delta\theta = 0.11°$ in arc, respectively. The scattered intensity measured with the CCD was calibrated for a variation of the solid angle subtended by each channel and for the detector sensitivity for each channel. For the latter calibration we used fluorescence intensity, having no θ-dependence, which is emitted from an aqueous solution of ethylene blue with the concentration of 10^{-5} mol/l when the solution is irradiated by He–Ne laser ($\lambda_0 = 632.8$nm). Through the calibration process, one can also confirm the effects of spherical aberration of the lens, if there are any.

3.1.8 Synchronization of Light Scattering Measurements and Optical Microscope Imaging with Shear-Strain Phase under Oscillatory Shear Flow

Under a dynamic shear mode in the strong shear regime, dissipative structures evolved may change with strain phase ϕ_s. In such a case, it is necessary to synchronize the data acquisition for both the scattering and the microscopy with ϕ_s. For this purpose, we use a computer as a synchronizer which takes the electronic signal of ϕ_s from the control unit of the rheometer. The computer sends a triggering signal for data acquisition for both the scattering and the optical microscopy at a specified ϕ_s and over the specified phase interval $\Delta\phi_s$. The detailed calibration and correction methods required for the synchronized data acquisition should be referred to the original reference [1].

3.2 Examples: Simultaneous Measurements of Stress, Shear SALS, and Shear Microscopy

In most of complex liquids to be dealt with here, the structures and the rheological properties of the liquids are generally very sensitive to the thermal and shear histories imposed on the liquids. Therefore, it is desired to conduct truly simultaneous measurements of the various properties and structures *in situ* and *at real time* for a given system. By so doing, one can avoid complications that are brought about by the fact that the measured various quantities reflect the different states of the systems due to the different thermal and/or shear histories. We shall demonstrate below examples of the simultaneous measurements.

3.2.1 Continuous Shear Mode: Shear-Induced Two-Phase to Single-Phase Liquid–Liquid Phase Transition

The system to be discussed here was described earlier in Sect. 2.1 and showed the disappearance of the strong scattering with increasing the shear rate above the critical shear rate $\dot{\gamma}_{c,\mathrm{single}}$ for the single-phase formation (from part a to part b in Fig. 2.2). Figure 3.5 presents the $\dot{\gamma}$ dependence of viscosity η measured by the apparatus described in Sect. 3.1 [1, 2]. The figure also includes steady-state shear viscosity η measured by a commercially available rheometer ARES-FS® (Rheometric Scientific Inc.) dedicated only to rheology experiments for comparison. It is clearly confirmed that the two data sets are identical within the error of $\pm 5\%$, indicating that the measurement of the rheological properties works well with this rheo-optical apparatus. Thus, this apparatus really enables us to measure simultaneously rheological properties, SALS patterns and OM images even for systems having small stress-levels, weak scattering intensity and weak OM image-contrast, respectively as characterized by the weak shear stress of the system investigated here.

Figure 3.6 presents steady-state light scattering patterns simultaneously measured for the same solution as that in Fig. 3.5 as a function of $\dot{\gamma}$ [1, 2]. Figure 3.7 presents the steady-state transmission optical microscope images [1, 2] measured simultaneously with the scattering patterns shown in Fig. 3.6 and the rheology shown in Fig. 3.5. It is important to confirm whether the obtained microscope images truly reflect the structures developed under the shear field. For this purpose, we calculated the 2D fast Fourier transformation (FFT) spectra from the OM images shown in Fig. 3.7. The results revealed that the $\dot{\gamma}$ dependence of the FFT spectrum is essentially consistent with that of the corresponding scattering patterns in Fig. 3.6 [1]. Hence, we conclude that the real-space images in Fig. 3.7 captured by the microscopy truly reflect the phase-separated structures under the shear flow.

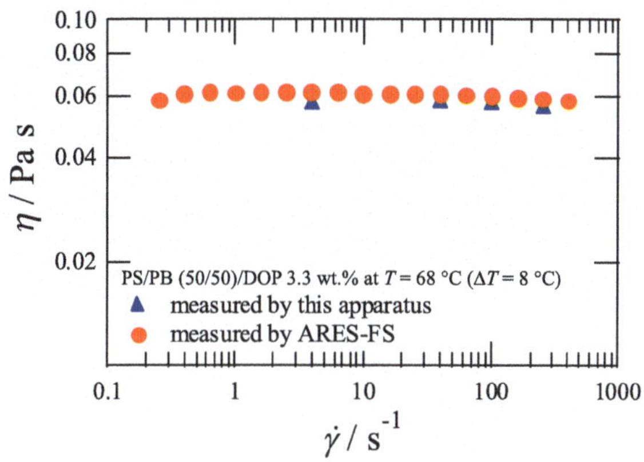

Fig. 3.5 Steady-state shear viscosity η of PS/PB (50:50)/DOP 3.3 wt.% measured by the rheo-optical apparatus described in Sect. 3.1. The figure also includes the data obtained with ARES-PS®. Based on Hashimoto [2]

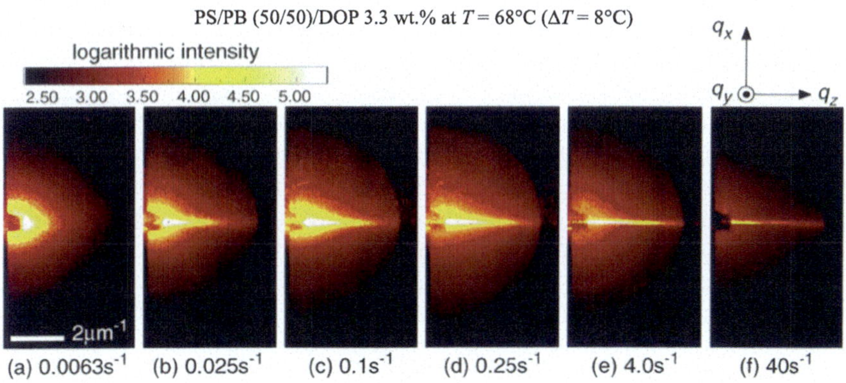

Fig. 3.6 Shear-SALS pattern of PS/PB (50:50)/DOP 3.3 wt.% at steady state measured simultaneously with the rheology (Fig. 3.5) and the optical microscopy (Fig. 3.7). Based on Hashimoto [2]

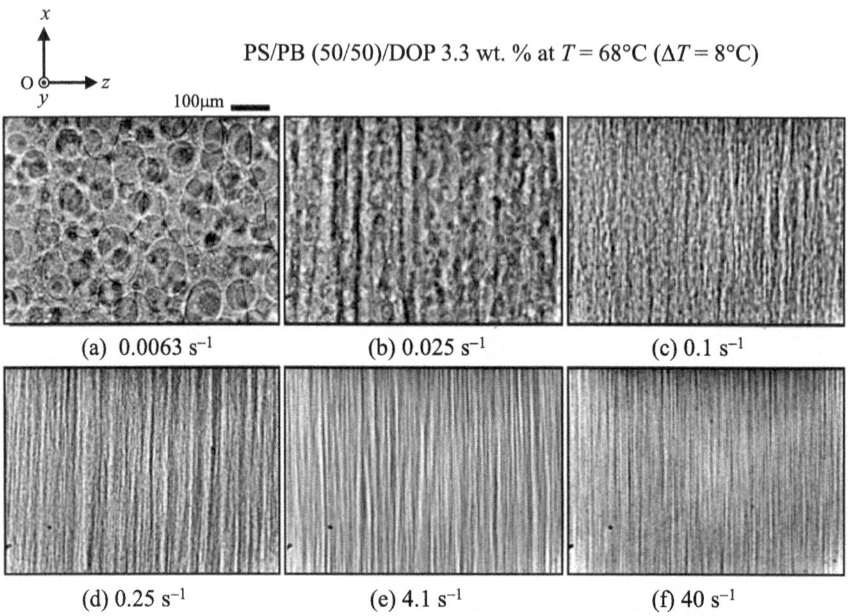

Fig. 3.7 Transmission optical micrographs showing steady-state structures of PS/PB (50:50)/DOP 3.3 wt.% obtained at various shear rates, which are simultaneously measured with the scattering patterns shown in Fig. 3.6 and the viscosity data shown in Fig. 3.5. Based on Hashimoto [2]

Figure 3.7 shows the large and nearly round droplets existing at the weak shear rate of $\dot{\gamma} = 0.0063\ \text{s}^{-1}$ (part a). They tend to become smaller droplets elongated and aligned parallel to the flow direction (e.g., part b at $\dot{\gamma} = 0.025\ \text{s}^{-1}$). They are percolated ($\dot{\gamma} = 0.1 - 0.25\ \text{s}^{-1}$) (parts c and d) and eventually transformed into stable strings with almost macroscopic dimension along the flow direction ($\dot{\gamma} = 4.0 - 40\ \text{s}^{-1}$) (parts e and f). The real-space images thus captured as a function of $\dot{\gamma}$ is consistent with the shear-rate dependence of the SALS, as will be detailed in Sect. 4.3. Although not shown in Fig. 3.6, when $\dot{\gamma}$ is increased above $\dot{\gamma}_{\text{c,single}}$, the streak-like scattering pattern oriented along q_z shown in Fig. 3.6f disappears as shown in Fig. 2.2b. The string-like structure elucidated by the OM image shown in Fig. 3.7f also disappears at $\dot{\gamma} > \dot{\gamma}_{\text{c,single}}$. Since the interfaces of the domains align parallel to the flow direction in the range of the shear rates from 0.25 to 40 s^{-1} as shown in Fig. 3.7d to Fig. 3.7f, the interface does not contribute to the shear viscosity, giving rise to the constant value of η in this range of the shear rates as shown in Fig. 3.5, i.e., η is predicted by a weighted average of a homogeneous solution of PS/DOP and that of PB/DOP with the same concentration as the blend solution.

3.2.2 *Oscillatory Shear Mode: Shear-Induced Concentration Fluctuations and Phase Separation*

The system to be discussed here was described also in Sect. 2.1. The single-phase solution of PS548/DOP 6.0 wt.% exhibited a strong butterfly-type scattering pattern with increasing $\dot{\gamma}$ when $\dot{\gamma}$ exceeds the critical shear rate $\dot{\gamma}_{c,x}$, as shown in the change of the scattering pattern from Fig. 2.2c to Fig. 2.2d for UHMWPS/DOP 2.0 wt.%, because of the shear-enhanced concentration fluctuations. Here, we extend the problem to the shear-enhanced concentration fluctuations under the dynamic shear mode.

Figure 3.8 presents a Lissajous curve, the shear stress σ against the shear strain γ, obtained with the rheo-optical apparatus described in Sect. 3.1 [1, 2]. The figure indicates a nonlinear rheological response. The reason for the nonlinear behavior is the large strain amplitude γ_0 of 4.2, which brings about shear-induced phase separation or concentration fluctuations, to be clarified in Chap. 5, and its relaxation, during a given cycle of strain. The structural changes in the solution in the given cycle should contribute to the nonlinear response of σ with γ.

Figure 3.9 shows the strain phase ϕ_s dependence of the scattering pattern [1, 2] measured simultaneous with the results shown in Fig. 3.8. At a first glance, we can clearly observe the butterfly patterns which show the characteristics of the shear-induced phase separation or concentration fluctuations. Thus, the dynamic shear flow under a given set of (γ_0, ω, T) creates the shear-induced structures as in the case of the steady shear flow. We also recognize that the intensity of the SALS pattern changes in harmony with ϕ_s. We calculated the integrated scattered intensity along

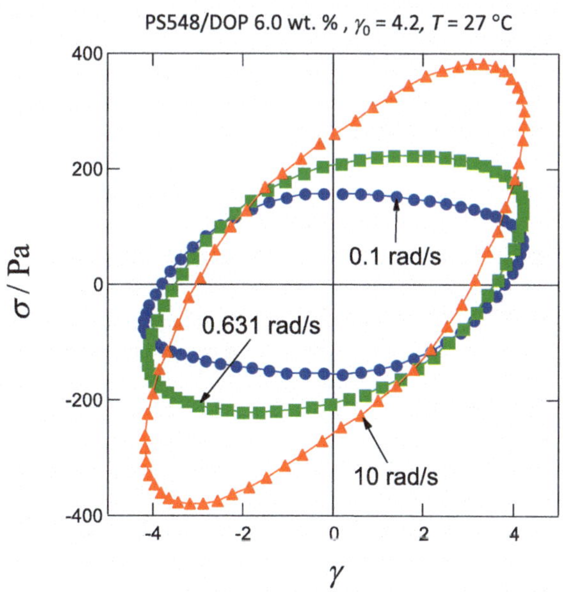

Fig. 3.8 Lissajous figures of shear stress σ versus strain γ for PS548/DOP 6.0 wt.% obtained at 27 °C at various ω and at a given strain amplitude $\gamma_0 = 4.2$ by using the shear-rheo-optical apparatus. Based on Hashimoto [2]

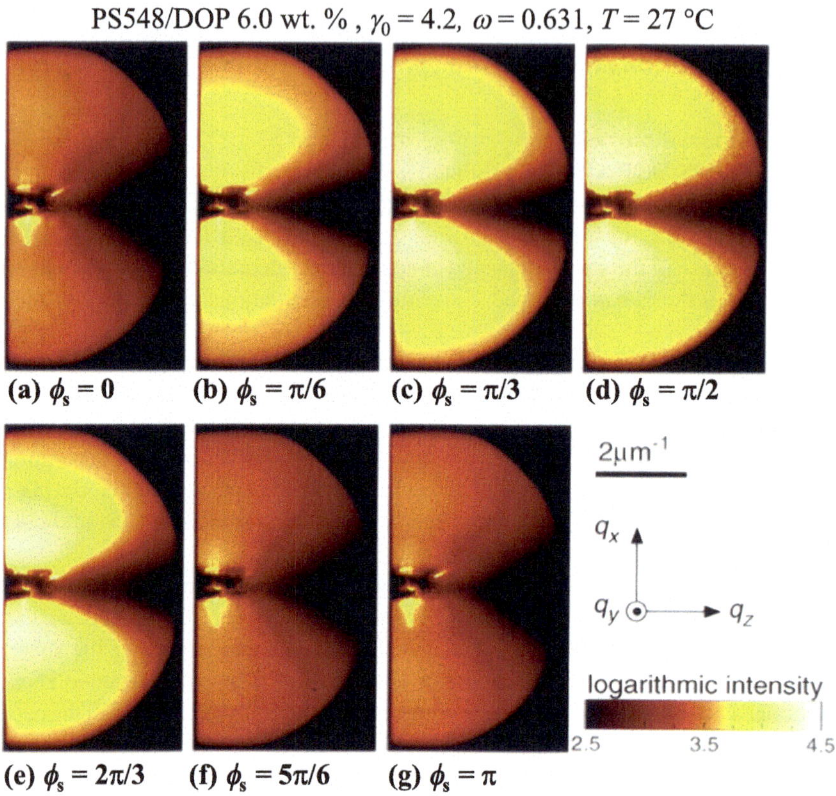

Fig. 3.9 SALS patterns at various strain phases ϕ_s for PS/DOP 6.0 wt.% under the dynamic shear mode of $\gamma_0 = 4.2$ and $\omega = 0.631$ rad/s at 27 °C Based on Hashimoto [2]

the q_x axis, $\mathcal{J}(q_x)$, to quantitatively analyze the ϕ_s dependence of scattering intensity, where $\mathcal{J}(q_x)$ is defined by.

$$\mathcal{J}(q_x) \equiv \int_{q_1}^{q_2} I(q_x, q_y = 0, q_z = 0)\, dq_x. \tag{3.2.1}$$

Here, $I\ (q_x, q_y = 0, q_z = 0)$ is the scattered intensity along Ox-axis measured over the lower and upper limit of $q_1 = 0.47\ \mu\text{m}^{-1}$ and $q_2 = 3.36\ \mu\text{m}^{-1}$.

Figure 3.10 presents $\mathcal{J}(q_x)$ as a function of ϕ_s, indicating clearly a nonlinear response of $\mathcal{J}(q_x)$ with ϕ_s [1, 2]. As ϕ_s increases, $\mathcal{J}(q_x)$ gradually increases in the region from $\phi_s = 0$ to $9\pi/16$ and decreases more quickly in the region from $\phi_s = 9\pi/16$ to $15\pi/16$. This indicates that the concentration fluctuations induced by the shear flow gradually grow with an increase of $|\gamma|$, and they decay quickly with the decrease of $|\gamma|$, although a small phase lag exists with respect to $|\gamma|$. The quick decrease of the scattered intensity indicates a quick elastic recoil of the stretched

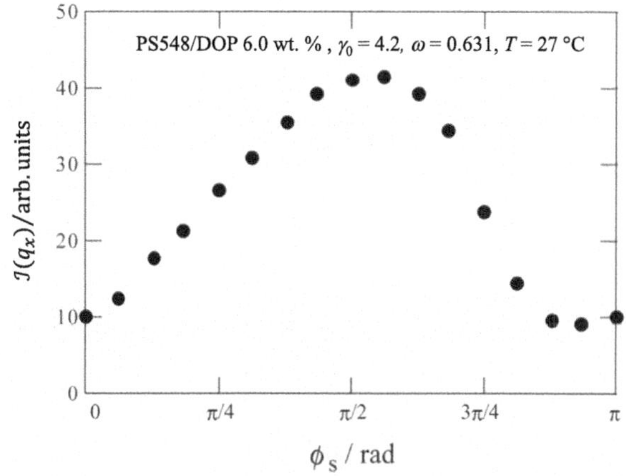

Fig. 3.10 Integrated scattering intensity along q_x, $\mathcal{J}(q_x)$, against ϕ_s under the same dynamic shear mode of $\gamma_0 = 4.2$ and $\omega = 0.631$ rad/s at 27 °C as that in Fig. 3.9. Based on Hashimoto [2]

entangled networks and a quick solvent back flow (counter-squeezing effect of solvents) driven by osmotic pressure as will be clarified in Chap. 5. Thus, the dynamic shear flow induces the oscillation of the concentration fluctuations in harmony with ϕ_s. It illuminates an intriguing physical problem on how the nonlinearity in σ is related to the nonlinearity in the scattering response. This problem has to be solved in future.

3.2.3 Remarks

Up to this point in this section, we described the in situ observation of scattering patterns and microscope images in Oxz plane both under the continuous and oscillatory shear mode. We like to point out an apparatus which enables the in situ observation of the scattering in Oxy plane under the oscillatory shear mode. The apparatus shown below uses a shear-sandwich mode for small-angle X-ray scattering (SAXS) [7]. The sample surface can be rotated by 90° around the x-axis for the observation of the SAXS pattern in Oxz plane.

Figure 3.11 presents a layout of the dynamic SAXS (DSAXS) system. The DSAXS system comprises a stage-mounted imaging-plate system and a hydraulic sample-deformation device. Details concerning the sample-deformation device have been given elsewhere [8].

The sample can be uniaxially elongated or subjected to the shear deformation. Figure 3.11 shows the latter case. In either case, the displacement of the actuator determines the macroscopic strain on the sample. The maximum load and maximum displacement are 490 N and ± 7.5 mm, respectively, and the operated linear frequency is between 0 and 60 Hz. The load on the sample is detected by a load cell, and the displacement of the actuator is measured using a linear variable differential transformer (LVDT). The temperature of the sample can be controlled at a constant

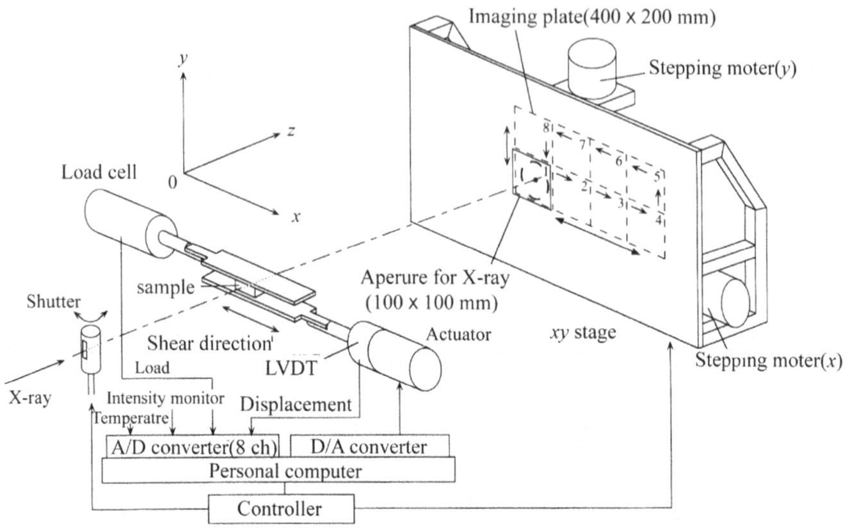

Fig. 3.11 Schematic diagram of the DSAXS system consisting of a stage-mounted imaging-plate detector for the 2D SAXS pattern and a hydraulic sample-deformation device, both of which are controlled by a personal computer. From Hashimoto [2]

value between 173 and 573 K by using a sample chamber made of a copper block with heating elements and a channel for the gas flow from liquid nitrogen.

The imaging plate as a two-dimensional detector for the time-resolved measurements is inserted into its cassette and is mounted on an *xy* stage which is driven by two stepping motors through an intelligent stepping-motor controller. The *xy* stage is made of one set of conventional translation stages which are placed perpendicularly to each other. There is a 100×100 mm^2 square aperture for the scattered *X*-rays in front of the imaging-plate stage. The imaging plate (400×200 mm^2) divided into eight sections (100×100 mm^2) is used for detecting the SAXS patterns. Each section is moved in turn to the position of the aperture to record the 2DSAXS pattern as shown in Fig. 3.12. The time required for the translational movement of the imaging-plate stage along O*x*-axis or O*y*-axis is *ca* 0.5 s/100 mm. Therefore, time-resolved measurements up to eight time slices are possible in a time scale shorter than 1 s per frame. The total number of frames for the two-dimensional SAXS patterns per one imaging plate can be increased, and the dead time between the successive measurements can be shortened accordingly by using a smaller aperture and a shorter sample-to-detector length at the expense of the spatial resolution of the *X*-ray scattering patterns. For the details of the controlled system, including the SAXS data acquisition with the strain phase by using a personal computer, reference is made to the original paper [7].

Figure 3.12 demonstrates an example of the synchrotron dynamic SAXS (SR-DSAXS) for a polystyrene-*block*-poly(ethylene-*alt*-propylene) copolymer having number-average molecular weight (M_n) of 3.4×10^4, heterogeneity index (M_w/M_n) of 1.3, and PS volume fraction in the copolymer of 0.103 [2, 6]. The copolymer has the spherical microdomains comprising PS block chains in the matrix of

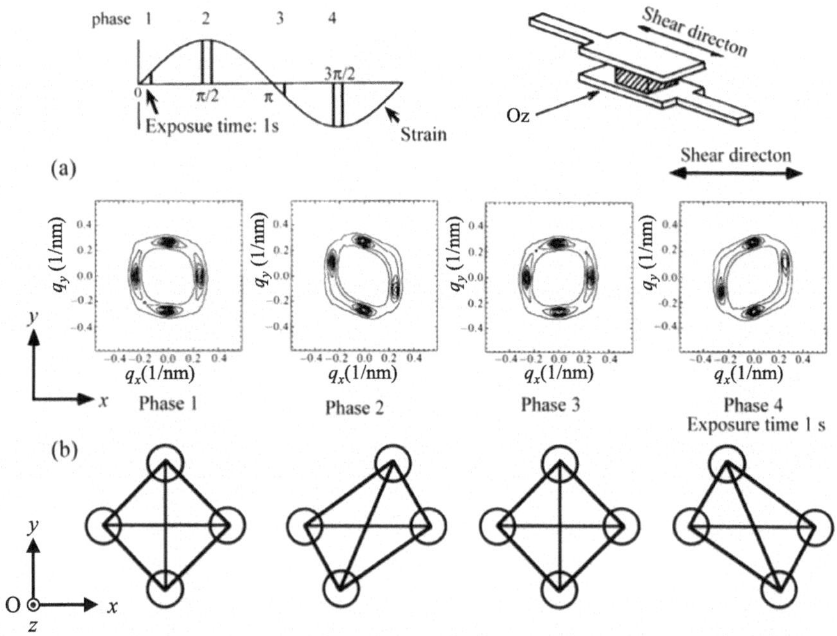

Fig. 3.12 Part **a** presents 2D SAXS patterns obtained at the four strain phases ϕ_s, defined by the top left corner of the figure, under the dynamic shear strain mode, presented by top right corner of the figure, where the propagation direction of the incident *X*-ray beam and shear direction are set along O*z*- and O*x*-axes, respectively. Part **b** presents the lattice-deformation model to explain the four diffraction spots from (110) and $(\bar{1}10)$ planes (**b**). The *X*-ray exposure time was 1 s for each pattern. The pattern at each strain phase is taken for the strain-phase interval of 0.03π and cumulated over a strain cycle of 110. From Hashimoto [2]

poly(ethylene-*alt*-propylene) (PEP) block chains. The amplitude of the dynamic shear strain is 50%, the static strain is 0%, and the angular frequency is 0.0944 rad s^{-1}. The measurements were carried out at room temperature where the PS spheres and the PEP matrix are in a glassy and rubbery state, respectively.

The SAXS profile before the shear deformation shows that the PS spherical microdomains, comprised of PS block chains, of average radius $R = 7.1$nm are in body-centered-cubic lattice (bcc) with the Bragg spacing of 21.2 nm and with the *g*-factor characterizing the paracrystalline lattice distortion of $0.09 \leq g \leq 0.13$ [8–10].

Upon imposing the oscillatory shear deformation, (110) and $(\bar{1}10)$ planes orient parallel and perpendicular to O*xz* plane, respectively, resulting in the four-point SAXS pattern, as shown in part (a) and part (b) at Phase 1 (the left edge of the figures), respectively. The degree of orientation of the lattice planes increases with the strain cycle. Figure 3.12a shows the SAXS patterns at the four strain phases ϕ_s of 0, π/2, π, and 3π/2, designated as phase 1 to phase 4, respectively. These SAXS patterns were taken with an exposure time of 1 s for the *X*-ray and at the 110th strain

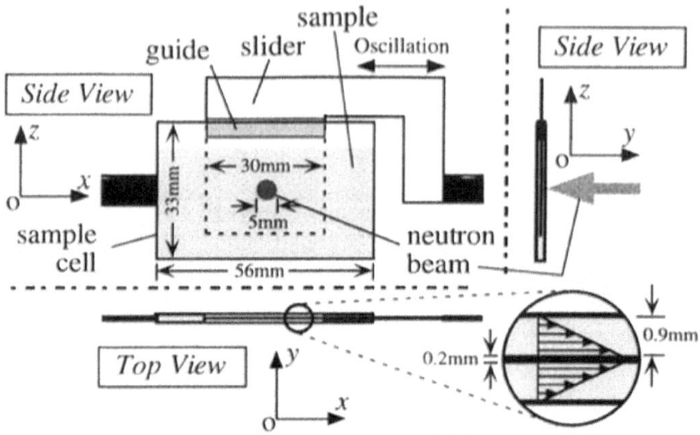

Fig. 3.13 Schematic diagram of the shear cell used for 2D SANS. The incident neutron beam is sent along the velocity gradient direction (along O*y*-axis), the velocity gradient exists in O*xz* plane, and O*z*-axis is neutral direction. The 2D detector is set normal to O*y*-axis in O*xz* plane. From [12], reprinted with permission from American Chemical Society

cycle after imposing the oscillatory strain, at which the degree of (110) orientation parallel to O*xz* plane reached a steady value. Figure 3.12b shows a model used to describe the response of the lattice deformation at each strain phase to explain the four diffraction spots appearing in the 2DSAXS patterns (part a). For more details, the reader is referred to the original papers [9, 11].

Figure 3.13 presents a schematic diagram of a sandwich-type shear cell used for SANS [12]. This shear cell imposes the oscillatory shear flow on the solution by moving the slider plate at the center of the fixed outer cell with the surfaces of the slider and the outer cell parallel to O*xz* plane. The motion of the slider along O*x*-axis is driven by the servo-controlled hydraulic deformation apparatus shown in Fig. 3.11. The cell is made of copper having a thickness of 0.2 mm. The incident neutron beam is irradiated along the velocity gradient direction parallel to O*y*-axis and the 2DSANS is observed in O*xz* plane as in the case of the 2D SALS patterns (Fig. 3.9). The phase-resolved data acquisition is performed using the same data acquisition system as described in Sect. 3.2.2.

Figure 3.14 presents a schematic diagram for the Couette shear cell made of quartz to impose the steady-state shear flow for the SANS experiments [13, 14]. The incident neutron beam is irradiated along the velocity gradient direction parallel to O*y*-axis, which enables us to observe the 2DSANS patterns in the same plane as the 2D SALS patterns, i.e., O*xz* plane (Figs. 1.4, 2.2, 3.6, and 3.9).

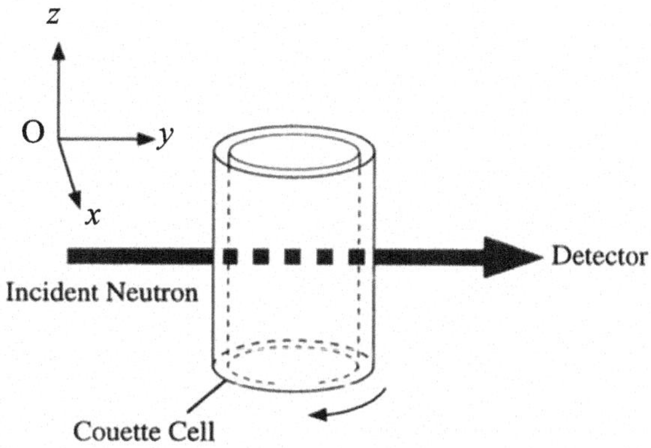

Fig. 3.14 Schematic diagram of the SANS set up under shear flow. The incident beam is sent along the velocity gradient direction (O*y*-axis). The 2D detector is set normal to O*y*-axis in O*xz* plane, and O*z*-axis is the neutral direction. From [14], reprinted with permission from American Chemical Society

References

This Chapter is partly based on Section 4 of Hashimoto, T. (2008) In Soft Matter Characterization; Borsali, R. and Pecora, R., (ed), Springer, New York, Chapter 8, pp. 378–462 [2]

1. Matsuzaka, K. and Hashimoto, T. (1999) *Rev. Sci. Instr.*, **70**, 2387.
2. Hashimoto, T. (2008) In *Soft Matter Characterization*, Borsali, R. and Pecora, R., (ed), Springer, New York, Chapter 8, pp. 378–462.
3. Hashimoto, T. (2005) *Bull. Chem. Soc. Jpn.*, **78** No.1, 1-39.
4. Hashimoto, T. Takebe, T., and Suehiro, S. (1986) *Polym. J.*, **18**, 123.
5. Kume, T., Asakawa, K., Moses, E., Matsuzaka, K., and Hashimoto, T. (1995) *Acta Polym.*, **46**, 79.
6. Kume, T. Hattori, T., and Hashimoto, T. (1997) *Macromolecules*, 30, 427.
7. Suehiro, S., Saijo, K., Seto, T., Sakamoto, N., Hashimoto, T., Ito, K., and Amemiya, Y. (1996) *J. Synchrtron Rad.*, **3**, 225.
8. Suehiro, S., Saijo, K., Ohta, Y., Hashimoto, T., and Kawai, H. (1986) *Anayl. Chim. Acta.*, **189**, 41.
9. Matsuoka, H., Tanaka, H., Iizuka, N., Hashimoto, T., and Ise, N. (1990) *Phys. Rev., B.*, **41**, 3854.
10. Okamoto, S., Saijo, K., and Hashimoto, T., (1994) *Macromolecules*, **27**, 3753.
11. Hashimoto, T. (2022) *Principles n Applications of X-ray, Light and Neutron Scattering*, Springer Nature, Singapore.
12. Saijo, K., Shin, G., and Hashimoto, T., Macromolecules, 46, 1549-1562 (2013).
13. Saito, S., Koizumi, S. Matsuzaka, K., Suehiro, S., and Hashimoto, T. (2000) *Macromolecules*, **33**, 2153.
14. Linder, P. ad Obrethür, R. (1985) *Colloid. Polym. Sci.*, **263**, 443.

Chapter 4
Flow-Induced Mixing: Dynamically Symmetric Mixtures

Abstract This chapter describes binary polymer mixtures with the dynamical symmetry described in Sect. 2.2 under external flow fields. The flow fields break the macroscopically phase-separated two-phase structure and self-organizes a series of characteristic dissipative structures as classified from Regime I to IV under increased flow rate, including flow-induced cluster-to-percolation transition in Regime III and stable strings aligned along flow direction in Regime IV. The flow eventually brings about the liquid–liquid phase transition ("bifurcation") from the two-phase (in Regime IV) to single-phase mixture (in Regime V). This chapter presents self-organization of order/disorder through evolution of a series of the dissipative structures. Specifically the following themes are discussed: Flow-induced phase transition from two-phase to single-phase with increased shear rate: Shear-induced shift of phase transition temperature: Time evolution of dissipative structures at the strong shear regime after shear drop from shear-induced single- phase state: Scaling law in the critical phenomena on space–time self-organization of dissipative structures under flow: Reversible changes of the steady-state structures with shear rate. This chapter also describes off-critical mixtures under flow as follows: Composition dependence of dissipative structures and flow-induced phase transition: Uniformity of droplet size in Regime II: Flow-induced phase transition in off-critical mixtures: Hysteresis in phase transition of off-critical mixtures: Shear-rate dependent "steady-state phase diagram" or "phase diagram at dynamic equilibrium".

4.1 Introduction

In this Chapter, we describe binary polymer mixtures with the dynamical symmetry described in Sect. 2.2 under external flow fields. The flow fields break the symmetry of the macroscopically phase-separated two-phase structure and subsequentry self-organize a series of shear-rate dependent characteristic dissipative structure under the increased flow rate. The flow eventually brings about the liquid-liquid phase transition ("bifurcation") from two-phase to single-phase mixture. This chapter

T. Hashimoto, *Polymer Phasing*, https://doi.org/10.1007/978-981-97-9648-9_4

presents the self-organization in nonequilibrium systems from dissipative structures to disordered, single-phase structure.

Let us consider macroscopically phase-separated two-phase systems composed of dynamically symmetric two components. The systems to be dealt with here have the phase diagram of the upper critical solution temperature (UCST). When the shear flow with a shear rate $\dot{\gamma}$ is applied to the systems under conditions where $\dot{\gamma} < \Gamma_{MD}$, the relaxation rate for deformation of polymer chains, the systems would not exhibit either molecular deformation or stress due to the deformation. Even in this case, if this $\dot{\gamma}$ is larger than the relaxation rate of the interfacial deformation, the interface is deformed and broken, giving rise to smaller domains with a larger interfacial area. The broken small domains tend to grow, driven by the thermodynamic force (or interfacial tension) inherent to the systems, but overgrown large domains tend to be burst, driven by the viscous force, i.e., the energy imposed to the systems. The burst and growth and its repetition control the steady-state domain size and spatial distribution of the domains under the shear, i.e., the flow-rate-dependent dissipative structure, as will be described later in this section.

If $\dot{\gamma}$ can be further increased up to $R(q_m)$, the growth rate of the dominant Fourier mode of concentration fluctuations at the early stage of spinodal decomposition, the domain size becomes smaller than the critical size comparable to the interface thickness. Under this condition, the shear-induced mixing may take place because there is no time available for the domain to grow at the given shear rate. It should be noted that a considerable degree of intermixing of the two components in the mixtures may occur in each domain prior to the shear-induced mixing. For the shear-induced single-phase formation to occur, $R(q_m)$ should be sufficiently small so that the applied $\dot{\gamma}$ can exceed $R(q_m)$. Mixtures composed of high-molecular weight polymers with the dynamical symmetry and with a small interfacial tension can fulfill this criterion. Under the condition of $R(q_m) < \dot{\gamma} < \Gamma_{MD}$, the shear flow effectively affects the domain structures without causing a significant molecular deformation to result in the shear-induced single-phase formation with increased shear rate. We shall further elaborate below the general picture described above.

4.2 Experimental Systems

The systems to be discussed in this chapter are the binary mixtures of polystyrene (PS) and polybutadiene (PB) in a neutral solvent of dioctyl phthalate (DOP) designated as PS/PB (X: Y)/ DOP Z wt. % as summarize in Table 4.1. Here, X and Y denote the weight faction (in percent) of PS and PB in the polymer mixtures, respectively, and Z denotes the total weight fraction of PS and PB in the solution. Any given quiescent solution at $\dot{\gamma} = 0$ shown in Table 4.1 macroscopically phase-separates into the PB-rich solution and PS-rich solution below its binodal temperature defined by $T_c(0)$ where $T_c(0)$ is estimated by extrapolating the cloud point temperatures measured as a function of heating rates to the zero-heating rate. The phase separation of PS/PB/DOP is predicted to give rise to the spatial distribution of thermally averaged local volume fraction of Kth segment (K = PS, or PB or DOP) as shown in Fig. 4.1a: the spatial distribution of the concentration of the neutral solvent, $\phi_{DOP}(r)$, is independent of r, constant ϕ_{DOP} everywhere in space as shown by the dot-dashed line, whereas $\phi_{PS}(r)$ and $\phi_{PB}(r)$ vary with r, as shown by the solid line and broken line, respectively, under the condition given below,

$$\phi_{DOP}(r) = \phi_{DOP} = (\text{constant}),\ \ \phi_{PS}(r) + \phi_{PB}(r) = 1 - \phi_{DOP}.$$

This condition is full field because of neutrality of the solvent and incompressity of the system.

From this point view, the phase separation in the ternary system PS/PB/DOP is predicted to be the phase separation between the PS solution and the PB solution as schematically shown in Fig. 4.1a or it is predicted to be identical to the phase separation in the *binary* system PS/PB with the reduced binary segmental interaction between PS and PB by the concentration of the neutral solvent ϕ_{DOP}. This is the so-called "**pseudo-binary approximation**", a further detail of which is specifically described below.

In the semidilute solution of PS and PB with the number of monomeric units per chain N_{PS} and N_{PB}, respectively, in the neutral solvent, the bear thermodynamic segmental interaction of PS and PB in melts, defined by $\chi_{PS/PB}$, is reduced to χ_{eff} due to the presence of the neutral solvent. In the context of pseudo-binary approximation, we can indicate that this ternary system is equivalent to the binary mixtures of PS and PB in the melt which are composed of the number of blob units per chain $(N_b)_{PS}$ and

Table 4.1 Characteristic of polymers and list of systems

Polymers	$M_n \times 10^{-5}$	M_w/M_n	
Polystyrene (PS)	2.05	1.05	
Polybutadiene (PB)	1.65	1.09	
Systems	**Total polymer concentration (wt.%)**	**Weight fraction of PS/PB (wt.%)**	**$T_c(0)$ (°C)**
PS/PB (80:20)/DOP	3.3	80/20	58.0
PS/PB (70:30)/DOP	3.3	70/30	63.0
PS/PB (60:40)/DOP	3.3	60/40	69.0
PS/PB (50:50)/DOP	3.3	50/50	76.0
PS/PB (50;50)/DOP	3.0	50/50	54.0
PS/PB (40:60)/DOP	3.3	40/60	78.0
PS/PB (30:70)/DOP	3.3	30/70	78.0
PS/PB (20:80)/DOP	3.3	20/80	68.0
PS/PB (10:90)/DOP	3.3	10/90	46.0

M_n and M_w are the number- and weight-average molecular weight, respectively.
DOP is a neutral solvent, close to Θ-solvent to PS and PB. $T_c(0)$ is the binodal point determined by extrapolating the cloud point temperatures measured as a function of heating rate to zero-heating rate.

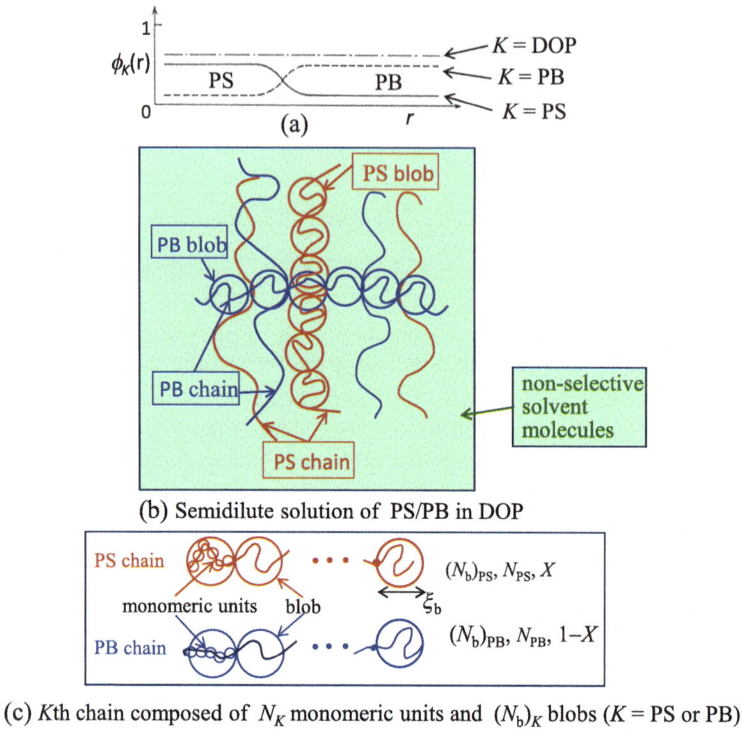

Fig. 4.1 Pseudo-binary approximation. In a semidilute solution of PS/PB in a neutral solvent of DOP, the thermodynamic interaction between DOP and PS segment is identical to that between DOP and PB segment. When PS and PB are phase-separated, DOP is equally partitioned into the PS and PB phases to result in the constant DOP concentration in space, as shown by the dash-dot line for $\phi_K(K = \text{DOP})$, and DOP primarily acts a diluent to decrease the effective segmental interactions between PS and PB, χ_{eff} (part a). The ternary mixture of PS/PB/DOP is identical to the binary mixture of PS/PB (part b) by replacing PS and PB chains with N_{PS} and N_{PB} monomeric units with those composed of $(N_b)_{\text{PS}}$ and $(N_b)_{\text{PB}}$ blobs, respectively (part c); χ_{eff} is also replaced by the inter-blob interaction χ_{blob} between the PS and PB blobs in the binary mixtures

$(N_b)_{\text{PB}}$, respectively, with the thermodynamic inter-blob interaction χ_{blob} between the PS blobs and the PB blobs as shown in part (b) and (c). Thus, the discussion extended herein for the ternary system PS/PB/DOP can be generally applied to that in the binary mixtures of polymers.

Pseudo-binary approximation. In the PS/PB/DOP solution, we focus on one of the chains of PS and that of PB composed of the blobs labeled by red and blue circles, respectively: many other PS and PB chains also composed of chains of the blobs are shown only by red and blue lines, respectively (Fig. 4.1b and c). The blobs in many other chains are not drawn just for the sake of simplicity. We may visualize each chain as a succession of units designated as "blobs" of size ξ_b which is defined as the mesh size between neighboring entanglement coupling points in the solution as illustrated by the blue blob chain oriented horizontally in the center of Fig. 4.1b.

Inside one blob, the chain does not interact with other chains, so that it is subjected to the excluded volume effect. However, the excluded volume effect is screened out in the chains outside the blobs. In the neutral solvent, the Flory–Huggins segmental interaction parameter between PS and solvent, $\chi_{PS/DOP}$, and PB and solvent, $\chi_{PB/DOP}$, are equal, close to that for Θ solvent, so that

$$\chi_{PS/DOP} \cong \chi_{PB/DOP} \cong 1/2. \tag{4.2.1}$$

Thus, the blob size for PS and PB is equal: PS and PB contain the same number of the monomeric units per blob, defined by g_b.

In the blob picture g_b is given by [1]

$$g_b \cong f_0^{-2}(f_0/\phi_P)^{1/(3\nu-1)}, \tag{4.2.2}$$

where ϕ_P is the total polymer concentration in the ternary system, and

$$f_0 = \left(1 - 2\,\chi_{PS/DOP}\right)/u^* = \left(1 - 2\,\chi_{PB/DOP}\right)/u^*. \tag{4.2.3}$$

Here, u^* is given by

$$u^*/(2\pi)^2 = \varepsilon/8 + O\left(\varepsilon^2\right), \quad \varepsilon = 4 - d, \tag{4.2.4}$$

and d is the space dimensionality. In the Θ solvent, $\nu = 1/2$ so that Eq. (4.2.2) gives

$$g_b \cong \phi_P^{-2}. \tag{4.2.5}$$

Thus, $g_b \cong 1000$ and 530 for the 3 wt. % and 3.3 wt. % PS/PB/DOP solution, respectively.

The number of the blobs per PS and PB chain defined by $(N_b)_{PS}$ and $(N_b)_{PB}$, respectively, is estimated by

$$(N_b)_K = N_K/g_b \ (K = \text{ PS or PB}), \tag{4.2.6}$$

where N_K is the number of the monomeric units per Kth chain. N_K calculated from the weight-average molecular weight is given by,

$$N_{PS} = 2.1 \times 10^3, N_{PB} = 5.8 \times 10^3 \tag{4.2.7}$$

$$(N_b)_{PS} \cong 1.9, (N_b)_{PB} \cong 5.2 \text{ for 3 wt. \% solution} \tag{4.2.8}$$

$$(N_b)_{PS} \cong 2.2, (N_b)_{PB} \cong 6.3 \text{ for 3.3 wt. \% solution.} \tag{4.2.9}$$

Heat of mixing of PS and PB per unit volume ΔH_{mix} is given by

$$\Delta H_{\text{mix}} = \begin{cases} \chi_{\text{eff}}\, \phi_{\text{PS}}\, \phi_{\text{PB}}/a^3, \ \phi_{\text{PS}} + \phi_{\text{PB}} = \phi_{\text{P}} \\ \chi_{\text{blob}} X(1-X)/\xi_{\text{b}}^3 \end{cases}, \tag{4.2.10}$$

where a is the segmental length or the length of monomeric units of PS and PB: ϕ_{PS} and ϕ_{PB} are volume fraction of PS and PB in the ternary system, respectively: X is the volume fraction of PS blob in the binary mixture of PS and PB in the blob picture shown in Fig. 4.1b, c. X is given by

$$X = \phi_{\text{PS}}/(\phi_{\text{PS}} + \phi_{\text{PB}}) = \phi_{\text{PS}}/\phi_{\text{P}}. \tag{4.2.11}$$

Note that

$$\xi_{\text{b}}^3\, \phi_{\text{P}} = g_{\text{b}}\, a^3. \tag{4.2.12}$$

Equations (4.2.10)–(4.2.12) lead to

$$\chi_{\text{eff}} = (g_{\text{b}}\, \phi_{\text{P}})^{-1} \chi_{\text{blob}} = \phi_{\text{P}}\, \chi_{\text{blob}}. \tag{4.2.13}$$

The second line of Eq. (4.2.13) is obtained by using Eq. (4.2.5) for the neutral Θ solvent.

χ_{blob} in the Θ solvent is given by [1]

$$\chi_{\text{blob}} \cong \chi_{\text{PS/PB}} \Big/ \left[\phi_{\text{P}} + \frac{\chi_{\text{PS/PB}}}{u_{12}^*}\right]. \tag{4.2.14}$$

The fixed-point value u_{12}^* is given by

$$u_{12}^*/(2\pi)^2 = \varepsilon/4 + O(\varepsilon^2) \cong 0.22. \tag{4.2.15}$$

Thus, in the context of the **pseudo-binary approximation**, we can predict the following:

(the ternary mixture of the PS/PB/DOP with the numbers of the monomeric units N_{PS} and N_{PB} per PS and PB chains, respectively, and with the effective interaction parameter χ_{eff} between the PS and PB monomeric units)
= (the binary mixture of the PS/PB in melt with their numbers of the blob units $(N_{\text{b}})_{\text{PS}}$ and $(N_{\text{b}})_{\text{PB}}$ per PS and PB chains, respectively). (4.2.16)

For the present systems, with the PS/PB (X:Y)/DOP Z wt. % having
i) any given PS/PB composition (X:Y) in DOP and
ii) any given total concentration Z of PS/PB Z wt. % in DOP, and
iii) the degree of polymerization N_{PS} and N_{PB} given by Eq. (4.2.7),
a) the numbers of the blobs $(N_{\text{b}})_{\text{PS}}$ and $(N_{\text{b}})_{\text{PB}}$ are given by Eq. (4.2.8) for 3.0 wt. % solution, and by Eq. 4.2.9 for 3.3 wt. % solution, respectively.

b) the thermodynamic interaction parameters χ_{eff} and χ_{blob} are given by Eqs. (4.2.13) and (4.2.14), respectively.

4.3 Flow-Induced Dissipative Structures

4.3.1 Experimental Systems and Shear Protocol

Systems. The system to be discussed here is the one designated PS/PB (50/50)/DOP 3.3 wt. % in Sect. 2.1 and in Table 4.1. DOP is a neutral solvent for PS and PB and weakens the repulsive segmental interactions between PS and PB. The system can be treated as **a pseudo-binary system** [1] where the phase separation between PS and PB occurs in the medium of DOP and the thermally averaged local volume fraction of DOP is constant everywhere in space as shown in Fig. 4.1a. In other words, the phase separation between the polymers and solvent molecules is insignificant: the phase separation is treated as the one between PS/DOP and PB/DOP solution. The pseudo-binary system is regarded to be equivalent to the binary systems in melt when the segments of polymers in the solution are replaced by the **blobs** [2]: The pseudo-binary systems of polymers A and B are equivalent to bulk binary systems of A and B having the number of the effective repeating units of $(N_{\mathrm{A}})_{\mathrm{b}}$ and $(N_{\mathrm{B}})_{\mathrm{b}}$, respectively. Here, $(N_{\mathrm{A}})_{\mathrm{b}}$ and $(N_{\mathrm{B}})_{\mathrm{B}}$ are number of blobs for A and B polymers, respectively: $(N_{\mathrm{A}})_{\mathrm{b}} = N_{\mathrm{A}}/g$ and $(N_{\mathrm{B}})_{\mathrm{b}} = (N_{\mathrm{B}})_{\mathrm{b}}/g$ where g is the number of segments in the blobs.

The system has the following characteristics [3, 4] at the given quench depth $\Delta T(0) \equiv T_{\mathrm{c}}(0) - T = 8$ °C: and $T_{\mathrm{c}}(0) = 76$ °C (the binodal temperature in quiescent state);

(1) the characteristic wave number $q_{\mathrm{m}}(0) = 4 \times 10^{-3}\ \mathrm{nm}^{-1}$
(2) the characteristic wavelength $\Lambda_{\mathrm{m}} = 2\pi/q_{\mathrm{m}}(0) = 1.5\ \mu\mathrm{m}$, which characterizes the Fourier modes of the concentration fluctuations with a maximum growth rate in the early stage of SD,
(3) the collective diffusivity $D_{\mathrm{app}} \cong 1.0 \times 10^{4}\ \mathrm{nm}^2/\ \mathrm{s}$
(4) characteristic time $\tau \equiv \left[q_{\mathrm{m}}^2(0) D_{\mathrm{app}}\right]^{-1} = 6$ s
(5) the characteristic interface thickness $t_{\mathrm{I}} \simeq 100$ nm and
(6) interfacial tension between the PS solution and the PB solution $\gamma_{\mathrm{int}} \cong 10^{-7}$ N/m

The large t_{I} value and the small γ_{int} value are indicative of the system being is in the very weak segregation limit. Thus, the system can attain various states ranging from a macroscopically phase-separated state to a single-phase state under the experimentally accessible range of shear rates (0.005 s^{-1} to 500 s^{-1}). It has $c/c^* \sim 2$ where c is the total polymer concentration in the solution, and the concentration c^* is the overlap concentration of PS and PB. It has the viscosity of 0.63 poise (0.063 Pa s), independent of the shear rates from 0.2 to 400 s^{-1} (Fig. 3.5).

Shear protocol: We applied the following shear history to the solution to attain the steady-state solution under the shear with a sufficient reproducibility. The solution in the shear cell was first brought into the single-phase solution above $T_c(0)$, e.g., at 79 °C, for PS/PB(50:50)/DOP 3.3 wt. % with $T_c(0) = 76$ °C. Then, the solution was cooled to $T = 68$ °C ($\Delta T_c(0) = T_c(0) - T = 8$ °C) for a sufficiently long time (24 h.) to attain the macroscopic phase separation into the two phases at $\dot{\gamma} = 0\ \mathrm{s}^{-1}$. The macroscopically phase-separated two-phase solution is then subjected to the shear flow of $\dot{\gamma} = 0.0033\ \mathrm{s}^{-1}$ (the lowest shear rate employed) at least for 4 h to achieve a steady state at $\dot{\gamma} = 0.0033\ \mathrm{s}^{-1}$. The shear rate was step-wisely increased to a next higher shear rate, e.g., 0.0063 s^{-1} at which the solution was kept for ca. 20 min. to attain the steady state: in the protocol of shear-rate versus time shown in Fig. 4.2a, time $t = 0$ is set at the time in which shear-rate is jumped from 0.0033 to 0.0063 st^{-1}. The solution was subsequently applied to the simultaneous measurement of the

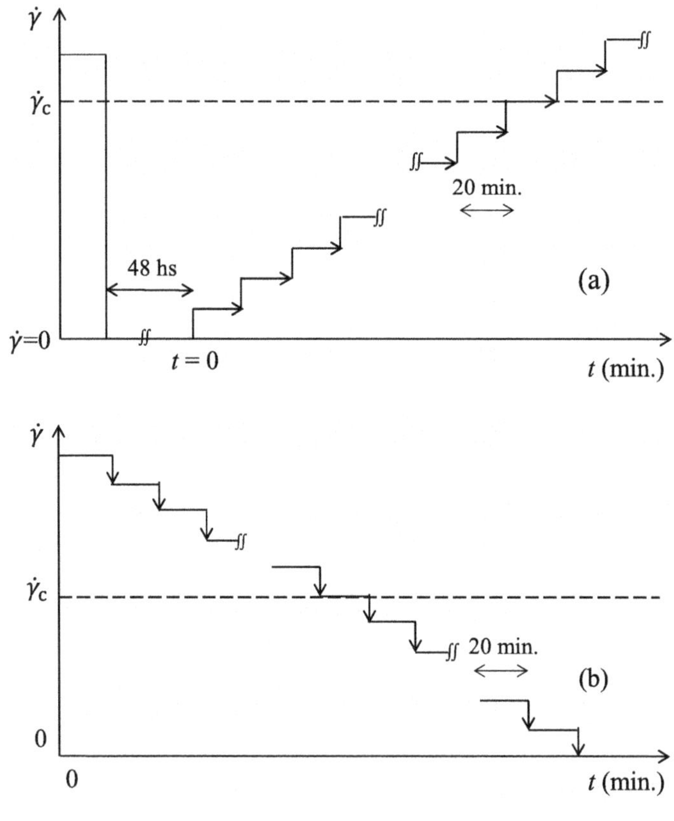

Fig. 4.2 Shear protocol in increased shear rate (**a**) and decreased shear rate with time (**b**). The protocol involves a stepwise increase or decrease of shear rate with 20 min. holding time at each shear rate to achieve the steady state shear, after which stress, SALS, OM, etc. are measured as a function of shear rate. The onset time $t = 0$ is set at the time when the solution is subjected to shear-jump from 0.0033 to 0.0063 s^{-1} after shearing at shear rate 0.0033 s^{-1} for 4h

shear-SALS and shear-OM. This process of the stepwise increase of $\dot{\gamma}$ with 20 min. holding time before the SALS and OM being measured is repeated to explore the shear-rate dependence of the steady-state SALS and OM, as indicated in Fig. 4.2a.

4.3.2 Real-Time, In Situ, Simultaneous Observation of SALS and OM as a Function of $\dot{\gamma}$

Figure 4.3 presents the SALS pattern for PS/PB (50:50)/DOP 3.0 wt. % solution with $T_c(0) = 54.0\,°C$ in the quiescent state (part a) and in the steady-state shear flow with increased $\dot{\gamma}$ at $T = 43.2$ °C, $\Delta T(0) = 10.8$ °C (part b to part g) [5, 6]. The SALS patterns were taken with the cooled CCD at the exposure time of 1/50 s. The flow direction, the incident beam propagation direction, and the neutral (vorticity) direction are along Ox, Oy, and Oz direction, respectively, in the Cartesian coordinate Oxyz, as indicated in the figure. The velocity gradient exists in Oxy plane, and the SALS pattern is detected in Oxz plane with the 2D detector.

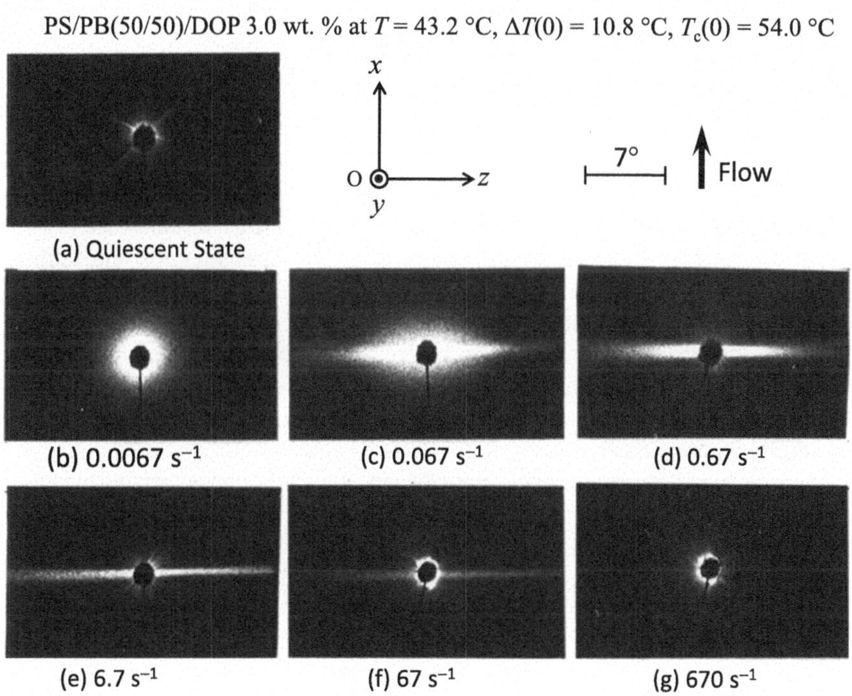

Fig. 4.3 Shear-rate dependence of the SALS pattern for PS/PB (50/50)/DOP 3.0 wt. % solution with $T_c(0) = 54.0\,°C$ at $T = 43.2\,°C$ under the increased shear protocol. Based on [5], reprinted by AIP Publishing with permission

In the quiescent state at $\dot{\gamma} = 0$, the solution is macroscopically phase-separated into the PB-rich homogeneous solution phase and the PS-rich homogeneous solution phase across the interface parallel to Oxz plane, so that the interface does not contribute to the SALS (part a). The scattering around the dark circle at the center of the SALS pattern (the incident beam stopper) is the artifact attributed to the tail part of the incident beam leaked out from the incident beam stopper.

At $\dot{\gamma} = 0.0067\ \mathrm{s}^{-1}$, the solution exhibits a strong round SALS pattern (part b), revealing that the macroscopic two-phase structure is broken into the domains rich in PS dispersed in the matrix rich in PB. The domains are expected to be spherically symmetric with respect to a statistically averaged shape, judged from the shape of the SALS pattern.

At $\dot{\gamma} = 0.067\ \mathrm{s}^{-1}$, the SALS pattern is composed of the two kinds of elemental scattering patterns superposed to each other: the elliptical pattern and the streak-like pattern, both of which are extended and compressed along Oz axis and Ox axis, respectively (part c). This reveals that the round domains in part (b) are elongated into the ellipsoidal domains whose major axis oriented along the flow direction (along Ox-axis), and some of them are further interconnected into the longer domains or the threads. Both the ellipsoidal domains and the threads have the statistically averaged orientation along the flow direction, giving rise to the elliptical and streak-like SALS pattern extended along the neutral (Oz) direction, respectively.

At $\dot{\gamma} = 0.67\ \mathrm{s}^{-1}$, the elliptical SALS pattern disappears to degenerate into the strong streak-like pattern oriented and further extended along Oz axis (part d). This indicates that all the ellipsoidal domains are elongated and interconnected into the threads oriented along the flow direction.

At $\dot{\gamma} = 6.7\ \mathrm{s}^{-1}$, the streak-like pattern is further elongated along Oz axis and narrowed along the flow direction (Ox axis) (part e). This reveals that (1) the threads are further elongated longitudinally along the flow direction and thinned laterally along Oz axis: (2) the threads have increased uniformity both laterally and longitudinally. We designated such threads as described above as "**strings**".

Upon further increasing $\dot{\gamma}$ from $6.7\ \mathrm{s}^{-1}$ to $67\ \mathrm{s}^{-1}$, the absolute scattered intensity of the streak-like pattern is remarkably reduced (part f). This reveals that an intermixing of the PS and PB chains occurs to reduce the scattering contrast between the strings and surrounding matrix.

At $\dot{\gamma} = 670\ \mathrm{s}^{-1}$, the streak-like pattern disappears. The scattering around the black circle is the artifact, same as that in part (a). The disappearance of the streak-like pattern reveals that the strings are dissolved, i.e., the solution becomes the homogeneous single-phase solution at the given shear rate. This phase transition is more quantitatively investigated in depth later in conjunction with Fig. 4.22 in Sect. 4.3.9.

Figure 4.4 presents the integrated scattered intensity from the same solution [5, 6] as that in Fig. 4.3. The integrated scattered intensity parallel and perpendicular to the flow direction as a function of $\dot{\gamma}$, defined as $\mathcal{J}_x(\dot{\gamma})$ and $\mathcal{J}_z(\dot{\gamma})$, respectively, are obtained by integrating the scattered intensity distribution $I(q_x, q_z; \dot{\gamma})$, measured by the 2D detector as shown in Figs. 3.3 and 3.6, with the narrow q range with respect to q_x and q_z as follows: $-\delta \leq q_x \leq +\delta$ for $\mathcal{J}_z$ and $-\delta \leq q_z \leq +\delta$ for $\mathcal{J}_x$, respectively,

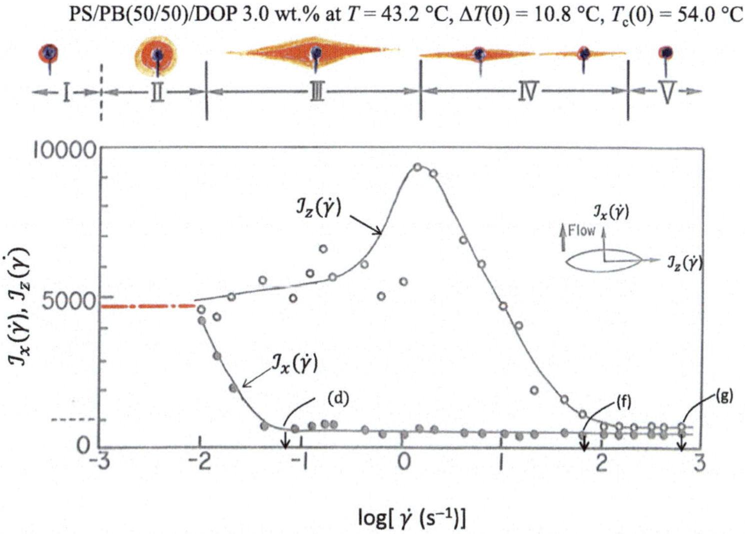

Fig. 4.4 Integrated SALS intensity along the flow direction $\mathcal{J}_x(\dot{\gamma})$ and the neutral direction $\mathcal{J}_z(\dot{\gamma})$ as a function of shear rate under the increased shear protocol for the PS/PB (50:50)/DOP 3.0 wt. % with $T_c(0) = 54.0\,°C$ at $T = 43.2\ °C$. The arrows marked (d), (f), and (g) present the shear rates where the SALS patterns (d), (f), and (g) in Fig. 4.3are taken. Based on [5], reprinted by AIP Publishing with permission

$$\mathcal{J}_z(\dot{\gamma}) \equiv \int_{-\delta}^{+\delta} dq_x \int_{q_1}^{\infty} dq_z I(q_x, q_z; \dot{\gamma}), \tag{4.3.1}$$

$$\mathcal{J}_x(\dot{\gamma}) \equiv \int_{-\delta}^{+\delta} dq_z \int_{q_1}^{\infty} dq_x I(q_x, q_z; \dot{\gamma}), \tag{4.3.2}$$

where $\delta = 3.1 \times 10^{-3}\ \text{nm}^{-1}$ and $q_1 = 2.0 \times 10^{-3}\text{nm}^{-1}$. Since the absolute scattered intensity was not measured, $I(q_x, q_z; \dot{\gamma})$ indicates only the relative intensity distribution as a function q_x and q_z at the given $\dot{\gamma}$. Thereby, $\mathcal{J}_x(\dot{\gamma})$ and $\mathcal{J}_z(\dot{\gamma})$ also indicate the relative integrated intensity as a function of $\dot{\gamma}$.

We can classify the shear-rate dependence of $\mathcal{J}_x(\dot{\gamma})$ and $\mathcal{J}_z(\dot{\gamma})$ in Fig. 4.4 into the five regimes: Regime I to Regime V as defined on the top of the figure along with the schematically illustrated characteristic SALS patterns representing each regime which are drawn based on Fig. 4.3.

In Regime I ($\dot{\gamma} \lesssim 10^{-3}\ \text{s}^{-1}$) where the shear-rate is close to zero,

$$\mathcal{J}_x = \mathcal{J}_z \sim 1000\,(\text{arb. unit}),$$

as indicated by the horizontal broken line shown in the edge of the left-hand side of the ordinate scale at the shear rate range smaller than 10^{-3} s^{-1}.

In Regime II (10^{-3} s^{-1} $< \dot{\gamma} <$ 10^{-2} s^{-1}), both $\mathcal{J}_x$ and $\mathcal{J}_z$ are abruptly increased to

$$\mathcal{J}_x = \mathcal{J}_z \sim 5000 \text{ (arb. unit)},$$

with increased $\dot{\gamma}$, as indicated by the horizontal dash-dot line shown by red color. The increase of the integrated intensity arises from the emergence of the strong, round SALS pattern as shown in Fig. 4.3 (b) and as schematically illustrated on the top of Fig. 4.4.

In Regime III $\left(10^{-2}\ \text{s}^{-1} < \dot{\gamma} < 10^{0}\ \text{s}^{-1}\right)$, $\mathcal{J}_z$ increases to a maximum with increased $\dot{\gamma}$ above 10^{-2} s^{-1}, whereas $\mathcal{J}_x$ decreases to a minimum intensity level. This is because the elliptical SALS pattern, which is superposed on the streak-like pattern in Fig. 4.3c, changes into the streak-like pattern with increased $\dot{\gamma}$, and this newly developed streak-like pattern is superposed on the as-existing elongated streak-like pattern. At $\dot{\gamma} \sim 1.0$ s^{-1} where the highest $\mathcal{J}_z$ is attained in Regime III, the width of the streak-like pattern along Ox axis may become minimum, which brings about the following two effects: (1) the whole strong SALS pattern falls into the narrow height of the window of $|q_x| < \delta$ for obtaining $\mathcal{J}_z$, giving rise to a maximum value for $\mathcal{J}_z$, and (2) the detector window of $q_1 < q_x < \infty$ for obtaining $\mathcal{J}_x$ covers only a weak scattering arising from thermal concentration fluctuations within the domain phase and the matrix phase, due to the strong SALS along q_x being degenerated in the q range at $q < q_1$, thereby giving rise to the minimum intensity level for $\mathcal{J}_x$.

In Regime IV (10^{0} s^{-1} $< \dot{\gamma} <$ 10^{2} s^{-1}), upon increasing $\dot{\gamma}$ above 1 s^{-1}, $\mathcal{J}_z$ decreases from the maximum value of ~ 10,000 to the value close to the minimum intensity level of ~ 500 in Regime V, whereas $\mathcal{J}_x$ keeps the minimum intensity level attained in Regime III. The decreased $\mathcal{J}_z$ is primarily due to the decreased absolute intensity of the streak-like SALS pattern with $\dot{\gamma}$, as shown in the pattern at $\dot{\gamma} = 67$ s^{-1} in Fig. 4.3f. The decreased absolute intensity reflects the enhanced intermixing of the PS and PB chains in the string phase and the matrix phase, which reduces the difference of the scattering contrast between the strings and the matrix.

In Regime V ($\dot{\gamma} >$ 10^{2} s^{-1}), $\mathcal{J}_z$ becomes the minimum intensity level of ~ 500 independent of shear rate, whereas $\mathcal{J}_x$ keeps the minimum intensity level, attained in Regime III and Regime IV. This is due to the disappearance of the streak-like SALS pattern which in turn reflects the dissolution of the strings. Thus, the solution is predicted to become a single-phase solution. Both $\mathcal{J}_z$ and $\mathcal{J}_x$ in this regime reflect the scattering from thermal concentration fluctuations which are anisotropic under the shear flow as will be clarified in conjunction with Fig. 4.16 later in Sect. 4.3.8.

Figure 4.5 represents typical steady-state SALS patterns and transmission optical microscope (OM) images measured simultaneously as a function of shear rate in the increased shear protocol [7] on the top half and bottom half, respectively, while Fig. 4.6 schematically represents the summary of the dissipative structures elucidated from a series of the studies as a function of shear rate in the increased shear protocol [8, 9].

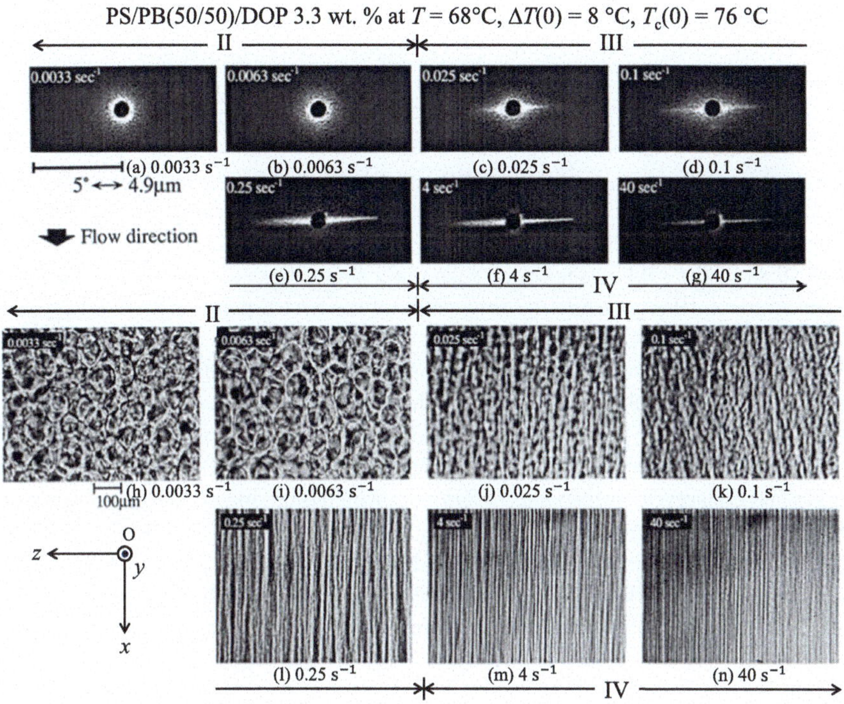

Fig. 4.5 Typical steady-state SALS patterns (top half) and OM images (bottom half) measured simultaneously as the function of shear rate under the increased shear protocol for PS/PB (50:50)/ DOP 3.3 wt. % with $T_c(0) = 76\,°C$ at $T = 68\,°C$

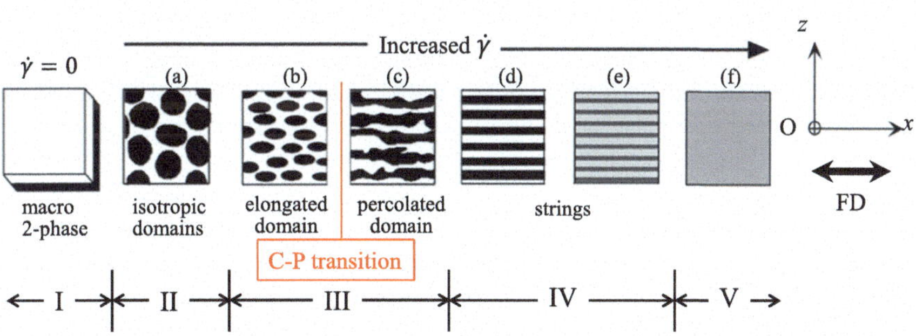

Fig. 4.6 Summary of the dissipative structures as a function of shear rate under the increased shear protocol for PS/PB/DOP (50:50) 3.0 wt. % with $T_c(0) = 54.0\,°C$ at $T = 43.2\,°C$ and 3.3 wt. % with $T_c(0) = 74\,°C$ at $T = 66\,°C$. Based on [8], permission with Oxford University Press

The experimental results shown in Fig. 4.5 were obtained for the PS/PB (50:50)/DOP 3.3 wt. % with $T_c(0) = 76$ °C. The shear flow is applied at $T = 68$ °C to the macroscopically phase-separated solutions of the PS and PB in DOP having its interface parallel to Oxz plane in quiescent state, as shown in the left edge in Fig. 4.6, with an incident laser beam irradiated along Oy axis as shown in the right edge of the same figure and Fig. 1.4 for the schematically illustrated experimental setup and the Cartesian coordinate system. The SALS patterns and OM images were recorded at steady states under the shear protocol with a stepwise increase of the shear rate. Snap shots of the OM images were taken at a shutter speed of 10^{-4} s and a focal depth of about 10 μm, while the snap shots of the SALS patterns were taken with the exposure time of 1/60 s.

At the very small shear rates smaller than 0.0033 s^{-1}, the macroscopic interface parallel to the Oxz plane (the plane of the paper) is conserved as schematically illustrated in the left edge of Fig. 4.6, although the shear flow induces the local, dynamic perturbations of the interface. The shear-rate range where the macroscopic interface is conserved corresponds to Regime I which is not included in Fig. 4.5. Upon increasing the shear rate, the macroscopic interface is broken, and the large nearly spherical droplets are formed as evidenced by Fig. 4.5h and i and as schematically illustrated in Fig. 4.6a because of the viscous drag outweighing the interfacial tension. Correspondingly, the SALS patterns which have the intensity distribution nearly independent of the azimuthal angle μ appear at the small angles as shown in Fig. 4.5 a and b. This shear-rate range corresponds to Regime II ($\dot{\gamma} \leq 0.0063$ s^{-1}) as indicated in the top of Fig. 4.5 and in Fig. 4.6a.

At the higher shear rates, the droplets are broken into an increased number of the smaller droplets elongated parallel to the flow direction as evidenced by Fig. 4.5j, so that the interfacial area is increased as schematically shown in Fig. 4.6b. The increased interfacial area is stabilized by the mechanical energy steadily imposed on the system. With the increased $\dot{\gamma}$, the droplets become smaller and deformed, and the elongated droplets tend to align parallel to the flow direction (Fig. 4.6b). The droplets are eventually interconnected to result in the formation of the percolated structure parallel to the flow direction as shown in some part of areas in Fig. 4.5j, in the increased areas in Fig. 4.5k, and in the whole area in Fig. 4.5l and as schematically illustrated in Fig. 4.6c. This is the so-called **flow-induced cluster-to-percolation transition phenomenon** [8, 9], as designated by **C-P transition** in Fig. 4.6, which develops an oriented percolated structure comprised of the solution rich in one polymer component in the matrix solution rich in another polymer component. The percolated structure is still unstable, exhibiting several random irregularities because of such dynamical processes as frequent interconnections, disconnections, branching, and exchanges of the domains through interactions with neighboring domains, as schematically illustrated in Fig. 4.6c. The scattering pattern in Regime III is composed the following two components, both oriented normal to the flow direction: (1) the streak-like SALS pattern arising from the interconnected or percolated elongated domains and (2) the elliptical SALS pattern arising from the isolated elongated domains free from the interconnection. Upon increasing $\dot{\gamma}$, the streak-like pattern (1) has higher intensity than the elliptical pattern (2) and is more elongated normal to the flow direction, whereas the elliptical pattern loses its intensity as shown by the change of the SALS pattern from Fig. 4.5c to 4.5e. This shear-rate range is defined as Regime III (0.025 s^{-1} < $\dot{\gamma}$ < 0.25 s^{-1}) for the 3.3 wt. % solution (Fig. 4.6b and 4.6c).

At $\dot{\gamma} \geq 0.25\ s^{-1}$ in Regime IV, the system forms the stable **string** structure [4, 10] with a macroscopic continuity of the order of 500 μm along its axis parallel to flow direction as shown in the OM images in Fig. 4.5m and 4.5n and as schematically illustrated in Fig. 4.6d and 4.6e. The corresponding SALS pattern is the sharp streak oriented normal to the flow direction as shown in Figs. 4.5f and 4.5g.

Described below are the analyses and discussion of the experimental results in Regime I to Regime V.

4.3.3 *Regime I: Macroscopically Phase-Separated Two-Phase*

The solution of the PS/PB (50:50)/DOP 3.0 wt. % with $T_c(0) = 54.0$ °C at $T = 43.2$ °C and 3.3 wt.% with $T_c(0) = 74$ °C at $T = 66.0$ °C exhibit the macroscopically phase-separated two-phase solution across the interface parallel to Oxz plane at $\dot{\gamma} = 0$ and at the very small $\dot{\gamma}(< 10^{-3}\ s^{-1})$ for the 3.0 wt.% solution as shown in Figs. 4.4, 4.6, and 4.7. The PB-rich solution phase and the PS-rich solution phase exist in the upper and lower part of the interface, and both phases are homogeneous as schematically illustrated in the bottom-left of Fig. 4.7. Thus, the OM image viewed along Oy axis is featureless, and the SALS arises from the thermal concentration fluctuations from the two homogeneous solution phases as shown in the top-left (schematic) and top-right of Fig. 4.7, i.e., the interface does not contribute to the SALS. Since the shear rate in Regime I belongs to the weak shear regime with respect to the thermal concentration fluctuations which satisfices $\dot{\gamma} < \Gamma_{\mathrm{conc.}}$ in Regime $[\dot{\gamma}_1]$ (Fig. 1.7b and Fig. 1.8d), the fluctuations along the neutral direction are not affected by the shear flow and thereby the SALS intensity along q_z is predicted by the Ornstein–Zernike scattering formula [11] for the quiescent solution,

$$I(q) = I(0)/\left(1 + \xi_T^2\, q^2\right), \tag{4.3.3}$$

where $q \equiv q_z$. $I(0)$ is the scattered intensity $I(q)$ at $q = 0$, and ξ_T is the thermal correlation length given by Eq. (2.5.3).

4.3.4 *Regime II: Round Droplets*

The macroscopically phase-separated two-phase solution in Regime I is burst into the droplets rich in the PS dispersed in the matrix solution rich in PB as demonstrated by

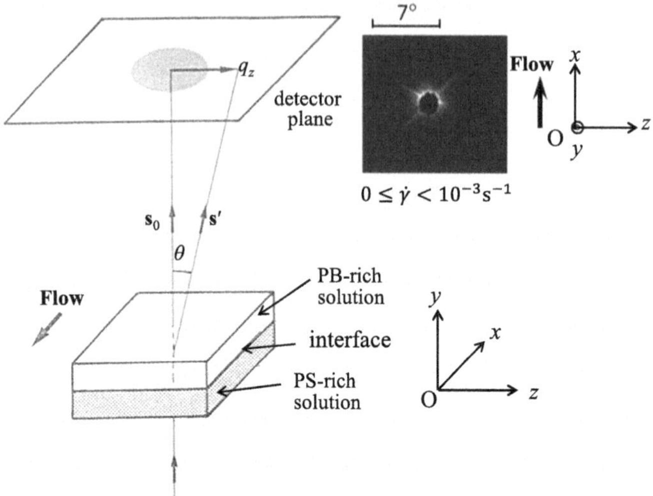

Fig. 4.7 2D SALS pattern from macrophase-separated PS/PB/DOP solution in Regime I. $\mathbf{s}_0$ and $\mathbf{s}'$ are unit vectors along the propagation direction of incident light beam and scattered light beam, respectively. q_z is component of scattering vector $\mathbf{q}$ along Oz axis

the enlarged OM image and the SALS pattern (inset) in Fig. 4.8 for PS/PB (50:50)/DOP 3.3wt. % at $T = 68$ °C and at $\dot{\gamma} = 0.0063\ \mathrm{s}^{-1}$. The radius of the droplet at a given $\dot{\gamma}$, defined as $R(\dot{\gamma})$, may be controlled by the dissipation of the applied energy imposed to the solution at the given $\dot{\gamma}$ to attain the dynamic equilibrium between the burst of the droplets and the growth of the droplets driven by thermodynamic driving force. The droplets tend to be deformed and burst into the smaller ones due to the disruptive force driven by the viscous drag imposed to the droplets by the applied flow field, whereas the burst droplets tend to grow via the coagulation (flow-induced diffusion and coalescence) driven by the interfacial tension as schematically illustrated in Fig. 4.9.

Thus, the cut-off size $R(\dot{\gamma})$ of the droplet at a given $\dot{\gamma}$ in Regime II is controlled by the balance of the disruptive force imposed to the droplets by the applied flow and the interfacial tension intrinsic to the emulsion of the droplets. Since the former and the latter are given as follows, respectively,

$$(\text{the disruptive energy}) \sim \eta\ \dot{\gamma}\ \{R(\dot{\gamma})\}^3, \tag{4.3.4}$$

and

$$(\text{the interfacial energy}) \sim \gamma_{\text{int}}\{R(\dot{\gamma})\}^2, \tag{4.3.5}$$

the cut-off size of the droplets at steady state at $\dot{\gamma}$ is given by

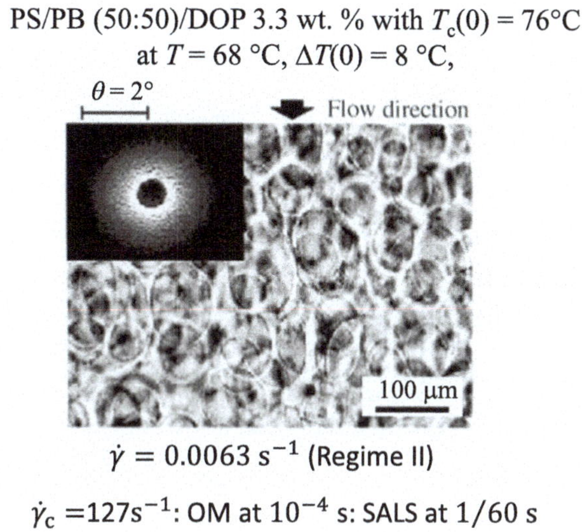

Fig. 4.8 2D SALS pattern (inset) and POM image at $\dot{\gamma} = 0.0063\ s^{-1}$ in Regime II for PS/PB (50:50)/DOP 3.3 wt. % with $T_c(0) = 76\,°C$ at = 68 °C. The flow direction is the vertical direction of the figure

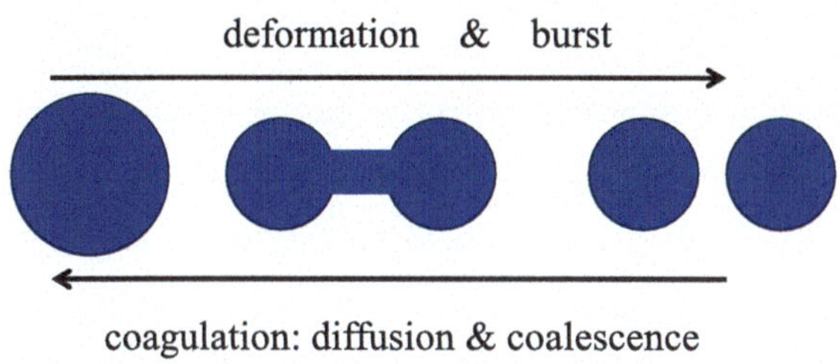

Fig. 4.9 Dynamic equilibrium between deformation and burst of droplets (from left to right) and the diffusion-coalescence (coagulation of droplets) (from right to left) in Regime II at the steady-state shear. The flow direction is the horizontal direction of the figure

$$R(\dot{\gamma}) \sim \frac{\gamma_{int}}{\eta\,\dot{\gamma}} \tag{4.3.6}$$

for the Newtonian fluids where η is the viscosity, and γ_{int} is the interfacial tension or the interfacial energy per unit area.

$R(\dot{\gamma})$ increases with decreased $\dot{\gamma}$. There is the crossover $R(\dot{\gamma})$, defined as $R_{grav}(\dot{\gamma})$, above which the droplets become unstable for the gravitational force, i.e., the gravitational force outweighs the interfacial tension of the droplets, so that the solution crossovers from Regime II to Regime I. Since

$$\text{(gravitational force)} \sim \{R(\dot{\gamma})\}^3 (\Delta\rho) g, \tag{4.3.7}$$

and

$$\text{(interfacial tension)} \sim \{R(\dot{\gamma})\}\gamma_{int}. \tag{4.3.8}$$

Since the two forces given by Eqs. 4.3.7 and 4.3.8 balance at $R(\dot{\gamma}) = R_{\text{grav}}$

$$R_{\text{grav}} \sim \left\{ \frac{\gamma_{\text{int}}}{(\Delta\rho)g} \right\}^{1/2}, \tag{4.3.9}$$

where $\Delta\rho$ is the difference of the mass-density between the droplets and matrix, and g is the gravitational acceleration. The critical shear rate $\dot{\gamma}_{\text{I-II}}$ for the crossover of the solution between Regime I and Regime II satisfies,

$$R_{\text{grav}}(\dot{\gamma} = \dot{\gamma}_{\text{I-II}}) = R(\dot{\gamma} = \dot{\gamma}_{\text{I-II}}), \tag{4.3.10}$$

where $R(\dot{\gamma})$ and $R_{\text{grav}}(\dot{\gamma})$ is given by Eq. (4.3.6) and Eq. (4.3.9), respectively. Hence,

$$\dot{\gamma}_{\text{I-II}} = \frac{(\gamma_{\text{int}}\ g\ \Delta\rho)^{1/2}}{\eta}. \tag{4.3.11}$$

The high viscosity of polymer systems makes $\dot{\gamma}_{\text{I-II}}$ and $R(\dot{\gamma})$ at $\dot{\gamma}_{\text{I-II}}$ small for the given set of the parameters (γ_{int}, $\Delta\rho$).

4.3.5 (A) Regime III: Cluster-to-Percolation Transition and Transition from Percolated Domains to Strings

In the lower shear-rate range in Regime III, the solution is composed of clusters of ellipsoidal droplets (the solution rich in the PS) dispersed in the matrix solution rich in the PB as schematically illustrated in Fig. 4.6b. In the higher shear-rate range in Regime III, the ellipsoidal particles tend to be interconnected in the long threads oriented along the flow direction as schematically illustrated in Fig. 4.6c and as demonstrated also by the SALS pattern (inset) and OM image in Fig. 4.10a. Thus, the solution undergoes the **flow-induced cluster-to-percolation transition** (C-P transition) with increased $\dot{\gamma}$ in Regime III: the clusters of the oriented ellipsoidal droplets tend to be transformed into the percolated domains or the long threads oriented along the flow direction.

The interconnection of the droplets and the burst of the interconnected droplets may occur in the dynamical equilibrium at the steady state: the former and the latter are driven by the thermodynamic force to reduce the interfacial energy and the $\dot{\gamma}$-dependent viscous drag, respectively. The fraction of the droplets decreases from 1 to 0, whereas that of the interconnected droplets increases from 0 to 1 with increased $\dot{\gamma}$ in Regime III. The elliptical SALS pattern superposed on the streak-like pattern at $\dot{\gamma} = 0.1\ \text{s}^{-1}$ (Fig. 4.10a) disappears and only the streak-like pattern with enhanced intensity is observed at $\dot{\gamma} = 0.025\ \text{s}^{-1}$ (Fig. 4.10b), indicating that all the ellipsoidal droplets are interconnected into the threads. The percolated domains at the lower shear-rate range have such dynamical random irregularities as frequent

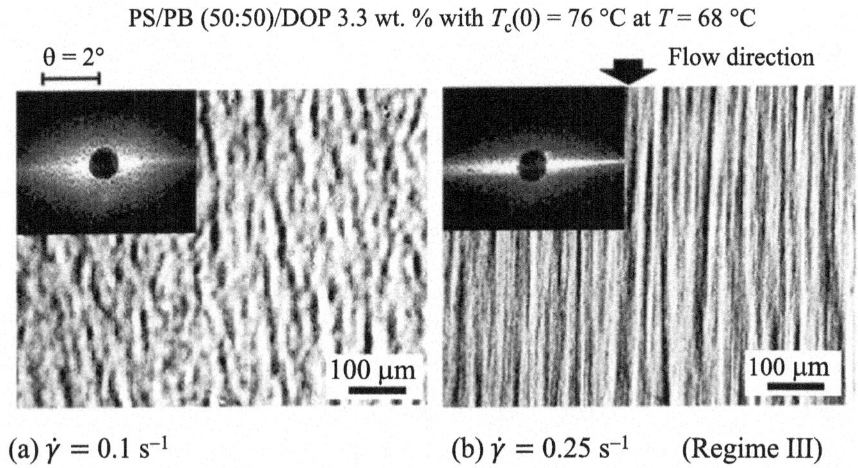

Fig. 4.10 2D SALS pattern (inset) and POM image at $\dot{\gamma} = 0.1\ \mathrm{s}^{-1}$ (**a**) and $0.25\ \mathrm{s}^{-1}$ (**b**) in Regime III for the PS/PB (50:50)/DOP 3.3 wt. % with $T_c(0) = 76\,°\mathrm{C}$ at $T = 68\ °\mathrm{C}$

break-up, branching, interconnection, and exchange of the domains, although the overall, time-averaged domain structure is stationary. The threads tend to have the suppressed irregularities with increased $\dot{\gamma}$, as shown by the changes in the OM image and SALS pattern from Fig. 4.10a and b, and eventually become the strings with high uniformity at the higher shear-rate part in Regime III.

4.3.6 (B) Regime III: SALS Analysis

The scattering intensity distribution $I_{\mathrm{domain}}(q_x, q_y, q_z)$ from the oriented two-phase systems composed of the domains dispersed in the matrix in Regime II to IV can be generally described by

$$I_{\mathrm{domain}}(q_x, q_y, q_z) = K\langle\eta^2\rangle \iiint_{-\infty}^{\infty} dxdydz\ \gamma(x, y, z)\exp\left[-\mathrm{i}\left(q_x x + q_y y + q_z z\right)\right], \tag{4.3.12}$$

where the proportionality constant K associated with the absolute scattered intensity, the mean square of scattering length density (SLD) fluctuations $\langle\eta^2\rangle$, and the correlation function for the oriented two-phase systems $\gamma(x, y, z)$ are given as follows by [12, 13],

$$K = \frac{I_{\mathrm{i}}}{R_{\mathrm{D}}^2} V_{\mathrm{ir}}, \tag{4.3.13}$$

$$\langle\eta^2\rangle = (p_A - p_B)^2\phi_A(1-\phi_A), \tag{4.3.14}$$

and

$$\gamma(x,y,z) = \exp\left[-\left(\frac{x^2}{\xi_x^2} + \frac{y^2}{\xi_y^2} + \frac{z^2}{\xi_z^2}\right)^{1/2}\right], \tag{4.3.15}$$

respectively. q_x, q_y, and q_z are the components of the scattering vector $\mathbf{q}$ along Ox, Oy, and Oz axis. ξ_x, ξ_y, and ξ_z are the correlation length of the systems composed of the domains dispersed in the matrix along Ox, Oy, and Oz axis, respectively. I_i is the incident beam intensity, R_D is the distance between the detector and the given system, V_{ir} is the volume of the system irradiated by the incident beam. p_K is the SLD of the Kth phase (K = A or B) given by [13],

$$p_K = (2\pi/\lambda)^2\alpha_K \sin\gamma, \tag{4.3.16}$$

where λ is the wavelength of the incident beam in the system, α_K is the polarizability per unit volume of the Kth phase, and $\sin\gamma$ is related to the polarization factor, $\sin^2\gamma = (1+\cos^2\theta)/2$ for unpolarized incident beam and $\sin^2\gamma = 1$ for the V_V scattering [13].

From Eqs. (4.3.12) to (4.3.15), we obtain [12],

$$I_{domain}(q_x,q_y,q_z) = 8\,\pi\,K\langle\eta^2\rangle\frac{\xi_x\,\xi_y\,\xi_z}{\left[1+\left(q_x^2\,\xi_x^2+q_y^2\,\xi_y^2+q_z^2\,\xi_z^2\right)\right]^2}. \tag{4.3.17}$$

Equation 4.3.17 is a generalized version of Deye-Bueche scattering equation extended to oriented two-phase systems. In the case when $\xi_x = \xi_y = \xi_z = \xi$, and $q_x = q_y = q_z = q$, Eq. 4.3.17 reduces to the Debye-Bueche scattering equation for the isotropic two-phase systems [14]. For the scattering normal to the flow direction, $q_x = q_y = 0$, so that

$$\begin{aligned} I_{domain}(q_z) &\equiv I_{domain}(q_x=0, q_y=0, q_z) \\ &= 8\,\pi K\langle\eta^2\rangle\frac{\xi_x\,\xi_y\,\xi_z}{\left(1+q_z^2\,\xi_z^2\right)^2}. \end{aligned} \tag{4.3.18}$$

For the uniaxially symmetric systems with respect to the flow direction (Ox axis),

$$\xi_x\,\xi_y\,\xi_z = \xi_x\,\xi_z^2 \tag{4.3.19}$$

Figure 4.11 presents also the experimental SALS intensity distribution $I_{exp}(q_z)$ (open circles) along the direction normal to the flow direction (Oz axis) for clusters of spherical droplets in Regime II (part a in Fig. 4.11) [12]. In this shear-rate regime,

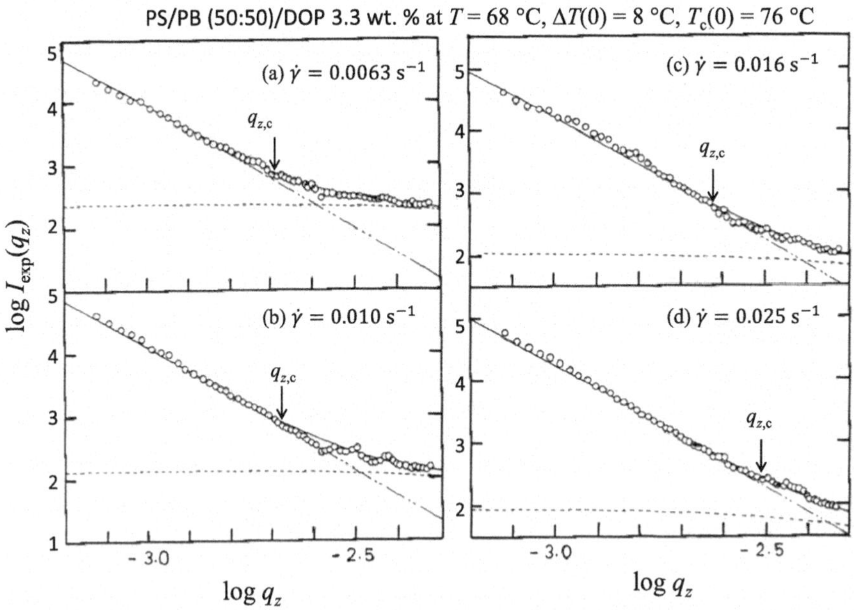

Fig. 4.11 Log $I_{\mathrm{exp}}(q_z)$ versus log q_z for the measured SALS intensity profile along the neutral direction q_z at $\dot{\gamma} = 0.0063\ \mathrm{s}^{-1}$ (**a**) and $0.010\ \mathrm{s}^{-1}$ (**b**) in Regime II and at $\dot{\gamma} = 0.016\ \mathrm{s}^{-1}$ (**c**) and $0.025\ \mathrm{s}^{-1}$ (**d**) in Regime III for PS/PB (50:50)/DOP 3.3 wt. % with $T_c(0) = 76\,°\mathrm{C}$ at $T = 68\ °\mathrm{C}$

the droplet size normal to the flow direction related to ξ_z is large relative to the q-scale covered by the SALS apparatus, so that $I_{\mathrm{exp}}(q_z)$ covers only the Porod' law [15] regime, i.e., $q_z\ \xi_z \gg 1$.

Thus, Eq. 4.3.18 at $q_z \to \infty$ leads to,

$$I_{\mathrm{domain}}(q_z) = 8\ \pi K\ \langle \eta^2 \rangle\ \xi_x\ \xi_z^{-2}\ q_z^{-4}. \tag{4.3.20}$$

The Porod's law of q_z^{-4} is confirmed for all the measured $I_{\mathrm{exp}}(q_z)$ in the small q_z range smaller than the critical value $q_{z,\mathrm{c}}$, as indicated by the arrow marked with $q_{z,\mathrm{c}}$, which increases with increased $\dot{\gamma}$: $q_{z,\mathrm{c}}$ is the q value above which $I_{\mathrm{domain}}(q_z)$ deviates from the Porod's law q_z^{-4} [12]. Here, the Porod's law is indicated by the straight dash-dot-dot lines in Fig. 4.11 [12]. The small q_z range covered satisfies the Porod's criterion of $1/\xi_z \ll q_z < q_{z,\mathrm{c}}$ i.e., $1 \ll q_z\xi_z < q_{z,\mathrm{c}}\xi_z$. The deviation of $I_{\mathrm{exp}}(q_z)$ (open circles) from the Porod's law (straight dash-dot-dot line) is presented by the dotted line in Fig. 4.11; this deviation can be attributed to the thermal concentration fluctuations within the domains and their matrix, defined hereafter as $I_{\mathrm{therm}}(q_z)$.

The scattering predicted by Eq. (4.3.20) assumes that the two phases are homogeneous with no thermal concentration fluctuations within them. However, the thermal concentration fluctuations exist in the two phases, that is, one phase is the matrix phase which is composed of PB-rich solution, and the other phase is the phase of dispersed domains which is composed of PS-rich solution. These thermal concentration fluctuations give rise to the scattering excess to that given by Eq. 4.3.20 at $q_z > q_{z,c}$, defined above as $I_{therm}(q_z)$.

The scattering from the thermal concentration fluctuations from the single phase solution in the direction normal to the flow direction, $I_{therm,s}(q_z)$, is proven both experimentally and theoretically to be described by the OZ scattering formula for solutions at equilibrium, free from the flow [12, 13, 16, 17], which is given by the equation presented below

$$I_{therm,s}(q_z) = K\ \phi_P\ v_0\ (\langle p_P \rangle - p_{solv})^2\ \frac{18\ \xi_T^2}{a^2} \frac{1}{1 + \xi_T^2\ q_z^2}, \tag{4.3.21}$$

where ϕ_P is the total volume concentration of the PS chain and the PB chain in the solution, v_0 is the volume of the solvent molecule and monomeric unit of the polymers, and a is the segment length of the polymers. ξ_T is the thermal correlation length for the thermal concentration fluctuations along Oz axis, $\langle p_P \rangle$ is the average SLD of the PS and PB chains, and p_{solv} is the SLD of the solvent. Here, the subscript "s" in $I_{therm,s}(q_z)$ signifies the scattering intensity from the thermal concentration fluctuations in the phase-separated solution where the two phases are coexisting under the shear flow. One may assume that the sum of the measured SALS intensity, $I_{thermal}(q)$, from the thermal concentration from the two-phase system, i.e., the matrix phase and the phase composed of the dispersed domains, is identical to $I_{thermal,s}(q_z)$ given by Eq. 4.3.21 for the single-phase solution with the same set of parameters ϕ_P and $< p_P >$ as that of the experimental system, for simplicity of the arguments without loosing physical significance.

Figure 4.11 presents the best fit of $I_{exp}(q_z)$ (shown by symbols) with the $I_{theor}(q_z)$ (shown by the straight lines). It should be noted that $I_{exp}(q_z)$ (symbols) and $I_{theor}(q_z)$ (straight line) are fitted very well, so that the symbols and straight line effectively overlap each other, making them indistinguishable. Here, $I_{theor}(q_z)$ is given by

$$I_{theor}(q_z) = I_{domain}(q_z) + I_{therm,s}(q_z). \tag{4.3.22}$$

$I_{domain}(q_z)$ and $I_{therm,s}(q_z)$ are shown by the dash-dot-dot-line and the dotted line, respectively. The smaller the $\dot{\gamma}$ is, the larger is the domain size, so that $I_{domain}(q_z)$ shifts left toward the small q_z: thereby, the relative contribution of $I_{therm,s}(q_z)$ appears to become more significant.

Figure 4.12 presents $q_z{}^4 I_{domain}(q_z)$ versus $\dot{\gamma}$ where $I_{domain}(q_z)$ is obtained at the six different shear rates by the best-fit of $I_{exp}(q_z)$ with $I_{theor}(q_z)$ [12]. The plot shows the straight line with the slope 1 in Regime II and in the slow shear-rate range of Regime III. The deviation from the slope 1 at the two points in the high shear-rate range may be due to the contribution of the streak-like SALS pattern superposed to the elliptical SALS pattern, i.e., the contribution of interconnected ellipsoidal particles in the higher shear-rate range of Regime III (see Fig. 6 in ref [12]). The results along with Eq. 4.3.20 indicate that

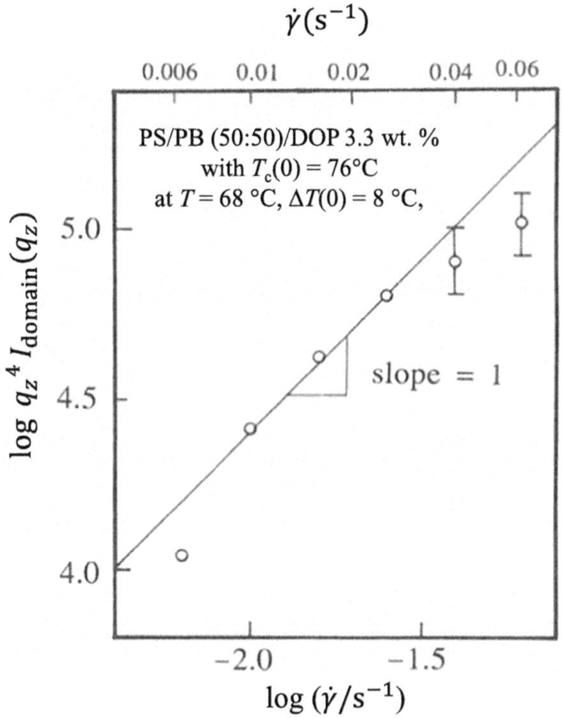

Fig. 4.12 $q_z^4 I_{\mathrm{domain}}(q_z)$ versus $\dot{\gamma}$ in the double logarithmic scale based on Eq. 4.3.20, where I_{domain} is evaluated by the best fit of $I_{\exp}(q_z)$ with $I_{\mathrm{theor}}(q_z)$ based on Eq. 4.3.22, as shown in Fig. 4.11. PS/PB (50:50)/DOP 3.3 wt. % with $T_c(0) = 76$ °C at $T = 68$ °C. Based on [12], reprinted by AIP Publishing with permission

$$\xi_x \xi_z^{-2} \sim \dot{\gamma}, \tag{4.3.23}$$

since $\langle \eta^2 \rangle$ is independent of $\dot{\gamma}$ in this shear-rate range. In Eq. 4.3.23, we can verify that ξ_x and ξ_z is proportional to the radius of the domain size along the Ox axis and Oz axis, R_x and R_z, respectively, with the proportionality constant independent of $\dot{\gamma}$. In the shear rate range where the ellipsoidal particles are isolated, both ξ_x and ξ_z should have the same shear-rate dependence,

$$\xi_K \sim \dot{\gamma}^{-n}, K = x \text{ or } z. \tag{4.3.24}$$

Thus, Eqs. 4.3.23 and 4.3.24 give the results consistent with the Taylor's law,

$$\xi_x \sim R_x \sim \dot{\gamma}^{-1} \text{ and } \xi_z \sim R_z \sim \dot{\gamma}^{-1}. \tag{4.3.25}$$

Since $I_{\mathrm{domain}}(q_x) \equiv I_{\mathrm{domain}}\left(q_x, q_y = q_z = 0\right)$ and at its limit at $q_x \to \infty$ are given below,

$$I_{\mathrm{domain}}(q_x) = 8\,\pi K \langle \eta^2 \rangle \frac{\xi_x \xi_z^2}{(1 + q_x^2 \xi_x^2)^2} \sim K \langle \eta^2 \rangle \xi_x^{-3} \xi_z^2 q_x^{-4}, \tag{4.3.26}$$

so that the second line in the rhs of Eq. 4.3.26 along with Eq. 4.3.20 predicts the relation given below

$$I_{\mathrm{doamin}}(q_x) < I_{\mathrm{domain}}(q_z), \tag{4.3.27}$$

at any given $q_x = q_y$ in the q-range where the Porod's law is valid. We can predict from Eqs. (4.3.20), (4.3.26), and (4.3.27) that

$$\xi_x > \xi_z, \tag{4.3.28}$$

although $I_{\mathrm{domain}}(q_x)$ could not be accurately estimated because of its weak intensity in the covered q_z range.

The flow-induced self-organization of the dissipative structures in Regime III is summarized as follows, as schematically illustrated in Fig. 4.13 from part (a) to (c) with the increased flow rate:

(a) Formation of a cluster of many oriented prolate-ellipsoidal domains, whose centers of mass are predicted to be randomly placed in space (part a): the revolution axis of the prolate-domains is kept preferentially oriented along Oz-axis under the flow,

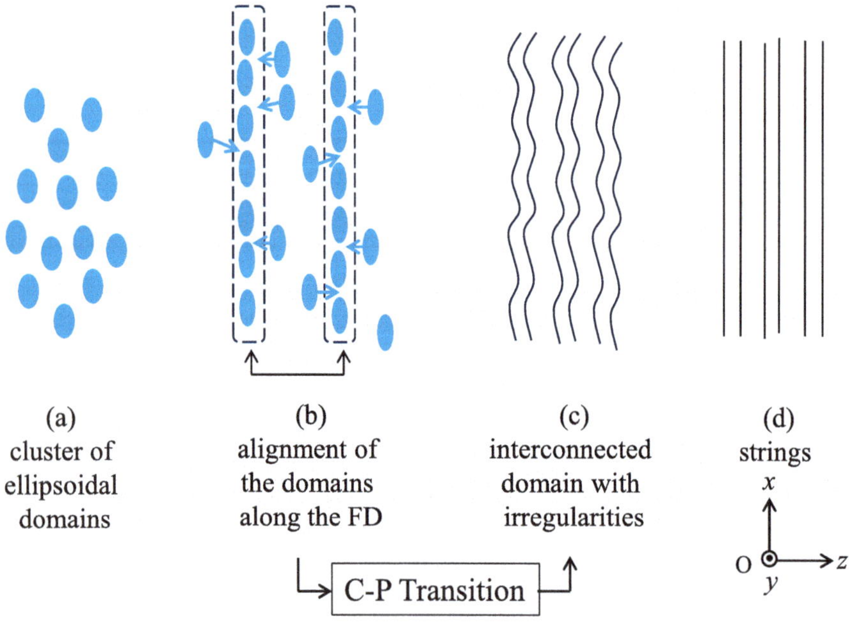

Fig. 4.13 Flow-induced self-organization of the dissipative structures in Regime III as schematically presented from part (a) to (c). The ellipsoids with the arrow in part (b) indicate the flow-induced incorporation of off-aligned domains into assemblies of the domains aligned along the flow direction FD (along Ox axis). Part (d) illustrates the strings in Regime IV

(b) Alignment of the mass centers of the domains in lines along the flow direction and incorporation of the off-aligned domains into the lines (part b),
(c) Interconnection of the aligned domains into the percolated domains along the flow direction (the C-P transition) (part c). The percolated domains have the irregulates as described in conjunction with Fig. 4.10a.
(d) Relaxation of the irregularities of the percolated domains to form uniform percolated domains along the flow direction, i.e., the so-called strings in Regime IV (part d).

4.3.7 Regime IV: Flow-Induced Dissolution of Strings

Figure 4.14 presents the SALS intensity distribution along q_z direction [12] for the PS/PB (50:50)/DOP 3.3 wt. % with $T_c(0) = 76$ °C at $T = 68$ °C in Regime IV (representatively at $\dot{\gamma} = 16, 40$, and $100\ \mathrm{s}^{-1}$) and in Regime V (representatively at $\dot{\gamma} = 253\ \mathrm{s}^{-1}$). The experimental scattering profile $I_{\mathrm{exp}}(q_z)$ (symbols) in Regime IV is best fitted with the theoretical scattering profile (solid line) from the long uniform cylindrical domains $I_{\mathrm{domain}}(q_z)$ given by the square Lorentzian (SQL) with $\xi_y = \xi_z$ (in Eqs. 4.3.18 and 4.3.19), which leads to the equation given below, to obtain ξ_z,

$$I_{\mathrm{domain}}(q_z) = \frac{I_{\mathrm{domain}}(0)}{(1+\xi_z^2\, q_z^2)^2}, \tag{4.3.29}$$

$$I_{\mathrm{domain}}(0) = 8\ \pi K\langle\eta^2\rangle\xi_x\ \xi_z^2. \tag{4.3.30}$$

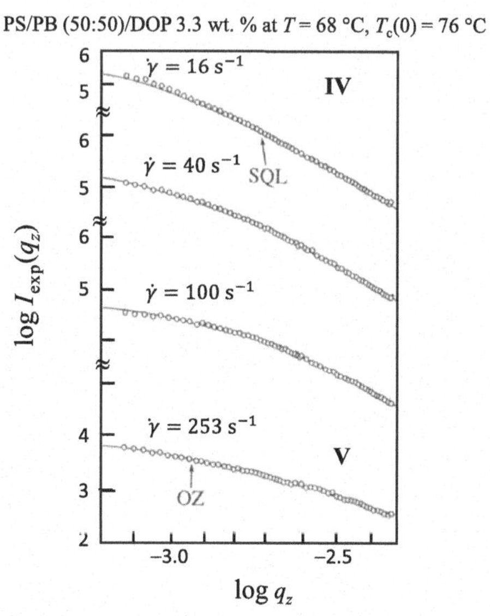

Fig. 4.14 $I_{\mathrm{exp}}(q_z)$ versus q_z at $\dot{\gamma} = 16, 40$, and $100\ \mathrm{s}^{-1}$ in Regime IV and that at $\dot{\gamma} = 253\ \mathrm{s}^{-1}$ in Regime V for the PS/PB (50:50)/DOP 3.3 wt. % with $T_c(0) = 76$ °C at $T = 68$ °C (symbols). The solid lines are the SQL function in Regime IV and OZ function in Regime V best-fitted with the experimental scattering function $I_{\mathrm{exp}}(q_z)$ shown by circles

Note that $I_{\text{domain}}(0)$ is defined as $I_{\text{domain}}(q_z)$ at $q_z = 0$.

Figure 4.15a presents the characteristic length ξ_z obtained by the best fit between $I_{\exp}(q_z)$ and $I_{\text{domain}}(q_z)$ in Regime IV ($0.8 \lesssim \log \dot{\gamma} \lesssim 2.1$) as well as in Regime III ($0.2 \lesssim \log \dot{\gamma} \lesssim 0.8$) [9, 18]. The data at $\dot{\gamma} \geq \dot{\gamma}_c = 127\ \text{s}^{-1}$ where $\dot{\gamma}_c$ is the critical shear rate for the flow-induced single-phase formation assessed by the increased $\dot{\gamma}$ is described later in Sect. 4.3.8. In Regime IV, ξ_z is predicted to be given by,

$$\xi_z \equiv \xi_{d,z} \sim \dot{\gamma}^{-n}, \quad n \simeq 1/4 \tag{4.3.31}$$

as shown by red straight line in Fig. 4.15a. On the other hand, the exponent n varies from 1/2 (the Regime III) to 1/3 (Regime II, $\log \dot{\gamma} < 0.2$) with decreased $\dot{\gamma}$.

The size of the dissipative structure in the steady state at the any given shear rates satisfying $\dot{\gamma} < \dot{\gamma}_c$ is generally controlled by the "**shear distortion time**" $t \sim \dot{\gamma}^{-1}$. This is due to the following reasons:

(1) In the time scale of $t < \dot{\gamma}^{-1}$, the domains grow driven by the thermodynamic driving force.
(2) In the time scale of $t \geq \dot{\gamma}^{-1}$, the overgrown domains are burst into smaller domains driven by shear deformation, whereas the burst smaller domains grow driven by the thermodynamic force.

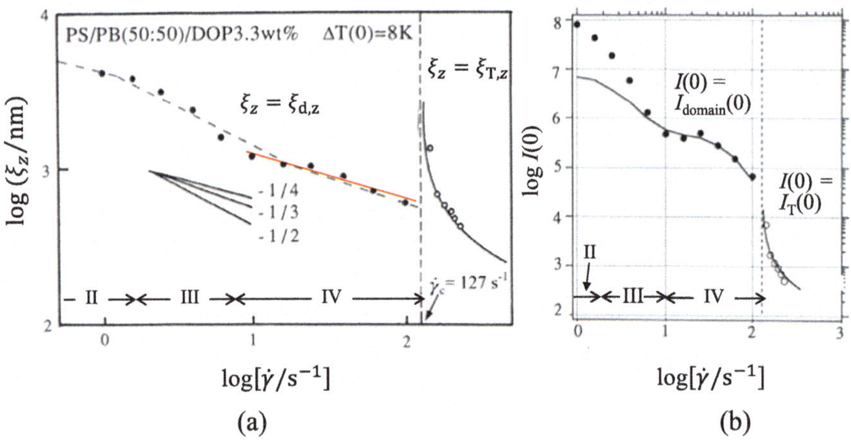

Fig. 4.15 Double logarithmic plot of the characteristic length ξ_z versus $\dot{\gamma}$ (part a) and $I(0) \equiv I(q_z = 0)$ versus $\dot{\gamma}$ (part b) for the PS/PB (50:50)/DOP 3.3 wt. % with $T_c(0) = 76\,°\text{C}$ at $T = 68\,°\text{C}$. In Regime III and IV at $\dot{\gamma} < \dot{\gamma}_c = 127\ \text{s}^{-1}$, the critical shear rate at the phase transition, ξ_z is the characteristic length of the domain (string), $\xi_{d,z}$, along the neutral direction (Oz-axis), and $I(0)$ is $I_{\text{domain}}(0)$. In Regime V at $\dot{\gamma} > \dot{\gamma}_c$, $\xi_z = \xi_{T,z}$ and $I(0) = I_T(0)$ for the thermal concentration fluctuations in the shear-induced single phase. The solid lines for $\xi_{T,z}$ and $I_T(0)$ in part (a) and part (b), respectively, are predicted from Eq. 4.3.51. The solid line for $I_{\text{domain}}(0)$ is predicted from Eqs. 4.3.32–4.3.34 and 4.5.3. The dashed line for $\xi_{d,z}$ is for visual guide. The shear-rate dependence of $\xi_{d,z}$ (**a**) and $I_{\text{domain}}(0)$ (**b**) at $\dot{\gamma} < \dot{\gamma}_c$ characterize the scaled structure factor below $\dot{\gamma}_c$ as will be shown in Fig. 4.22

Thus, $\dot{\gamma}^{-1}$ provides the time for the domains to grow into the domains at the steady state controlled by $\dot{\gamma}$. The principle of equivalence between $\dot{\gamma}^{-1}$ and t in the self-organization of the dissipative structures will be validated later in Sect. 4.8. ξ_z above $\dot{\gamma}_c$, defined as $\xi_{T,z}$, is discussed in Sect. 4.3.8.

Figure 4.15b presents $I_{\text{domain}}(0) \equiv I_{\text{domain}}(q_z)$ at $q_z = 0$ versus $\dot{\gamma}$ for Regime IV and Regime III evaluated by the best fitting between $I_{\text{exp}}(q_z)$ (symbols) and $I_{\text{domain}}(q_z)$ given by Eqs. 4.3.29 and 4.3.30 (solid line). In these regimes, ξ_x is anticipated to be large and constant independent of $\dot{\gamma}$. In Eq. 4.3.30, $\langle \eta^2 \rangle$ is proportional to $\left[\Delta\phi_p(\dot{\gamma})\right]^2$, where $\Delta\phi_p(\dot{\gamma})$ is the difference in the total polymer concentration, in the PS/PB DOP solutions, between the elongated domains and their matrix solution at the any given $\dot{\gamma} < \dot{\gamma}_c$. Thus, we obtained from Eq. (4.3.30),

$$I_{\text{domain}}(0) \sim \left[\Delta\phi_p(\dot{\gamma})\right]^2 \xi_z^2. \tag{4.3.32}$$

Here, it should be noted that ξ_z depends on $\dot{\gamma}$ in Regime IV as presented in Fig. 4.15a, whereas ξ_x is anticipated to be constant, independent of $\dot{\gamma}$ in Regime IV where macroscopically long, uniform strings are formed along Ox-axis, as schematically presented in Fig. 4.13d. The mean-field theory predicts that

$$\left[\Delta\phi_p(\dot{\gamma})\right]^2 \sim \Delta T(\dot{\gamma}), \tag{4.3.33}$$

in the flow field at the shear rate $\dot{\gamma}$, where $\Delta T(\dot{\gamma})$ is the quench depth of the system at a given working T under the shear flow at the shear rate $\dot{\gamma}$, defined by the first line of the equation given below

$$\Delta T(\dot{\gamma}) \equiv T_c(\dot{\gamma}) - T = \Delta T(0) - \Delta T_c(\dot{\gamma}), \tag{4.3.34}$$

with $T_c(\dot{\gamma})$ being the critical temperature of the system at $\dot{\gamma}$ (see Fig. 4.23b to be shown later). $\Delta T(\dot{\gamma})$ is given also by the second line of Eq. 4.3.34 where $\Delta T(0)$ is the quench depth $\Delta T(\dot{\gamma})$ at $\dot{\gamma} = 0$. $\Delta T_c(\dot{\gamma})$ is the drop of the critical temperature due to the shear flow at the shear rate $\dot{\gamma}$ defined by

$$\Delta T_c(\dot{\gamma}) \equiv T_c(0) - T_c(\dot{\gamma}). \tag{4.3.35}$$

$\Delta T_c(\dot{\gamma})$ is given later by Eq. 4.5.3 in Sect. 4.5.

Figure 4.15b presents $I_{\text{domain}}(0)$ theoretically predicted based on Eqs. 4.3.32–4.3.34, and 4.5.3 to be given later in Sect. 4.5 (solid line) and $I_{\text{domain}}(0)$ experimentally evaluated (filled circles) using the so-called Debye-Bueche plot of $I_{\text{domain}}(q_z)^{-1/2}$ versus $q_z{}^2$ at the any given shear rates in Regime III and IV: the plot yields the linear relationship between the two quantities, and the linear extrapolation of the plot to $q_z{}^2 = 0$ gives $I_{\text{domain}}(0)^{-1/2}$, as a function of the shear rate, at the intercept with the axis of $I_{\text{domain}}(q_z)^{-1/2}$ at $q_z{}^2 = 0$. The increased deviation of the experimental results from the predicted results with decreased $\dot{\gamma}$ in Regime III is due to the increased contribution of the strong elliptical SALS pattern superposed to the streak pattern with decreased $\dot{\gamma}$. The data on $I_T(0)$ at $\dot{\gamma} > \dot{\gamma}_c$ in Fig. 4.15b are discussed later in Sect. 4.3.8.

Upon increasing $\dot{\gamma}$ in Regime IV, the composition of PS and PB in the string and matrix phase deviate from the equilibrium one at $\dot{\gamma} = 0$ in such a way that the intermixing of the PS and the PB increases so that the composition difference between the two phases decreases as predicted by Eqs. 4.3.32–4.3.34 and as manifested by the fact that the streak-like SALS pattern loses its intensity as shown in part (f) and (g) in Fig. 4.5. These trends of the decreased diameter and the increased intermixing of the strings are characteristic of Regime IV, shown schematically in the change from Fig. 4.6d to 4.6e. At $\dot{\gamma} > \dot{\gamma}_c$, the system is brought into the single phase as will be further discussed in Sect. 4.3.8. This shear-rate range is defined as Regime V.

Similar changes in the dissipative structures with $\dot{\gamma}$ were also reported by Han and co-workers. They found the change from droplets to string-like structure for the ternary system of the PS/PB/DOP [19] and the binary system for low-vinyl PB/low-vinyl Polyisoprene (PI) mixture [20]. Furthermore, by using the low molecular weight polymer mixtures of the PB/PI, they found the string-like structure is eventually dissolved to result in a homogeneous phase with the increased flow rate [21, 22].

4.3.8 Regime V: Flow-Induced Single Phase

Onuki presented the linearized dynamic equation for the thermal composition fluctuations in the single phase $\delta\phi(\mathbf{q}, t)$ in q-space [17],

$$\frac{\partial}{\partial t}\delta\phi(\mathbf{q}, t) = \dot{\gamma}\; q_x \frac{\partial}{\partial q_y}\delta\phi(\mathbf{q}, t) - \Gamma(\mathbf{q})\;\delta\phi(\mathbf{q}, t) + \zeta(\mathbf{q}, t). \tag{4.3.36}$$

It is important to note here that the systems under the considerations have the dynamic symmetry, so that the stress associated with the stress-diffusion coupling is not involved in the dynamic equation (see Sects. 2.2 and 5.7.1). The fluctuations are convected by the shear and simultaneously dissipated thermally with the decay rate $\Gamma(\mathbf{q})$ as described by the first and second term in the rhs of Eq. 4.3.36, respectively. The decay rate in the single-phase state is given by

$$\Gamma(q) = \Lambda(0) q^2 \left(r_R + q^2\right), \tag{4.3.37a}$$

where r_R is defined by

$$r_R \equiv r_0/\left(2\; C_g\right) = \xi_T^{-2}, \tag{4.3.37b}$$

for the dynamically symmetric systems, where $\Lambda(0)$ is the Onsager transport coefficient in the small q limit, and ξ_T is the thermal correlation length for the thermal concentration fluctuations. The structure factor $F_T(\mathbf{q}, t)$ for the thermal concentration fluctuations is given by

$$F_{\mathrm{T}}(\mathbf{q}, t) \sim \delta\phi(q, t)^2. \tag{4.3.38}$$

At the steady state,

$$\frac{\partial F_{\mathrm{T}}(\mathbf{q}, t)}{\partial t} = 0, \tag{4.3.39}$$

so that the steady-state structure factor $F_{\mathrm{T}}(\mathbf{q})$ satisfies

$$\left(2\Gamma(\mathbf{q}) - \dot{\gamma}\, q_x \frac{\partial}{\partial q_y}\right) F_{\mathrm{T}}(\mathbf{q}) = 2\,\Lambda(0)\, q^2. \tag{4.3.40}$$

The rhs of Eq. 4.3.40 arises from the thermal noise term.

The effect of shear is significant with respect to the thermal composition (or concentration) fluctuations under the conditions of $r_{\mathrm{R}} \cong 0$ and $q \ll q_{\mathrm{c}}$, where q_{c} is determined by $\Lambda(0) {q_{\mathrm{c}}}^4 \equiv \dot{\gamma}$. In this regime (the strong shear regime), $F_{\mathrm{T}}(\mathbf{q})$ is given by the following approximate equation [17],

$$F_{\mathrm{T}}(\mathbf{q}) = \left(r_{\mathrm{R}} + c\, q_{\mathrm{c}}^{8/5}\, q_x^{2/5} + q^2\right)^{-1}, \tag{4.3.41}$$

where $q^2 = {q_x}^2 + {q_y}^2 + {q_z}^2$, and $c \cong 0.76$. $F_{\mathrm{T}}(q_z)$ and $F_{\mathrm{T}}(q_x)$ defined as follows,

$$F_{\mathrm{T}}(q_z) \equiv F_{\mathrm{T}}\left(q_x = q_y = 0, q_z\right), F_{\mathrm{T}}(q_x) \equiv F_{\mathrm{T}}\left(q_x, q_y = q_z = 0\right), \tag{4.3.42}$$

are given by

$$F_{\mathrm{T}}(q_z) = (r_{\mathrm{R}} + q_z^2)^{-1}, \tag{4.3.43}$$

$$F_{\mathrm{T}}(q_x) = (r_{\mathrm{R}} + q_x^2 + c q_{\mathrm{c}}^{8/5}\, q_x^{2/5})^{-1}. \tag{4.3.44}$$

Thus, the structure factor normal to the flow (neutral) direction $F_{\mathrm{T}}(q_z)$ is unaffected by shear and given by OZ scattering function. On the other hand, the structure factor along the flow direction $F_{\mathrm{T}}(q_x)$ is suppressed by the shear due to the third term in the rhs of Eq. (4.3.44), which depends on the shear-rate dependent q_{c}. Equation (4.3.44) predicts that the higher is the shear rate, the larger the value q_{c} is, and thereby, the larger the suppression of $F_{\mathrm{T}}(q_x)$ and $\delta\phi(q_x) \equiv \delta\phi(q_x, q_y = 0, q_z = 0)$ are at a given q_x relative to OZ function along q_x; the large is q_x, the larger the suppression of the $F_{\mathrm{T}}(q_x)$ and $\delta\phi(q_x)$ are at a given $q_{\mathrm{c}}(\dot{\gamma})$.

Figure 4.16 presents $I_{\mathrm{T}}(q_x)^{-1}$ versus ${q_x}^2$ and $I_{\mathrm{T}}(q_z)^{-1}$ versus ${q_z}^2$ for the PS/PB (50:50)/DOP 3.0 wt. % at $T = 43.2$ °C ($\Delta T = 10.8$ °C) and at $\dot{\gamma} = 670\ \mathrm{s}^{-1}$($\dot{\gamma}_{\mathrm{c}} \cong 100\ \mathrm{s}^{-1}$), where $I_{\mathrm{T}}(q_x)$ and $I_{\mathrm{T}}(q_z)$ shown by open and solid circles, respectively, are the measured scattered intensity distributions along q_x and q_z, respectively, which are proportional to the structure factor $F_{\mathrm{T}}(q_x)$ and $F_{\mathrm{T}}(q_z)$, respectively. The results reveal that $I_{\mathrm{T}}(q_z)$ obeys OZ scattering function (shown by solid straight lines), unaffected by

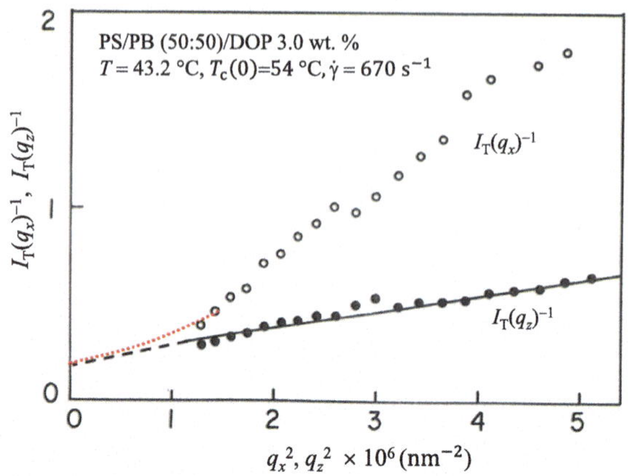

Fig. 4.16 OZ plot of $I_\mathrm{T}(q_x)^{-1}$ versus q_x^2 along the flow direction (Ox axis) (unfilled circles) and $I_\mathrm{T}(q_x)^{-1}$ versus q_z^2 along the neutral direction (along Oz axis) (filled circles) for the thermal concentration fluctuations in the shear-induced single phase at $\dot{\gamma} = 670\ \mathrm{s}^{-1}$ for the PS/PB (50:50) 3.0 wt. % with $T_\mathrm{c}(0) = 54.0\,°\mathrm{C}$ and $\dot{\gamma}_\mathrm{c} = 165\ \mathrm{s}^{-1}$ at $T = 43.2$ °C. The solid line is the best fit of the experimental results with the OZ equation. The dotted line is a guide for eye for $I_\mathrm{T}(q_x)^{-1}$ at the low q range of q_x^2

the shear flow, whereas $I_\mathrm{T}(q_x)$ does not obey OZ scattering: the suppression of $I_\mathrm{T}(q_x)$ from the OZ scattering intensity at the given shear rate increases with increased q_x, whose trend is predicted by the theoretical expression (Eq. 4.3.44).

Figure 4.17 presents $I_\mathrm{T}(q_z)^{-1}$ versus $q_z{}^2$ for the PS/PB (50:50)/DOP 3.3 wt. % at the various shear rates and at a given fixed temperature $T = 68$ °C ($\Delta T(0) = 6$ °C) [3]. All the data show the linear relationship between $I_\mathrm{T}(q_z)^{-1}$ and $q_z{}^2$ as shown by the straight line best-fitted with the experimental data (unfilled circles), revealing that the scattering intensity $I_\mathrm{T}(q_z)$ along the neutral direction (Oz axis) under the shear flow is described by the OZ scattering theory. $I_\mathrm{T}(q_z)$ is given from Eq. (4.3.43) by,

$$I_\mathrm{T}(q_z) = I_\mathrm{T}(0)/\left(1 + \xi_{\mathrm{T},z}^2\, q_z^2\right), \tag{4.3.45}$$

where $\xi_{\mathrm{T},z}$ is the corelation length of the thermal concentration fluctuations under the shear flow along the neutral direction (along Oz axis), and $I_\mathrm{T}(0) = I_\mathrm{T}(q_z)$ at $q_z = 0$ given by,

$$I_\mathrm{T}(0) = (\mathrm{const.})\xi_{\mathrm{T},z}^2,\ \xi_{\mathrm{T},z}^2 = 2\ C_\mathrm{g}/\mathrm{r}_0. \tag{4.3.46}$$

From the OZ plot of $I_\mathrm{T}(q_z)^{-1}$ versus $q_z{}^2$ shown in Fig. 4.17, $I_\mathrm{T}(0)^{-1}$ and $\xi_{T,z}{}^2$ are experimentally determined by

$$I_\mathrm{T}(0)^{-1} = (\text{the intercept of the straight line at } q_z^2 = 0) \tag{4.3.47}$$

and

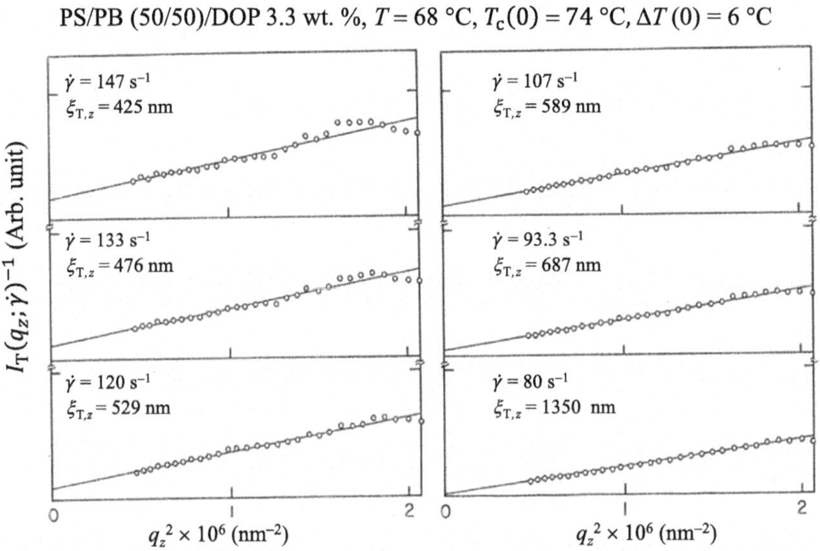

Fig. 4.17 OZ plot for the SALS intensity distributions $I_T(q_z)$ along the neutral direction at the different shear rates shown in the legends in the shear-induced single-phase solutionsfor the PS/PB (50:50)/DOP 3.3 wt. % with $T_c(0) = 74\,°\mathrm{C}$ and $\dot{\gamma}_c = 72\ \mathrm{s}^{-1}$at $T = 68\,°\mathrm{C}$. Based on [3], reprinted by AIP Publishing with permission

$$\xi_{T,z}^2 = \frac{\text{(the slope of the straight line in the OZ plot)}}{I_T(0)^{-1}}, \tag{4.3.48}$$

respectively. The $\xi_{T,z}$ value determined experimentally, as indicated in the figure legend in Fig. 4.17, is revealed to increase with decreased $\dot{\gamma}$. On the other hand, the value $I_T(0)^{-1}$is revealed to decrease with decreased $\dot{\gamma}$. Both trends reveal that the system approaches toward the spinodal point with the decreased $\dot{\gamma}$.

In the mean-field theory without shear flow, Eq. 4.3.46 predicts the equation given below

$$I_T(0)^{-1} \text{ and } \xi_T^{-2} \sim (\chi_s - \chi) \sim \frac{1}{T_s} - \frac{1}{T} \sim T - T_s \tag{4.3.49}$$

where χ_s and χ are the effective thermodynamic interaction parameter (χ_{eff}) between the PS and the PB in the presence of the neutral solvent at the spinodal temperature T_s and at a given temperature T in the single phase ($T > T_s$), respectively. The last line in Eq. (4.3.49) assumes that T is close to T_s. Under the shear flow, T_s depends on $\dot{\gamma}$, so that $T - T_s$ in Eq. 4.3.49 should be replaced by,

$$T - T_s(\dot{\gamma}) \equiv \Delta T(\dot{\gamma}). \tag{4.3.50}$$

If so, the depth of the superheating at a given T in the single phase above $T_s(\dot{\gamma})$, $\Delta T(\dot{\gamma})$, also depends on $\dot{\gamma}$. If the decrease of $\dot{\gamma}$ causes the increase of $T_s(\dot{\gamma})$ and thereby the decrease of $\Delta T(\dot{\gamma})$, both $I_T(0)^{-1}$ and $\xi_{T,z}^{-2}$ decrease with the decrease of $\dot{\gamma}$, as found experimentally shown in Fig. 4.17.

Figure 4.18 presents $\xi_{T,z}{}^{-2}$ versus $\dot{\gamma}^{1/2}$ and $I_T(0)^{-1}$ versus $\dot{\gamma}^{1/2}$ for the PS/PB (50:50)/DOP 3.3 wt. % at a given $T = 68$ °C ($\Delta T(0) = 8$ °C) [3]. The experimental results (symbols) are found to reveal that both $\xi_{T,z}{}^{-2}$ and $I_T(0)^{-1}$ decrease linearly with decreased $\dot{\gamma}^{1/2}$ to reach zero at the shear rate defined by the critical shear rate $\dot{\gamma}_c = 72\ \mathrm{s}^{-1}$ at $\Delta T(0) = 8$ °C, as indicated by the solid lines, which lead to the equations given below,

$$I_T(0)^{-1} \text{ and } \xi_{T,z}^{-2} \sim \dot{\gamma}^{1/2} - \dot{\gamma}_c^{1/2}. \tag{4.3.51}$$

This law experimentally found is predictable based on the mean-field theory under the shear flow, which is discussed in Sect. 4.4.

The solid lines fitted with the experimental data $\xi_z = \xi_{T,z}$ and $I(0) = I_T(0)$ at $\dot{\gamma} > \dot{\gamma}_c = 127\ \mathrm{s}^{-1}$ (indicated by symbols) in Fig. 4.15a, b are the one predicted by Eq. 4.3.51. Both $\xi_{T,z}$ and $I_T(0)$ reveal the shear-induced critical divergence as $\dot{\gamma} \to \dot{\gamma}_c$.

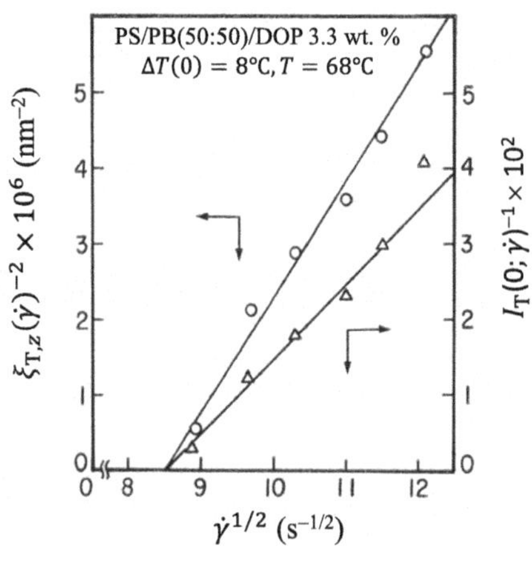

Fig. 4.18 Plot of $\xi_{T,z}(\dot{\gamma})^{-2}$ vs $\dot{\gamma}^{1/2}$ and $I_T(0;\dot{\gamma})^{-1}$ versus $\dot{\gamma}^{1/2}$ for the $\dot{\gamma}$-dependence of the thermal correlation length $\xi_{T,z}^{-2}(\dot{\gamma})$, and the SALS intensity at $q_{z,} = 0$, $I_T(0;\dot{\gamma})$, for the thermal concentration fluctuations in the shear-induced single-phase solution for the PS/PB(50:50)/DOP 3.3 wt. % with $T_c(0) = 76$ °C at $T = 68$ °C. $\xi_{T,z}(\dot{\gamma})$ and $I_T(0;\dot{\gamma})$ are evaluated from the OZ plots shown in Fig. 4.17. Based on [3], reprinted by AIP Publishing with permission

4.3.9 Stability of Strings in Regime IV

Figure 4.19 highlights the string structures as observed by OM and SALS for the PS/PB (50/50)/DOP 3.3 wt. % with $T_c(0) = 76$ °C at $T = 68$ °C ($\Delta T(0) = 8$ °C) and at $\dot{\gamma} = 40\ s^{-1}$ in Regime IV. The characteristic time τ_{conc} of the thermal composition fluctuations of the solution, $\tau_{conc} = 1/\Gamma_{conc} \cong 6$ s, where Γ_{conc} is evaluated by the growth rate of the SALS intensity $R(q_m(0))$ at the peak wave number $q_m(0)$ in the early stage spinodal decomposition after cassation of the shear in the single phase at $T = 68$ °C. Here, we note that $R(q_m(0)) = \Gamma_{conc} \cong 1/6\ s^{-1}$. Thus, the strong shear with respect to the composition fluctuations is predicted to be in the shear-rate range of $\dot{\gamma} > 0.2\ s^{-1}$: therefore, the shear rate range for the Regime IV (at $0.4 < \dot{\gamma} < 100\ s^{-1}$) belongs to the strong shear regime.

The strings at $\dot{\gamma} = 40\ s^{-1}$ have a high uniformity and spatial continuity along the flow direction (FD): the continuity extends from the top to the bottom of the OM images. Thus, the strings are macroscopically long with the length $\gtrsim 500$ μm. Their lateral length $\xi_z \equiv \xi_{d,z} \sim 800$ nm at $\dot{\gamma} = 40\ s^{-1}$ and goes down to ~ 500 nm at the largest $\dot{\gamma}$ measured in Regime IV as shown in Fig. 4.15a. The high spatial continuity along the FD as observed by OM image is confirmed also by a very small spread of the streak-like SALS pattern along the FD as shown in the inset to Fig. 4.19.

Tomotika [23] predicted the instability of an infinitely long cylindrical thread of the Newtonian fluid of viscosity η' in the matrix of another Newtonian fluid with viscosity η (Fig. 4.20a). The two fluids are assumed to be immiscible. Tomotika's linear stability analysis predicts that the growth of the amplitude of the undulation (lateral perturbation), ξ_{un}, of the thread with time, as defined in Fig. 4.20 (b), is given by

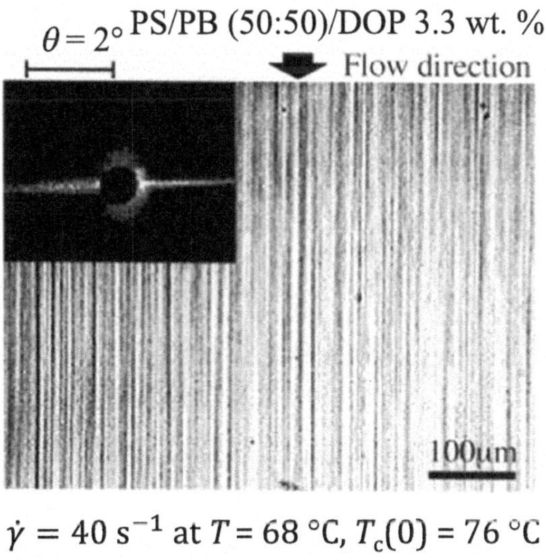

Fig. 4.19 SALS pattern (inset) and OM image at $\dot{\gamma} = 40\ s^{-1}$ and at $T = 68$°C representative in the Regime IV for the PS/PB (50:50)/DOP 3.3 wt. % with $T_c(0) = 76$°C and $\dot{\gamma}_c = 127\ s^{-1}$

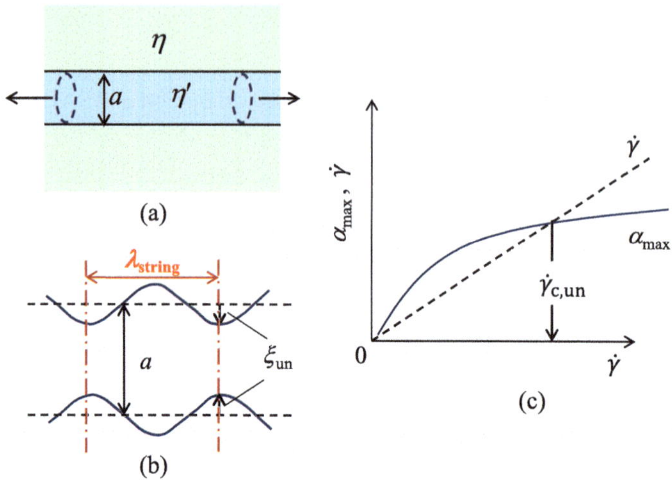

Fig. 4.20 Flow-induced stability of the string structure in Regime IV (part a) against the interface undulation (part b). λ_{string} is the wavelength of the undulation of the strings developed under the flow at the any given shear rates satisfying $\dot{\gamma} < \dot{\gamma}_{c,un}$ (part c)

$$\xi_{un} = \xi_0 \exp(\alpha_{max}\, t) \tag{4.3.52}$$

with the growth rate α_{max} given by

$$\alpha_{max} = \left[\gamma_{int}/(2\, \eta'\, a)\right] G(\eta'/\eta), \tag{4.3.53}$$

where γ_{int} is the interfacial tension of the thread, a is diameter of the thread, and G is the universal function of the viscosity ratio $(\eta\prime/\eta)$.

The system discussed here in this chapter has a very weak interfacial tension. The droplets developed in Regime II are elongated and aligned parallel to the flow and eventually interconnected along the flow direction via the shear-induced cluster-to-percolation transition as discussed earlier in Sect. 4.3.5. The percolated domain with the lateral size $\xi_{d,z}$ may correspond to the infinitely long cylindrical domain with the diameter a in Eqs. 4.3.52 and 4.3.53. In Regime IV, $\xi_{d,z}$ depends on shear rate as given by Eq. 4.3.31 and shown by Fig. 4.15a. The diameter a and the growth rate α_{max} in Eq. 4.3.53 may depend on $\dot{\gamma}$ as given by.

$$\alpha_{max} \sim a^{-1} \sim \xi_{d,z}^{-1} \sim \dot{\gamma}^{n}\ (n \cong 1/4). \tag{4.3.54}$$

Thus, the rate α_{max} in our case increases with increased $\dot{\gamma}$ due to the thinning of the strings with the shear rate as schematically shown in Fig. 4.20c.

In the case where $\dot{\gamma} > \alpha_{max}$, i.e., $\dot{\gamma} > \dot{\gamma}_{c,un}$, the critical shear rate for the undulation where $\dot{\gamma}_{c,\,un}$ is defined by

$$\dot{\gamma} \equiv \dot{\gamma}_{c,un},\ \text{where}\ \dot{\gamma} = \alpha_{max}. \tag{4.3.55}$$

the thread is stabilized under the shear against the instability driven by interfacial tension, i.e., against the undulation as schematically illustrated in Fig. 4.20b, which results in the straight cylindrical thread as illustrated in Fig. 4.20a or the straight, uniform string as shown in Fig. 4.19.

On the other hand, in the case where $\dot{\gamma} < \alpha_{max}$, the straight thread is unstable against the instability so that the undulation grows as predicted by Eq. 4.3.52. The experimental system under consideration herewith contains the many threads, whose volume fraction is equal to that of their matrix phase. In such a many-thread system, the instability may bring about the cooperative unduation of the threads as schematically illustrated by Fig. 4.13c. The instability may bring about also the random irregularities in the threads and/or the interconnection of the threads to form percolated domains oriented along the FD as shown in Figs. 4.10a in Regime III.

The instability is enhanced by the thermodynamic driving force inherent to the system at the given temperature which is independent of $\dot{\gamma}$, whereas it is supressed by the $\dot{\gamma}$-dependent shear force. The dynamic balance of these two forces controls the stability of the strings as dissipative structures self-organized in the solutions in the steady state.

The strings developed at the given shear rates become unstable after cessation of the shear. The lateral disturbance may grow at the rate given by Eq. 4.3.53, giving rise to undulating string with a wavelength λ_{string} and eventually splitting into a series of droplets aligned along the flow directions. Figure 4.21 presents typical time evolution of the SALS patterns (parts d and f) and the OM images (pats c and e) simultaneously taken *in situ* and *at real time* after the cessation of shear [9, 10]. As shown in the OM images, the stable strings under the steady flow in part (a) are split into a series of spheres aligned along the original flow direction in the form of "**pearl-necklace**" (the string composed of the droplets appeared as dark spheres) as shown in part (c). The droplets diffuse and coalesce into larger droplets and a memory of the spatial alignment of the droplets along the flow gradually fades away, the trend of which may be seen in part (e). Correspondingly, a diffuse butterfly-type SALS pattern appears above and below the sharp streak-like scattering pattern arising from the droplets aligned along the string in the flow direction as shown in part (d). The diffuse scattering has a maximum intensity at the magnitude of the scattering vector $q_{x,m}$ along the original flow direction, reflecting the inter-droplets spacing originating from λ_{string}.

The sharp streak is still conserved as a memory of the original string structure and due to the pearl-necklace alignment of the droplets. As time elapses, an average size of the droplets increases, giving rise to a low q shift of the position $q_{x,m}$ of the diffuse scattering. Simultaneously the streak loses its intensity, as the pearl-necklace alignment fades away, i.e., the correlation length of the pearl-necklace decays, specifically along the flow direction. It is quite intriguing to note the following two points: (i) the diffuse scattering pattern is analogous to the butterfly type scattering patterns shown in Figs. 2.2d and 3.9 developed in the shear-induced demixed systems with the dynamic asymmetry; (ii) **a combination of the butterfly scattering pattern and streak-like scattering pattern** is analogous to the scattering pattern due to the **string-like structure** developed for shear-induced demixed systems, both of which

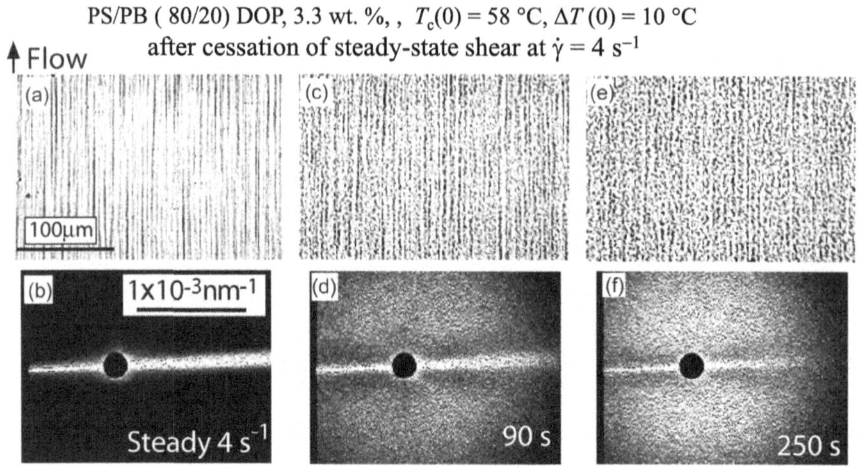

Fig. 4.21 OM images (**a**, **c**, **e**) and corresponding SALS patterns (**b**, **d**, **f**) for the PS/PB (80:20)DOP 3.3 wt. % solution with $T_c(0) = 58\,°C$ at $\Delta T(0) = 10\,°C$. Parts **a** and **b** were obtained under steady-state shear flow at $\dot{\gamma} = 4s^{-1}$: **c** and **d** were obtained at 90 s, and **e** and **f** were obtained at 250 s after cessation of the shear flow at $\dot{\gamma} = 4\ s^{-1}$. Based on [10]

will be extensively discussed in Chaps. 5 and 6 regarding the shear-induced phase separation of the solution from the single-phase solution to the two-phase solution.

The transformation of the uniform string into the pearl-necklace pattern of the burst droplets after the cessation of the shear from the shear-rate regime where the stable string is formed was extensively explored and analyzed by C.C. Han and co-workers [24] and J. Mewis and his co-workers [25–31] for polymer mixtures.

4.3.10 Remarks on Sect. 4.3

Table 4.2 presents the relation between the two kinds of the classifications on the flow-induced structures. The classifications of Regime [I] to Regime [IV] shown in the top row (a) are concerned with the classification of the dissipative structures self-organized with increased flow rate. The classification of Regime $\left[\dot{\gamma}_{\mathrm{I}}\right]$ and Regime $\left[\dot{\gamma}_{\mathrm{II}}\right]$ shown in the bottom row (b) is concerned with the classification of the characteristic shear-rate range. In the dynamically symmetric systems, the dissipative structures which belong to Regime [I] are self-organized under the shear-rate range defined as Regime $\left[\dot{\gamma}_{\mathrm{I}}\right]$ satisfying $\dot{\gamma} < \Gamma_{\mathrm{conc}}$, whereas the dissipative structures belong to Regime [II] to Regime [V] are self-organized under the shear-rate range defined as Regime $\left[\dot{\gamma}_{\mathrm{II}}\right]$ satisfying $\Gamma_{\mathrm{conc}} < \dot{\gamma} < \Gamma_{\mathrm{MD}}$.

Table 4.2 (a) Classification of Regime [I] to [IV] based on shear-induced structures (top row) and (b) the classification of Regime [$\dot{\gamma}_{\rm I}$] and, [$\dot{\gamma}_{\rm II}$] based on the effects of shear on CF and MD (bottom row) for dynamically symmetric systems

(a) I	II	III	IV	V
(b) [$\dot{\gamma}_{\rm I}$]	[$\dot{\gamma}_{\rm II}$]			

[$\dot{\gamma}_{\rm I}$] : $\dot{\gamma} < \Gamma_{\rm conc}$,   [$\dot{\gamma}_{\rm II}$] :$< \Gamma_{\rm conc} < \dot{\gamma} < \Gamma_{\rm MD}$
CF: Concentration fluctuations
MD: Molecular deformation

4.4 Flow-Induced Phase Transition: from Two-Phase to Single-Phase Phase Transition with Increased $\dot{\gamma}$

The shear-induced phase transition was determined from both the two-phase state (in Regime IV) [3, 5] and the single-phase state (in Regime V) [3] by analyzing a series of the SALS profiles $I(q_z)$ along the neutral axis (Oz-axis) measured at steady state as a function of shear rates under the increased and decreased shear protocol, respectively. Han and co-workers also reported the shear-induced phase transition of the polymer mixtures having low molecular weights, based on their studies of the effects of the shear flow not only for the single-phase mixtures in terms of the scattering from thermal concentration fluctuations [32–35] but also for the two-phase mixtures [21, 22] in terms of scattering from phase-separated domain structures. Higgins and co-workers also reported the shear-induced mixing of the polymer mixtures from the two-phase states [36, 37].

A qualitative observation of the disappearance and appearance of the streak-like SALS pattern with the increased and decreased share rate, respectively, does not provide the sufficient evidence for the assessment of the critical shear rate $\dot{\gamma}_{\rm c}$ for the phase transition from the two-phase to single-phase and single-phase to two-phase, respectively. This is because the disappearance and appearance of the streak-like SALS pattern may possibly occur when the scattering objects just come outside and inside the windows of the SALS observations, respectively, with respect to the q-range and with respect to the minimum scattered intensity that can be detected with a given SALS apparatus employed. One must indicate quantitatively the change in scattering functions between the two-phase state in Regime IV and the single-phase state in Regime V to unequivocally assess $\dot{\gamma}_{\rm c}$.

In the two-phase state, the measured SALS intensity profiles along the neutral direction of the flow (along Oz axis) $I(q_z; \dot{\gamma})$, defined by $I_{\rm domain}(q_z; \dot{\gamma})$, were characterized by the **squared Lorentzian functions (SQL)** in Eq. 4.3.29 where $\xi_{\rm z}(\dot{\gamma})$ is the correlation length of the domains along the q_z axis, defined as $\xi_{\rm d,z}(\dot{\gamma})$, in Fig. 4.15a. It depends on the lateral size of the strings: the larger the size is, the larger is the value of $\xi_{\rm d,z}$. The squared Loretzian function along q_z over the q_z range covered in our experiments suggests that the system has a random two-phase structure [14, 38]. $I_{\rm domain}(q_z - 0; \dot{\gamma})$ and $\xi_{\rm d,z}(\dot{\gamma})$ depend on $\dot{\gamma}$. Hence, the plots of log $I_{\rm domain}(q_z; \dot{\gamma})$ versus log q_z give a series of different curves when $\dot{\gamma}$ is varied. However, when $I(q_z; \dot{\gamma})$ and q_z are reduced by the shear-rate dependent

$I_{\text{domain}}(q_z = 0; \dot{\gamma})$ and $\xi_{d,z}(\dot{\gamma})$, respectively, the shear-rate independent master curve of the so-called **universal scaled structure factor** $F(Q_s)$ should be obtained as follows:

$$F(Q_s) = \left[1 + Q_s^2\right]^{-2} \tag{4.4.1}$$

with

$$F(Q_s) \equiv I_{\text{domain}}(q_z; \dot{\gamma})/I_{\text{domain}}(q_z = 0; \dot{\gamma}),\ Q_s \equiv q_z \xi_{d,z}(\dot{\gamma}). \tag{4.4.2}$$

Figure 4.22 presents the master curve of $F(Q_s)$ in the double logarithmic scale, log $F(Q_s)$ versus log Q_s, obtained at the quench depth of $\Delta T(0) = 8\,°\text{C}$ (T = 68 °C) for the quiescent solution and at the various shear rates from 1.0 to 100 s^{-1} (see the universal scaled structure factor of the SQL labeled Regime IV) [9, 18].

In the single-phase state, the measured profiles $I(q_z; \dot{\gamma})$, defined as $I_T(q_z; \dot{\gamma})$, were characterized by the **Lorentzian functions** (the **Ornstein-Zernicke equation, OZ**) [11] given by Eq. 4.3.45 where $I_T(0; \dot{\gamma})$ is $I_T(q_z = 0; \dot{\gamma})$ in the single-phase state, and $\xi_{T,z}(\dot{\gamma})$ is the thermal correlation length along q_z for the thermal concentration fluctuations under shear. $I_T(q_z = 0; \dot{\gamma})$ and $\xi_{T,z}(\dot{\gamma})$ depend on shear rate: both $I_T(q_z = 0; \dot{\gamma})$ and $\xi_{T,z}(\dot{\gamma})$ increase with decreased $\dot{\gamma}$ toward the critical shear rate $\dot{\gamma}_c$ for the shear-induced phase transition. However, when $I_T(q_z; \dot{\gamma})$ and q_z are scaled with $I_T(q_z = 0; \dot{\gamma})$ and $\xi_{T,z}(\dot{\gamma})$, respectively, the shear-rate independent master curve (universal scaled structure factor of OZ) $F(Q_s)$ should be obtained as follows:

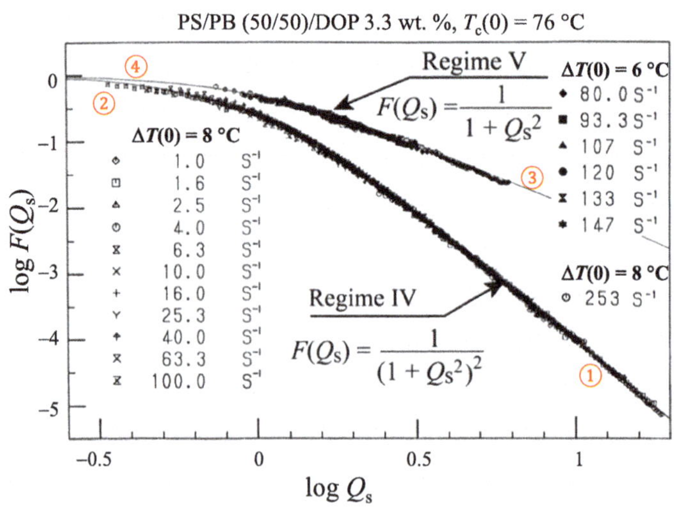

Fig. 4.22 Scaled structure factors $F(Q_s)$ versus reduced magnitude of scattering vector Q_s for the PS/PB (50:50) 3.3 wt. % solution under the shear flow in Regime IV at $\Delta T(0) = 8\,°\text{C}$ at the varying shear rates from 1.0 to 100.0 s^{-1} and Regime V at $\Delta T(0) = 6\,°\text{C}$ at the varying shear rates from 80.0 to 147 s^{-1} and $\Delta T(0) = 8\,°\text{C}$ at 253 s^{-1}

$$F(Q_s) = \left[1 + Q_s^2\right]^{-1}, \quad (4.4.3)$$

with

$$F(Q_s) \equiv I_T(q_z; \dot{\gamma})/I_T(q_z = 0; \dot{\gamma}),\ Q_s \equiv q_z \xi_{T,z}(\dot{\gamma}). \quad (4.4.4)$$

Figure 4.22 presents the master curve obtained for the $I_T(q_z : \dot{\gamma})$ at $\Delta T(0) = 6°C$ (T = 70 °C) and at a range of shear rates from 80.0 to 147 s^{-1} as well as that at $\Delta T(0) = 8°C$ (T = 76 °C) and at $\dot{\gamma} = 253\ s^{-1}$(see the universal scaled structure factor of the OZ function labeled Regime V).

Thus, it is revealed that the shear-induced phase transition dramatically changes the nature of the scattering function $I(q_z)$ along the neutral axis. The shear-induced phase transition changes the scaled structure factor of the systems $F(Q_s)$ from the **scaled SQL** to the **scaled Lorentzian** or vice versa, both universal with $\dot{\gamma}$ and T. This change in the universal scaled structure factor as presented in Fig. 4.22 and the change in $\xi_{d,z}$, $\xi_{T,z}$, $I_{domain}(q_z = 0)$, and $I_T(0)$ as presented in Fig. 4.15, both accompanied by the change in $\dot{\gamma}$ across the critical shear rate $\dot{\gamma}_c$ for the phase transition, enable us to determine unequivocally the shear-induced phase transition.

For the given fixed q-range of observation with our shear-rheo-optical apparatus, the observed profile $I_{domain}(q_z)$ undergoes a unique shift over the master curve with the increased $\dot{\gamma}$ from Regime IV to V as follows: the profile $I_{domain}(q_z)$ starts from the tail part of the SQL master curve at the large Q_s (labeled ①) toward the head part of the SQL at the small Q_s (labeled ②), because $\xi_{d,z}$ decreases with the increased $\dot{\gamma}$; Upon further increase of $\dot{\gamma}$, the phase transition from the two-phase to the single-phase state occurs, so that the profile discontinuously shifts to the tail part of the OZ mater curve at the large Q_s (labeled ③), which is followed by the shifting toward the head part of the OZ at the small Q_s (labeled ④), because $\xi_{T,z}$ decreases with increased $\dot{\gamma}$.

The shear-rate dependence of $\xi_{d,z}$ and $\xi_{T,z}$ are shown earlier in Fig. 4.15a for the two-phase solution and the single-phase solution, respectively, and that of $I_{domain}(q_z = 0)$ and $I_T(0)$ in Fig. 4.15b [18]. With the increased shear rate in the two-phase state (the data points shown by filled circles in Fig. 4.15a), the power-law exponent n in $\xi_{d,z}(\dot{\gamma})$ given by Eq. 4.3.31 appears to change from 1/4 to1/2 (from Regime II to Regime III, respectively) and eventually to 1/4 as highlighted by the red solid line (in Regime IV). Upon increasing $\dot{\gamma}$ across the critical shear rate of $\dot{\gamma}_c = 127\ s^{-1}$ for the shear-induced two-phase to single-phase transition, $\xi_{z,d}$ increases discontinuously to $\xi_{T,z}$, which is followed by the monotonic decrease with increased $\dot{\gamma}$ in the single-phase region (open circles in Fig. 4.15a). As shown in Fig. 4.15b, $I_{domain}(0)$ decreases with $\dot{\gamma}$ in the two-phase state (solid circles). Upon further increasing $\dot{\gamma}$ above the critical shear rate, $I_{domain}(0)$ increases discontinuously to $I_T(0)$, which is followed by the monotonic decrease with increased $\dot{\gamma}$ in the single-phase region (open circles in Fig. 4.15b).

The temperature dependence of $\xi_{T,z}$ and $I_T(0)$ at $T > T_c(0)$ in the quiescent single-phase systems is predicted by

$$\xi_T^{-2} \sim \Delta T(0), I(0)^{-1} \sim \Delta T(0), \tag{4.4.5}$$

in the context of the mean-field theory, where $\Delta T(0)$ is defined by

$$\Delta T(0) \equiv T - T_c(0) \tag{4.4.6}$$

with $T_c(0)$ being the phase transition temperature at $\dot{\gamma} = 0$. If the shear flow lowers the phase transition temperature from $T_c(0)$ at equilibrium to $T_c(\dot{\gamma})$ at dynamic equilibrium (steady state) under the flow at a given $\dot{\gamma}(> \dot{\gamma}_c)$, as shown in Fig. 4.23(a), ξ_T and $I(0)^{-1}$ at a given T under the flow-induced single-phase state, defined as $\xi_{T,z}$ and $I_T(0)^{-1}$, respectively, are predicted to be given by replacing $\Delta T(0)$ in Eq. 4.4.6 with $\Delta T(\dot{\gamma})$ as follows: (see Fig. 4.23a)

$$\Delta T(\dot{\gamma}) \equiv T - T_c(\dot{\gamma}) = \Delta T_c(\dot{\gamma}) - \Delta T(0), \tag{4.4.7}$$

where $\Delta T_c(\dot{\gamma})$ is the drop of critical temperature due to shear defined as follows:

$$\Delta T_c(\dot{\gamma}) \equiv T_c(0) - T_c(\dot{\gamma}). \tag{4.4.8}$$

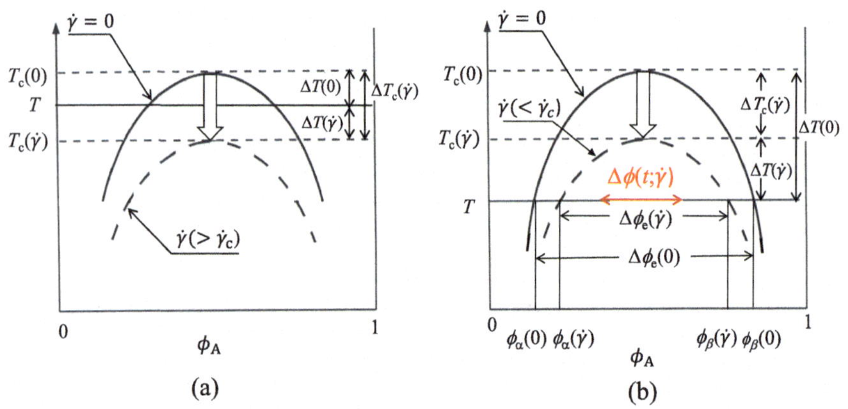

Fig. 4.23 Phase diagram of a binary mixture of A and B at the equilibrium in the absence of shear flow ($\dot{\gamma} = 0$, solid line) and at the dynamical equilibrium (or at the steady state) under the flow at a given $\dot{\gamma}$ (broken line). The flow at shear rate $\dot{\gamma}$ drops the critical temperature of the mixture from $T_c(0)$ at $\dot{\gamma} = 0$ to $T_c(\dot{\gamma})$ at a given $\dot{\gamma}$, as indicated by the large open arrow, pointing down. Part (a) and part (b) represents the case of $\dot{\gamma} > \dot{\gamma}_c$ and $\dot{\gamma} < \dot{\gamma}_c$, respectively. In part (a) the mixture at a given T exists in the shear-induced single-phase state with a superheating of $\Delta T\dot{\gamma}$with respect to the shear-rate dependent critical temperature $T_c(\dot{\gamma})$, while in part (b), the mixture at a given T exists in the shear-induced two-phase state with a quench depth of $\Delta T(\dot{\gamma})$ with respect to $T_c(\dot{\gamma})$. The critical shear-rate $\dot{\gamma}_c$ for the phase transition between two-phase and single-phase is attained, when $T_c(\dot{\gamma})$ reaches to the given working temperature T by lowering $\dot{\gamma}$ for the shear-indued single-phase mixture in part a and by raising $\dot{\gamma}$ for the shear-induced two-phase mixture in part b

Thus, the mean-field theory under the shear flow predicts that

$$\xi_{T,z}^{-2} \sim \Delta T(\dot{\gamma}), I_T(0)^{-1} \sim \Delta T(\dot{\gamma}), \tag{4.4.9}$$

where $\Delta T(\dot{\gamma})$ is given by Eq. 4.4.7. The experimentally established relationship between $\Delta T_c(\dot{\gamma})$ and $\dot{\gamma}$ is given by

$$\Delta T_c(\dot{\gamma})/T_c(0) = (\text{const})\dot{\gamma}^{1/2}, \tag{4.4.10}$$

as will be described in the next section, Sect. 4.5 (in Eqs. 4.5.2 and 4.5.3), where the proportionality constant, (const), has the dimension of square root of time. In Fig. 4.23a, one should note that

$$\Delta T_c(\dot{\gamma}) = \Delta T(0) \text{ at } \dot{\gamma} = \dot{\gamma}_c \tag{4.4.11}$$

for the system at a given $T < T_c(0)$ to attain the critical shear rate $\dot{\gamma}_c$ with the increased $\dot{\gamma}$. In our system, the critical temperature drops $\Delta T_c(\dot{\gamma})$ increases with increased $\dot{\gamma}$, so that the system at a given working temperature $T(< T_c(0))$ attains $\dot{\gamma}_c$ when $\Delta T_c(\dot{\gamma})$ is increased to $\Delta T(0)$ with the increased $\dot{\gamma}$ imposed to the single-phase solution (Fig. 4.23 above) Equations 4.4.10 and 4.4.11 give

$$\Delta T(0) = (\text{const.})T_c(0)\dot{\gamma}_c^{1/2}. \tag{4.4.12}$$

From Eqs. 4.4.7, 4.4.10 and 4.4.12 one obtains

$$\Delta T(\dot{\gamma}) = \Delta T(0)\left[(\dot{\gamma}/\dot{\gamma}_c)^{1/2} - 1\right] \text{ at a given } T > T_c(\dot{\gamma}). \tag{4.4.13}$$

Combining Eqs. 4.4.9 and 4.4.13, one obtains

$$\xi_{T,z} \sim \left(\dot{\gamma}^{1/2} - \dot{\gamma}_c^{1/2}\right)^{-1/2}, I_T(0) \sim \left(\dot{\gamma}^{1/2} - \dot{\gamma}_c^{1/2}\right) \text{ at given } T > T_c(\dot{\gamma}), \tag{4.4.14}$$

where $\dot{\gamma} \geq \gamma_c$.

The solid line in Fig. 4.15a (for $_{Tz}$) represents the best fit of Eq. 4.4.14 with the experimental data $\xi_{T,z}$(unfilled circles): the thermal correlation length $\xi_{T,z}$ tends to increase and diverges as the shear rate $\dot{\gamma}$ decreases to the critical shear rate $\dot{\gamma}_c$ shown by the vertical dotted line. The solid line in Fig. 4.15b for $I_T(0)$ represents the best-fit of Eq. 4.4.14 with the experimental data $I_T(0)$ (unfilled circles): $I_T(0)$ also tends to increase and diverge as $\dot{\gamma}$ decreases to the $\dot{\gamma}_c$ shown by the vertical dotted line. Thus, the $\dot{\gamma}$-dependence of $I_T(0)$ and $\xi_{T,z}$ in Regime V as well as that of $I_{\text{domain}}(0)$ in Regime IV can be consistently interpreted on the basis of the apparent downward shift of the phase boundary with the increased $\dot{\gamma}$ as will be shown in Figs. 4.23a, b and 4.24b as illustrated by the large open arrow, pointing down.

It may be worth noting here that the following intriguing phenomenon has been reported with respect to the shear-induced phase transition: that is, the given polymer mixtures exhibit not only the shear-induced mixing but also the shear-induced demixing. This phenomenon is first reported for the mixtures of PS and poly(vinyl methyl ether) by Winter and co-workers [39, 40]. Higgins and co-workers [41–43] and Chopra et al. [44] also reported both the shear-induced mixing and demixing for the given polymer mixtures reported by them.

The concept of the apparent downward shift of the phase boundary with shear rate can be applied also to the scaling analysis of the coarsening process of the dissipative structures with time via the **SD under the shear flow** [45, 46]. For example, after a shear-rate drop from Regime V (the shear-induced single-phase state) to Regime IV (the shear-induced two-phase state), we can **extend** the following **scaling hypothesis proposed in the quiescent state to that under shear flow**. In the quiescent state, the coarsening of the phase-separated domains via the SD can be characterized by the time-changes in the magnitude of the scattering vector $q_{\mathrm{m}}(t;\,\Delta T)$ at the maximum scattered intensity $I_{\mathrm{m}}(t;\,\Delta T)$ at the various quench depth ΔT. When the quantities $q_{\mathrm{m}}(t;\,\Delta T)$, $I_{\mathrm{m}}(t;\,\Delta T)$ and t are scaled, respectively, with the temperature-dependent characteristic wave number $q_{\mathrm{m}}(0;\,\Delta T)$ and time $t_{\mathrm{c}}(\Delta T)$ of the system, the scaled quantities $Q_{\mathrm{m}}(\tau)$ and $\tilde{I}_{\mathrm{m}}(\tau)$ defined by

$$Q_{\mathrm{m}}(\tau) \equiv q_{\mathrm{m}}(t;\,\Delta T)/q_{\mathrm{m}}(0;\,\Delta T), \tilde{I}_{\mathrm{m}}(\tau) \equiv I_{\mathrm{m}}(t;\,\Delta T\,)\, q_{\mathrm{m}}(t;\,\Delta T)^3/\Delta\phi_{\mathrm{e}}(\Delta T)^2, \tag{4.4.15}$$

with the reduced time τ defined by

$$\tau \equiv t/t_{\mathrm{c}}(\Delta T), t_{\mathrm{c}}(\Delta T)^{-1} \equiv D_{\mathrm{app}}(\Delta T) q_{\mathrm{m}}(0;\,\Delta T)^2, \tag{4.4.16}$$

become universal with ΔT. ΔT changes only the spatial scale and the time scale via $q_{\mathrm{m}}(0;\,\Delta T)$ and $t_{\mathrm{c}}(\Delta T)$, respectively, but it does not change the coarsening mechanism of the systems to be considered here (**scaling hypothesis**).

Note that the quench depth ΔT for the solution in the quiescent state is defined as $\Delta T(0) \equiv T_{\mathrm{c}}(0) - T$ in Fig. 4.23.Under the shear flow, however, ΔT depends on $\dot{\gamma}$, so that ΔT should be replaced by $\Delta T(\dot{\gamma}) \equiv T_{\mathrm{c}}(\dot{\gamma}) - T$ for the solution in the two-phase state under the flow. In Eq. 4.4.15, $\Delta\phi_{\mathrm{e}}(\Delta T)$ is the composition difference of the one component of the polymers in between the two phases separated at the thermal equilibrium, so that $\Delta\phi_{\mathrm{e}}(\Delta T) = \Delta\phi_{\mathrm{e}}(\Delta T(0))$ which is redefined as $\Delta\phi_{\mathrm{e}}(0)$ in Fig. 4.23b for brevity. Under the shear flow $\Delta\phi_{\mathrm{e}}(\Delta T)$ is the composition difference of one component in between the two phases separated at the dynamic equilibrium (steady state), so that the $\Delta\phi_{\mathrm{e}}(\Delta T)$ should be replaced by $\Delta\phi_{\mathrm{e}}(\Delta T(\dot{\gamma}))$ which is redefined as $\Delta\phi_{\mathrm{e}}(\dot{\gamma})$ in Fig. 4.23b for brevity. The quantities q_{m} and I_{m} in the quiescent state should correspond to those q_{mz} and I_{mz} along the neutral axis under the shear flow. Then, the scaled quantities $\widehat{Q}_{\mathrm{m}}(\hat{\tau})$ and $\widehat{I}_{\mathrm{m}}(\hat{\tau})$ under the shear flow, which respectively correspond to the $Q_m(\tau)$ and $\tilde{I}_{\mathrm{m}}(\tau)$ in the quiescent state, have been shown to give a universal behavior independent of $\dot{\gamma}$ where $\hat{\tau}$ is the reduced time of the given system under the shear flow defined by $\hat{\tau} \equiv t/t_{\mathrm{c}}(\Delta T(\dot{\gamma}))$ [18]. Thus, **shear rate also changes the space–time scale** of mixtures through the shear-rate dependent $\Delta T(\dot{\gamma})$, as the **temperature** $\boldsymbol{T}$ or the **quench depth** $\boldsymbol{\Delta T(0) \equiv (T_c(0) - T)}$ **changes the space-time scale of the mixtures** in quiescent state **in the critical phenomena**. More specific discussion is given later in Sect. 4.7.

4.5 Shear-Induced Shift of Phase-Transition Temperature: Small Molecules versus Polymers

In the preceding section, we experimentally elucidated the shear-induced phase transition from the two-phase to the single-phase with the increased shear rate. The critical shear rate $\dot{\gamma}_c$ for the transition is assessed by observing the shear-rate dependence of the parameters which characterizes the structure factor along the q_z axis: these parameters are $\xi_z(=\xi_{d,z})$ and $I_{z,\mathrm{domain}}(0)$ of the SQL function (Eqs. 4.3.29 and 4.3.30) in the two-phase state and/or the integrated scattered intensity in the two-phase state $\mathcal{J}_z(\dot{\gamma})$ (Eq. 4.3.1, Fig. 4.4): $\xi_{T,z}$ and $I_T(0)$ of the OZ function in the single-phase state (Eq. 4.3.45, Figs. 4.15, 4.17).

The phase-separated system at a given temperature T below $T_c(0)$ in the quiescent state, as shown in Fig. 4.24a, is brought into a single-phase system under the shear flow with the increased $\dot{\gamma}$ such that $\dot{\gamma}$ satisfies $\dot{\gamma} \geq \dot{\gamma}_c$, the critical shear rate for the single-phase formation at the given T which depends on $\Delta T(0) \equiv T_c(0) - T$ [8, 9]. If one can measure $\dot{\gamma}_c$ as a function of T below $T_c(0)$, one can determine T versus $\dot{\gamma}_c(T)$, as schematically shown in Fig. 4.24b. Figure 4.24b also predicts a critical temperature drop from $T_c(0)$ to $T_c(\dot{\gamma})$ by the amount of $\Delta T_c(\dot{\gamma}) \equiv T_c(0) - T_c(\dot{\gamma})$, under a given shear rate of $\dot{\gamma}$ (see Fig. 4.23 also). The value $\Delta T_c(\dot{\gamma})$ assessed experimentally is of the order of 1 mK for small-molecule systems [47] and 10 K = 10^4 mK for polymer systems over the shear-rate range experimentally covered [3, 45]. The shear-rate dependence of the critical temperature drop $\Delta T_c(\dot{\gamma})$ was found to be described by Eq. 4.5.1 below for the binary mixture of the small molecules first reported by Beysens et al. [47] and by Eqs. 4.5.2 and 4.5.3 for the dynamically symmetric polymer systems by Hashimoto and co-workers. [3, 45]

$$\frac{\Delta T_c(\dot{\gamma})}{T_c(0)} = (5.9 \pm 0.7) \times 10^{-7} \dot{\gamma}^{0.53 \pm 0.3}, \tag{4.5.1}$$

$$\frac{\Delta T_c(\dot{\gamma})}{T_c(0)} = (2.6 \pm 0.6) \times 10^{-3} \dot{\gamma}^{0.50 \pm 0.02}, \quad \text{for PS/PB (50 : 50)/DOP 3.0 wt. \%} \tag{4.5.2}$$

$$= (1.3 \pm 0.1) \times 10^{-3} \dot{\gamma}^{0.50 \pm 0.02}, \quad \text{for PS/PB (50 : 50)/DOP 3.3 wt. \%} \tag{4.5.3}$$

Figure 4.25 presents $\Delta T_c(\dot{\gamma})/T_c(0)$ versus $\dot{\gamma}$ for the 3.0 wt. % DOP solution of PS/PB(50:50) and for the 3.3 wt. % DOP solution of PS/PB(20:80), PS/PB(50:50), PS/PB(60:40),and PS/PB(70:30), respectively, in the increased shear-rate protocol [3, 6, 45]. The rigorous assessment of the phase transition point under the shear

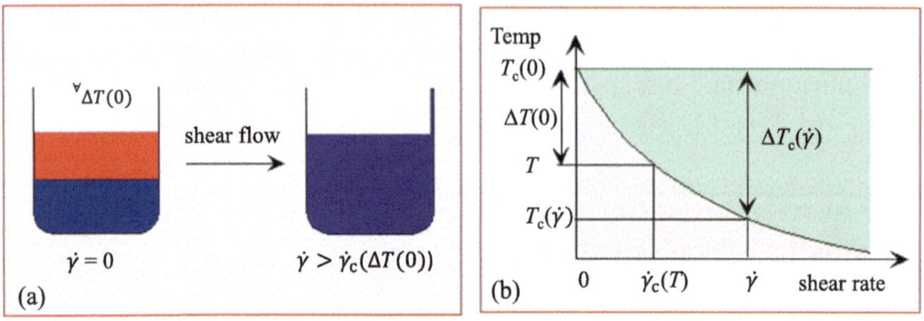

Fig. 4.24 (Part a) shear-induced two-phase to single-phase transition with increased shear rate (from left to right) at a given quench depth $\Delta T(0)$ at $\dot{\gamma} = 0$ is schematically shown to occur at shear rates greater than critical shear rate $\dot{\gamma}_c(\Delta T(0))$ for a dynamically symmetric system at a given T below the critical temperature $T_c(0)$ or $T(0) \equiv T_c(0) - T$ in the quiescent state. (Part b) shear-induced drop of the critical temperature $T_c(\dot{\gamma})$ at a given shear rate $\dot{\gamma}$: the shear flow at a given $\dot{\gamma}$ decreases the critical temperature from $T_c(0)$ to $T_c(\dot{\gamma})$. At a given $T < T_c(0)$, the increased shear rate from 0 shear rate brings the two-phase (colored red) to single-phase (colored green) transition at $\dot{\gamma}_c(T)$. Based on Hashimoto, 2008 In *Soft Matter Characterization*; Borsali, R., Pecora, R., Eds., Springer. New York; Chapter 8, pp. 378–462 [9]

flow was discussed earlier in Sect. 4.4. The results shown in Fig. 4.25 reveal the following predictions on the shear-rate dependence of the reduced temperature drop $\Delta T_c(\dot{\gamma})/T_c(0)$:

(1) The reduced temperature drop $\Delta T_c(\dot{\gamma})/T_c(0)$ is given by a power law of $\dot{\gamma}$, $\dot{\gamma}^n$, with the exponent $n = 1/2$ independent of the compositions of the polymers.
(2) The prefactor of the power law depends on the polymer concentration, but the exponent of the power law is independent of the composition of the PS/PB mixtures.

These laws found for the polymer systems are quite similar to that found for the small-molecule system except for the very large difference in the prefactor of the

Fig. 4.25 Drop of critical temperature $\Delta T_c(\dot{\gamma})$ reduced with $T_c(0)$, $\Delta T_c(\dot{\gamma})/T_c(0)$, as a function of $\dot{\gamma}$ under the increased $\dot{\gamma}$ for the PS/PB (20:80), (50:50), (60:40), and (70:30)/DOP at 3.3 wt. % and for the PS/PB (50:50)/ DOP 3.0 wt. %

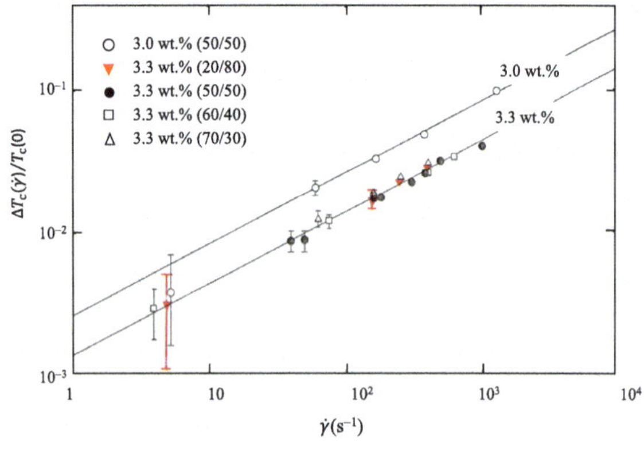

shear-rate dependent term $\dot{\gamma}^n$. The prefactor is the characteristic time for concentration fluctuations τ_s raised to the power 0.53 for the small-molecule system and the characteristic time for concentration fluctuations τ_p raised to the power 0.5 for the polymer systems. Since $\tau_p \gg \tau_s$ [8, 48, 49], it is quite reasonable, though striking, that the prefactor for the polymer system is very much larger than that of the small-molecule system **by a factor of about 10^4. Polymer systems are much more sensitive to the shear flow than the small molecule systems**. This is quite natural from the following viewpoints: (i) Γ_{conc} for the polymers $\left(\tau_p{}^{-1}\right)$ is much smaller than Γ_{conc} for the small molecules $\left(\tau_s{}^{-1}\right)$; (ii) the shear flow is universally effective when $\dot{\gamma} > \Gamma_{conc}$ as discussed earlier in Sect. 1.4.

The general principle introduced in Sect. 1.4 can be universally applied to the small molecules and polymers. The principles can be further extended to predict the difference in the critical-temperature drop, $\Delta T_c(\dot{\gamma})/T_c(0)$, for the mixtures of the small molecules and the polymers, as discussed above. The discussions regarding the difference in the critical-temperature drop of the small molecules and polymers can be further extended as given below.

If $\dot{\gamma} < \Gamma_{conc}$, the phase separation proceeds and the phase-seperated domains grow over the time of the order of $\dot{\gamma}^{-1}$. In the time scale longer than this, the shear deforms and bursts the overgrown domains into the small domains, but the burst small domains tend to grow again, driven by the thermodynamic driving force for the growth, giving rise to the shear-rate dependent steady-state domain structures, i.e., **shear-rate dependent domain structures (dissipative structures) at the dynamic equilibrium between the growth and the burst under the flow**. However, in the case when $\dot{\gamma} > \Gamma_{conc}$, there is no time available for the concentration fluctuations to grow into the domains. Thus, the system would be forced to stay in its single-phase state. The critical shear rate $\dot{\gamma}_c$ is the one which satisfies

$$\dot{\gamma}_c \sim \Gamma_{conc}. \tag{4.5.4}$$

This concept and the physics of the critical phenomenon given by

$$\Gamma_{conc} \sim \Delta T(0)^n \tag{4.5.5}$$

predict that

$$\dot{\gamma}_c \sim \Gamma_{conc} \sim \Delta T(0)^n, \tag{4.5.6}$$

and

$$\Delta T_c(\dot{\gamma}) \sim \dot{\gamma}^{1/n}. \tag{4.5.7}$$

Equation 4.5.6 predicts the critical shear rate as a function of quench depth $\Delta T(0) = T_c(0) - T$ at a given working T in the quiescent systems at $\dot{\gamma} = 0$, while Eq. 4.5.7 predicts the critical temperature drop $\Delta T_c(\dot{\gamma})$ as a function of shear rate, based on the relationships among these quantities as shown in the Fig. 4.24b. If the systems under consideration belong to the 3d Ising universality class, the exponent n in Eq. 4.5.5 is equal to 1.93. [50, 51]. If the systems belong to mean-field universality class, the value of n is equal to 2 [50]. The value of $n = 1.93$ or 2 together with Eq. 4.5.7 predicts Eq. 4.5.1 or Eqs. 4.5.2 and 4.5.3, respectively.

The critical temperature drops $\Delta T(\dot{\gamma})/T_c(0)$ with $\dot{\gamma}$ in the critical fluids of the small molecules are theoretically predicted by Onuki and Kawasaki [52–55]. At the strong shear flow, the shear rate $\dot{\gamma}$ satisfies that

$$\tau_\xi \sim \dot{\gamma}^{-1}, \tag{4.5.8}$$

where $\tau_\xi \sim {\Gamma_{\text{conc}}}^{-1}$ is the characteristic time for the concentration fluctuations of the mixtures of the small molecules or the characteristic time for the polymers with the thermal correlation length ξ (Fig. 1.8).

For the fluids of small molecules, the concentration fluctuations with correlation length ξ are convected by the local velocity field, so that the characteristic time for the concentration fluctuations τ_ξ is given by [54].

$$\tau_\xi = \frac{6\,\pi\eta\xi^3}{k_B T} \tag{4.5.9}$$

where η is the viscosity of the fluids. For the critical fluids, ξ is given by

$$\xi = \xi_{\text{small}}\,\varepsilon_T^{-\nu} \tag{4.5.10}$$

where ξ_{small} is the microscopic length of the small molecules, i.e., the size of the small molecules, and ε_T is the reduced quench depth in the quiescent system defined by

$$\varepsilon_T \equiv \frac{T_c(0) - T}{T_c(0)} = \frac{\Delta T(0)}{T_c(0)}, \tag{4.5.11}$$

and v is the critical exponent. Under the shear flow at the shear rate $\dot{\gamma}$ close to the critical shear rate $\dot{\gamma}_c$, the quench depth $\Delta T(0)$ at $\dot{\gamma} = 0$ (quiescent state) is close to $\Delta T_c(\dot{\gamma})$, as shown in Fig. 4.23a, b.Thereby, ε_T under the shear flow at the given shear rate is given by

$$\varepsilon_T = \frac{\Delta T_c(\dot{\gamma})}{T_c(0)}. \tag{4.5.12}$$

From Eqs. 4.5.8–4.5.10 and 4.5.12, one obtains that

$$\frac{\Delta T_c(\dot{\gamma})}{T_c(0)} \sim \tau_{\xi_{0,\text{small}}}^{1/3\nu}\,\dot{\gamma}^{1/3\nu}, \tag{4.5.13}$$

where $\tau_{\xi_{\text{small}}}$ is the microscopic time for small molecules defined by

$$\tau_{\xi_{\text{small}}} \equiv \frac{6\,\pi\eta\,\xi_{\text{small}}{}^{3}}{k_{\text{B}}T}. \tag{4.5.14}$$

The prefactor of Eq. 4.5.13 is calculated by Onuki and Kawasaki based on the renormalization group theory,

$$\frac{\Delta T_{\text{c}}(\dot{\gamma})}{T_{\text{c}}(0)} = 0.0832\,\varepsilon\,\tau_{\xi_{0,\text{small}}}^{1/3\nu}\,\dot{\gamma}^{1/3\nu}, \tag{4.5.15}$$

where $\varepsilon = 4 - d$ with d being the space dimensionality. When the critical fluids belong to the 3d Ising universality class,

$$\nu = 0.63. \tag{4.5.16}$$

Thus, the experimental result of Beysens et al. given by Eq. 4.5.1 is consistent with the theoretical prediction of Onuki and Kawasaki given by Eqs. 4.5.15 and 4.5.16.

For the fluids of the polymer molecules, the concentration fluctuations with ξ are controlled by individual molecular diffusion process, so that τ_ξ is given by

$$\tau_\xi = \xi^2/D_{\text{app}} = \xi^2/(D_{\text{c}}\varepsilon_{\text{T}}). \tag{4.5.17}$$

If ξ is given by the mean-field theory,

$$\xi = \xi_{\text{polym}}\,\varepsilon_{\text{T}}^{-\nu} \text{ with } \nu = 1/2, \tag{4.5.18}$$

where D_{app} is the collective diffusion constant for the concentration fluctuations: D_{c} is the self-diffusion constant of the polymers: ξ_{polym} is the characteristic length of the polymer chains. From Eqs. 4.5.8, 4.5.12, 4.5.17, 4.5.18, one obtains

$$\frac{\Delta T_{\text{c}}(\dot{\gamma})}{T_{\text{c}}(0)} \sim \left(\tau_{\xi_{\text{polym}}}\right)^{\frac{1}{2\nu+1}}\,\dot{\gamma}^{\frac{1}{2\nu+1}}, \tag{4.5.19}$$

where $\tau_{\xi_{\text{polym}}}$ is the microscopic time for the polymer chains.

$$\tau_{\xi_{\text{polym}}} \equiv \xi_{\text{polym}}^2\,/D_{\text{c}}. \tag{4.5.20}$$

In the context of the mean-field theory ($\nu = 1/2$), Eq. 4.5.19 reduces to

$$\Delta T_{\text{c}}(\dot{\gamma})/T_{\text{c}}(0) \sim \tau_{\xi_{\text{polym}}}^{1/2}\,\dot{\gamma}^{1/2}. \tag{4.5.21}$$

The predicted result given by Eq. 4.5.21 is consistent with the experimental results of Eqs. 4.5.2 and 4.5.3.

The large difference in the prefactor of the experimental result for the small molecules (Eq. 4.5.1) and the experimental results for the polymer molecules

(Eqs. 4.5.2 and 4.5.3) is elucidated to be due to the large difference in the microscopic time $\tau_{\xi_{\text{small}}}$ for the small molecules and $\tau_{\xi_{\text{polym}}}$ for the large molecules.

Han and co-workers reported the relationship between the critical temperature drop and shear rate for the mixtures of the deuterated PS and non-deuterated PB with the small molecular weights by means of the dynamic light scattering and SANS with and without the shear flow for the single-phase mixture [33, 35]. They found that their result is consistent with the renormalization group theory proposed by Onuki-Kawasaki, because the molecular-weight of the polymers employed are small enough to obey the non-mean-field behavior described by the 3d Ising universality class.

Hashimoto and co-workers [56–58] reported that the spinodally-decomposed, dynamically symmetric polymer mixture composed of PB and SBR, both having a large molecular weight of $M_n \cong 10^5$, is brought about into the single-phase state near the critical point by applying the repeated uniaxial compression at 25 °C designated by **Baker's transformation**. The mixture was estimated to have $T_c(0) \geq 170$ °C in the quiescent state. The Baker's transformation effectively induces the shear flow of the small shear rate such that $\dot{\gamma} < \Gamma_{\text{MD}}$ to the mixture where the polymer chains under the shear flow are in relaxed state in terms of the deformation and orientation of the chains. This small shear rate, however, satisfies the condition of $\dot{\gamma} > \Gamma_{\text{conc}}$, so that the mixture can be brought about into the shear-induced single-phase state by the Baker's transformation [56, 57]. Since Γ_{conc} is so small for this mixture, and hence $\tau_{\xi_{\text{polym}}}{}^{1/2}$ (the prefactor in Eq. 4.5.21) is so large that the small $\dot{\gamma}$ is elucidated to be sufficient to drop the critical temperature by $\Delta T_c(\dot{\gamma}) \geq 145$ °C.

In this chapter, the shear-induced phase transition is theoretically interpreted based on the basic kinetic equation, simply because this approach is believed to be the most fundamental one to the problems. However, we would like to point out that there is also a thermodynamic approach. This themodynamic approach phenomenologically accounts for the effects of the applied flow to the systems on the flow-induced demixing and mixing as follows. That is, the net Gibb's free energy of the mixtures under the flow is given by the sum of the free energy in the quiescent state and the mechanical energy stored by the applied flow [59–67]. It is natural, however, that the thermodynamic approach does not predict the dynamics of the phase transition process and the process of evolution of the shear-induced dissipative structures, although these important physical processes can be predicted based on the kinetic equation presented in Chap. 5 here in this Book (Eq. 5.71 in Sect. 5.7.1).

4.6 Phase Separation of Shear-Induced Single-Phase Mixture after Shear Drop to $\dot{\gamma} = 0$

Figure 4.26 presents the relaxation of the steady-state structures evolved at $\dot{\gamma} = 0.67$ s^{-1} (part a in Regime III), 67 s^{-1} (part b in Regime IV), and 670 s^{-1} (part c in Regime V) with time after the cessation of the shear to $\dot{\gamma} = 0$, as observed by

PS/PB (50:50)/DOP 3.0 wt. % at $T = 43.2$ °C, $\Delta T(0) = 10.8$ °C

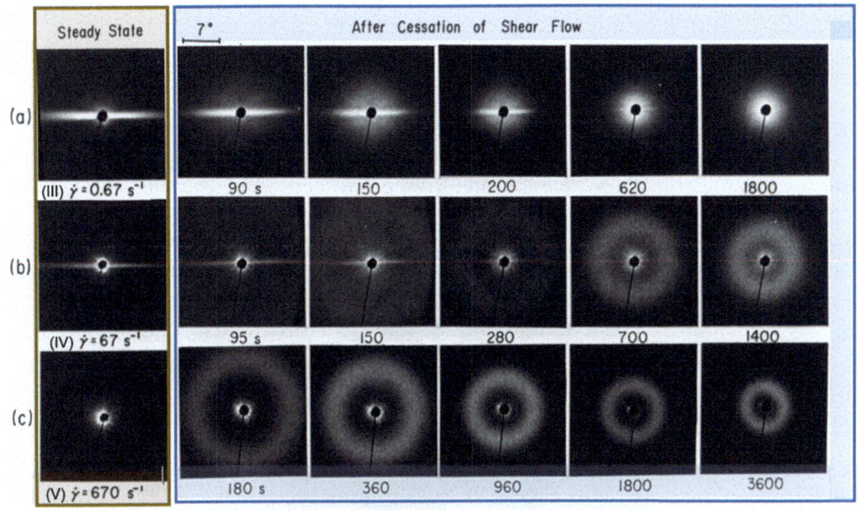

Fig. 4.26 SALS patterns under steady-state shear at $\dot{\gamma} = 0.67$ s^{-1} in Regime III part (a), 67 s^{-1} in Regime IV part (b), and 670 s^{-1} in Regime V part (c) shown in the left edge of the figure and the time evolution of the SALS patterns after the cessation of the steady-state shear flow at $\dot{\gamma} = 0.67$ s^{-1} (**a**), 67 s^{-1} (**b**), and 670 s^{-1} (**c**). The time after the cessation is indicated underneath each pattern in unit of seconds. The original flow direction is vertical. PS/PB (50:50)/DOP 3.0 wt. % with $T_c(0) = 54$°C at $T = 43.2$°C $\dot{\gamma}_c = 165$ s^{-1}. Based on [46]

the TRSALS for the PS/PB (50:50)/DOP, 3.0 wt. % at $T = 43.2$ °C ($\Delta T(0) = 10.8$ °C) [46]. The steady-state SALS patterns shown in the left edge in parts (a) to (c) correspond to those shown in parts (d), (f), and (g) in Fig. 4.3, respectively, and their integrated intensities $\mathcal{J}_x(\dot{\gamma})$ and $\mathcal{J}_z(\dot{\gamma})$ are taken at the shear rates labeled by the arrows marked with (d), (f) and (g) in Fig. 4.4.

The TRSALS patterns in part (a) presents the decreasing intensity of the streak-like scattering pattern, followed by the increased intensity of the diffuse elliptical-to-circular pattern with the elapse of time after the shear cessation. The time evolution of the SALS patterns after the cessation of shear indicates that the elongated long dissipative structure, oriented along FD, stabilized under the given shear rate is destabilized after the shear drop to $\dot{\gamma} = 0$. The destabilization driven by the interfacial tension breaks the long-elongated domains into the ellipsoidal domains with the decreased aspect ratio (the major axes of both domains being oriented along FD) and eventually to the spherical domains with time up to 1800s. Thus, the time evolution of the dissipative structure after the cessation involves "percolation-to-cluster" transition (PCT) from the elongated domains to the round droplets to lower the interfacial free energy of the dissipative structures in the given systems, which is quite opposite to the evolution of the steady-state dissipative structure through "cluster-to-percolation" transition (CPT) with the increased $\dot{\gamma}$ from the quiescent state, as shown earlier in Fig. 4.5. In other words, the self-organization of the dissipative structure at the given $\dot{\gamma}$ in Regime III into the equilibrium structure at $\dot{\gamma} = 0$ occurs according to the following pathway in reference to Fig. 4.5: (d / k) and (c / j) in Regime III → (b / i) and (a / h) in Regime II, where (x/y) designates the SALS pattern in part (x) and the OM image in part (y) in Fig. 4.5. This pathway is quite opposite to that of the self-organization of the dissipative structures with the increased $\dot{\gamma}$: (a/ h) → (b / i) → (c / j) → (d / k).

The TRSALS patterns after the cessation in part (b) in Fig. 4.26 present the decreased intensity of the streak-like SALS patterns with time after the cessation of the shear flow from Regime IV, similarly to the case of part (a), but the case of part (b) is followed by the process different from part (a), i.e., the increased intensity of the new pattern with the circular ring scattering pattern with its maximum intensity I_{max} at $q = q_{max}$, where I_{max} increases and q_{max} decreases with time. The time evolution of the SALS pattern at the late stage after cessation in part (b), i.e., 150 to 1400 s, is quite different from that in part (a), i.e., 150 to 1800s with respect to the self-organization of dissipative structures, as elucidated below.

The steady-state dissipative structure at $\dot{\gamma} = 67\ s^{-1}$ in Regime IV is the string structure self-organized at the shear rate $\dot{\gamma}$ quite close to $\dot{\gamma}_c$: thereby, the string phase and its matrix phase have a very small composition (or concentration) difference with respect to the PS and PB chains, as evidenced by the very weak integrated SALS intensity $\mathcal{J}_z(\dot{\gamma})$ at the shear rate marked by the arrow with (f) in Fig. 4.4 and as schematically illustrated in Fig. 4.6e in Sect. 4.3.2. In other words, a high degree of the intermixing of the PS and PB chains occurs both in the string phase and matrix phase, so that the composition or concentration difference between the two phases must be very small under the steady-state shear flow at the given shear rate $67\ s^{-1}$. After the cessation of the shear flow, the two phases developed in the steady state at $\dot{\gamma} = 67\ s^{-1}$ are subjected to the following two kinds of the instability.

(1) Each phase oriented along the flow direction (FD) at $\dot{\gamma} = 67\ s^{-1}$ as illustrated by Fig. 4.6e and 4.13d is brought into thermodynamic instability due to a large quench depth $\Delta T(0) = T_c(0) - T = 10.8$ °C with respect to the equilibrium phase diagram at $\dot{\gamma} = 0$. Note that the dynamic phase diagram at $\dot{\gamma} = 67\ s^{-1}$ with the dropped critical temperature $T_c(\dot{\gamma})$ with respect to $T_c(0)$ at $\dot{\gamma} = 0$ by the amount of $T_c(\dot{\gamma}) = T_c(0) - T_c(\dot{\gamma})$ becomes irrelevant after the cessation of shear from $\dot{\gamma} = 67\ s^{-1}$ to $\dot{\gamma} = 0\ s^{-1}$ (see Fig. 4.23b). Therefore, each phase with nearly the same composition of the PS and PB chains and the same volume fraction at $\dot{\gamma} = 67\ s^{-1}$ as illustrated in Fig. 4.6e in Sect. 4.3.2 and 4.13d in Sect. 4.3.6 (B) must undergo the spinodal decomposition (SD).

(2) The oriented string phase with the large extension along the FD, which is stable at the steady-state shear flow at $\dot{\gamma} = 67\ s^{-1}$, is subjected to the interfacial-tension-driven instability in the quiescent at $\dot{\gamma} = 0$, after the cessation of shear. The instability causes the growth of the amplitude of the undulation of the interface, and the interface is eventually burst. The growth of the undulation and resultant burst of the string phase would cause the memory of the anisotropic composition fluctuations of the PS and PB chains, initially existed along the neutral direction (along Oz axis) before the shear cessation, to decay, which results in isotropic composition fluctuations with the small composition difference.

The interfacial instability (2) described above causes the streak-like SALS pattern to disappear first with time after the cessation of the shear. The thermodynamic instability (1) subsequently causes the SD in the medium with isotropic composition fluctuations, which develops the SALS pattern of the so-called **spinodal ring** as clearly observed at 150 s: coarsening of the SD structures causes the I_{max} and q_{max} of the spinodal ring to increase and decrease with time, respectively as observed in the time span from 150 s to 1400 s.

The steady-state TRSALS patterns in prat (c) after the cessation of the given $\dot{\gamma}$ in the Regime V presents the evolution of the circular ring scattering having its maximum intensity I_{max} at $q = q_{max}$ with I_{max} and q_{max} increased and decreased with the time after the shear cessation, respectively as observed in the time span from 180 s to 3600 s. The time evolution of the SALS pattern is identical to that for the quiescent solution (the equilibrium solution at $\dot{\gamma} = 0$) which is quenched with the quench depth of $\Delta T = T_c(0) - T = 10.8$ °C ($T_c(0) = 54$ and $T = 43.2$ °C). Thus, the anisotropic SALS intensity distribution from the thermal concentration fluctuations in the shear-induced single-phase solution at $\dot{\gamma} = 670\ \mathrm{s}^{-1}$ as observed in Fig. 4.16 in Sect. 4.3.8, which validates the anisotropic concentration fluctuations in the shear-induced single-phase state at the given shear rate, is rapidly relaxed after the shear cessation to result in the isotropic thermal concentration fluctuations. The rapid relaxation of the anisotropic concentration fluctuations is subsequently followed by the SD, which is equivalent to the SD in the quiescent solution.

Analysis of the early stage SD. The Cahn's theory [68] in the early stage SD predicts the scattered intensity $I(q, t)$ exponentially increases with t and its growth rate $R(q)$ depends on q,

$$I(q, t) = I(q, 0)\exp\left[2R(q)t\right], \tag{4.6.1}$$

where $I(q, 0)$ is the scattered intensity $I(q, t)$ at $t = 0$, and $R(q)$ is the growth rate of the q-Fourier mode of the concentration fluctuations in the early stage SD given by

$$R(q) = D_{app}q^2\left(1 - q^2/q_c^2\right). \tag{4.6.2}$$

D_{app} is the collective diffusivity given by

$$D_{app} = D_c\varepsilon_T \tag{4.6.3}$$

with D_c is the self-diffusivity of the polymer chains in the mixture.

In Eq. 4.6.2, q_c is the characteristic wave number of the dominant Fourier modes of thermal concentration fluctuations given by

$$R(q) = 0 \text{ at } q = q_c,\ \ R(q) \gtrless 0 \text{ at } q \lessgtr q_c. \tag{4.6.4}$$

Hence, the thermal concentration fluctuations with the wave number q satisfying $q < q_c$ can grow with t, whereas those satisfying $q > q_c$ decay with time. This argument in the q-space (Fourier space) as given by Eq. 4.6.4 can be translated by that in real-space as given below by Eq. 4.6.6 as follows. Here in Eq. 4.6.4, the critical wave number q_c is related to the correlation length ξ of the thermal concentration fluctuations of the systems given by

$$q_c \sim 1/\xi. \tag{4.6.5}$$

Since q is related to the wavelength l of the Fourier modes of the thermal concentration fluctuations as given by, $q \sim 1/l$, Eq. 4.6.4 is rewritten by

$$R(l) \gtrless 0 \text{ at } l \gtrless \xi. \tag{4.6.6}$$

Hence, Fourier modes of the thermal concentration fluctuations with the wavelength l larger than the correlation length $\xi = 1/q_c$ can grow with time, whereas those with $l < \xi$, cannot grow but rather decay with time.

Figure 4.27 presents the plot of $\ln I(q, t)$ versus t [3] based on Eq. 4.6.1. In the early stage of the SD, $\ln I(q, t)$ increases linearly with t, from which the q-dependent growth rate $R(q)$ is evaluated. The deviation from the linear increase of $\ln I(q, t)$ with t is due to the coarsening of the SD structures in the later stage SD [51, 69, 70].

Figure 4.28 presents the plot of the evaluated $R(q)$ versus q^2 and $R(q)/q^2$ versus q^2 [3] based on Eq. 4.6.2. The growth rate $R(q)$ versus q has the maximum at $q = q_m(0) \simeq 3.5 \times 10^{-3}\text{nm}^{-1}$ which is the wave number of the q-Fourier mode of the concentration fluctuations which grows at the maximum growth rate in the early stage SD. $R(q)/q^2$ versus q^2 has the linear relationship. The collective diffusivity of the concentration fluctuations in the mixture $D_{app} \simeq 8.8 \times 10^3\text{nm}^2/\text{ s}$ is assessed from the intercept of the linear extrapolation of the plot of $R(q)/q^2$ versus q^2 to $q^2 = 0$, while $q_c(0)^2 \simeq 2.4 \times 10^{-5}\text{nm}^{-2}$ is assessed from the linear extrapolation of the plot $R(q)/q^2$ versus q^2 to $R(q)/q^2 = 0$. Here, note that "0" in $q_m(0)$ and $q_c(0)$ designates q_m and q_c estimated from the early state SD. The evaluated $q_m(0)$ and $q_c(0)$ satisfy the relation given by Eq. 4.6.2,

$$q_m(0)^2 = q_c^2/2. \tag{4.6.7}$$

The characteristic time for the concentration fluctuations, τ_{conc}, is determined by the diffusion law on the concentration fluctuations,

$$\tau_{conc} = \xi^2/D_{app} = \left(q_c^2 D_{app}\right)^{-1} = \left(2q_m(0)^2 D_{app}\right)^{-1}. \tag{4.6.8}$$

For the PS/PB (50:50)/DOP 3.0 wt. % with $T_c(0) = 54.0°\text{C}$ at $T = 48.0°\text{C}$, $\tau_{conc} = 4.6$ s. The strong shear with the concentration fluctuations is attained under the condition given by

$$\dot{\gamma} > 1/\tau_{conc} \equiv \dot{\gamma}_{c,conc}, \tag{4.6.9}$$

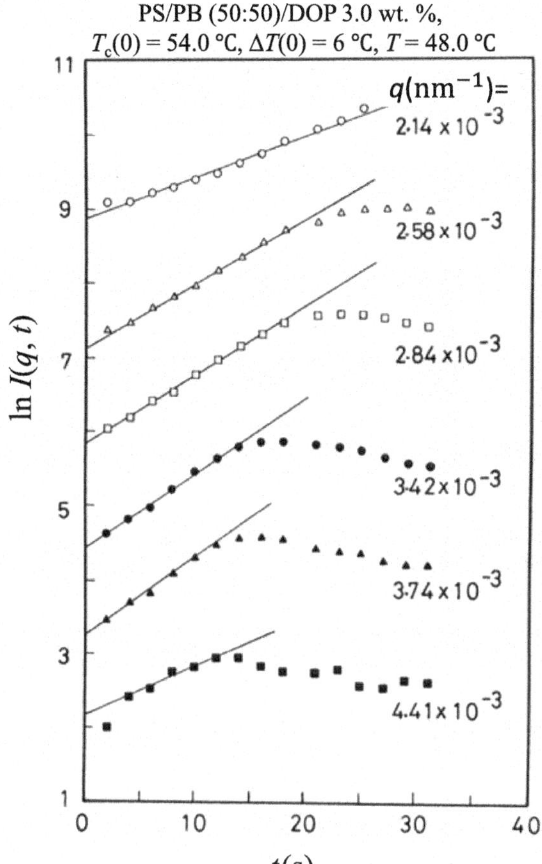

Fig. 4.27 ln $I(q, t)$ versus t at the various q's as indicated in the figure legend to determine $R(q)$ in the early stage SD induced by the cessation of the steady-state shear at $\dot{\gamma} = 670\ \mathrm{s}^{-1} (> \dot{\gamma}_c = 165\ \mathrm{s}^{-1})$ in the shear-induced single-phase solution of the PS/PB (50:50)/DOP 3.0 wt. % with $T_c(0) = 54$°C at $T = 48.0$°C. Based on [3], reprinted by AIP Publishing with permission

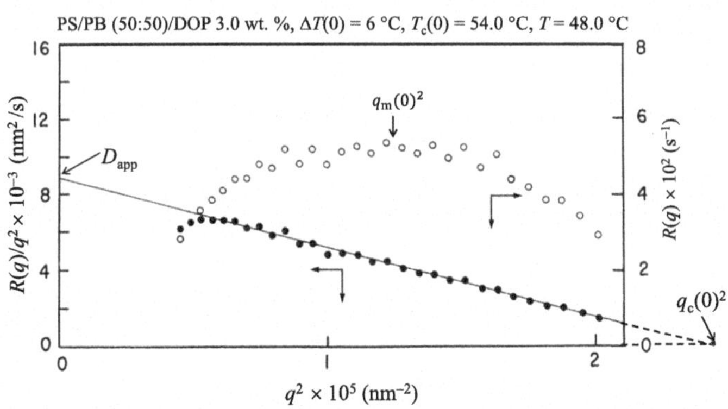

Fig. 4.28 Growth rate $R(q)$ of the q-Fourier mode of the concentration fluctuations plotted as a function of q^2 in the early stage SD (right ordinate) and the plot of $R(q)/q^2$ versus q^2 (left ordinate). PS/PB (50:50)/DOP 3.0 wt. % with $T_c(0) = 54$ °C at $T = 48.0$ °C

Table 4.3 Characteristic parameters* for PS/PB(50:50)DOP solution

Polymer concentrations wt.%	$T_c(0)$ °C	$\Delta T(0)$ °C	$q_{max}(0)$ nm^{-1}	D_{app} nm^2/s	$\dot{\gamma}_{c,conc}$ s^{-1}	$\dot{\gamma}_c$ s^{-1}
3.0	54.0	6.0	3.5×10^{-3}	8.8×10^3	0.22	60
3.0	54.0	10.8	4.7×10^{-3}	1.6×10^4	7.1	165
3.3	76.0	8.0	4.0×10^{-3}	1.0×10^4	0.32	127
3.3	76.0	6.0	$3.5 \times 10^{-3(a)}$	$7.5 \times 10^{3(b)}$	$0.18^{(c)}$	72
3.3	76.0	4.0	$2.8 \times 10^{-3(a)}$	$5.0 \times 10^{3(b)}$	$0.08^{(c)}$	2.5

*The data marked by (a), (b), and (c) are estimated from $q_{max}(0)$, D_{app}, and $\dot{\gamma}_{c,conc}$ for the 3.3 wt.% solution at $\Delta T(0) = 8$ °C, respectively, based on the mean-field theory for the respective quantity: (a) $q_{max}(0) \sim \Delta T^{1/2}$, (b) $D_{app} \sim \Delta T$, and (c) $\dot{\gamma}_{c,\,conc} \sim 1/\tau_c \sim R_m(0) \sim \Delta T^2$. τ_c : characteristic time for the concentration fluctuations with the correlation length ξ, $R_m(0)$: $R(q)$ at $q = q_m(0)$ in the early stage SD at the given $\dot{\gamma}$

where $\dot{\gamma}_{c,conc}$ is the critical shear rate for the strong shear, $\dot{\gamma}_{c,strong}$, with respect to the concentration fluctuations. $\dot{\gamma}_{c,strong}$ evaluated for this solution is 0.22 s^{-1}.

Similarly, for the same solution at the large quench depth, $\Delta T(0) = 10.8$°C, $q_m(0)$, D_{app}, and $\dot{\gamma}_{c,conc}$ can be evaluated as follows,

$q_m(0) = 4.7 \times 10^{-3}$ nm^{-1}, $D_{app} = 1.6 \times 10^4$ nm^2/ s, $\dot{\gamma}_{c,conc} = 7.1$ s^{-1} (Table 4.3).

The similar analysis conducted for the PS/PB (50:50)/DOP 3.3 wt. % at $\Delta T(0) = 8$°C and $T_c(0) = 76$°C yielded the following results: $q_m(0) = 4.0 \times 10^{-3}$ nm^{-1}, $D_{app} = 1.0 \times 10^4$nm^2/ s, $\dot{\gamma}_{c,conc} = 0.32$ s^{-1}. These characteristic parameters are summarized in Table 4.3.

4.7 Time Evolution of Dissipative Structures at a Given Shear Rate after Shear Drop from Shear-Induced Single Phase

In this section, we describe the time evolution of the dissipative structure at a given shear rate $\dot{\gamma} \neq 0$ after the shear drop from the shear-induced single phase attained at $\dot{\gamma} > \dot{\gamma}_c$.

4.7.1 Shear Drop to the Weak Shear Regime

Figure 4.29 presents the time evolution SALS pattern after the shear drop to $\dot{\gamma} = 0.0067$ s^{-1} in Regime II (part a) and to $\dot{\gamma} = 0.067$ s^{-1} in Regime III (part b) from the shear-induced single-phase state at $\dot{\gamma} = 670$ s^{-1} (the SALS pattern at the top

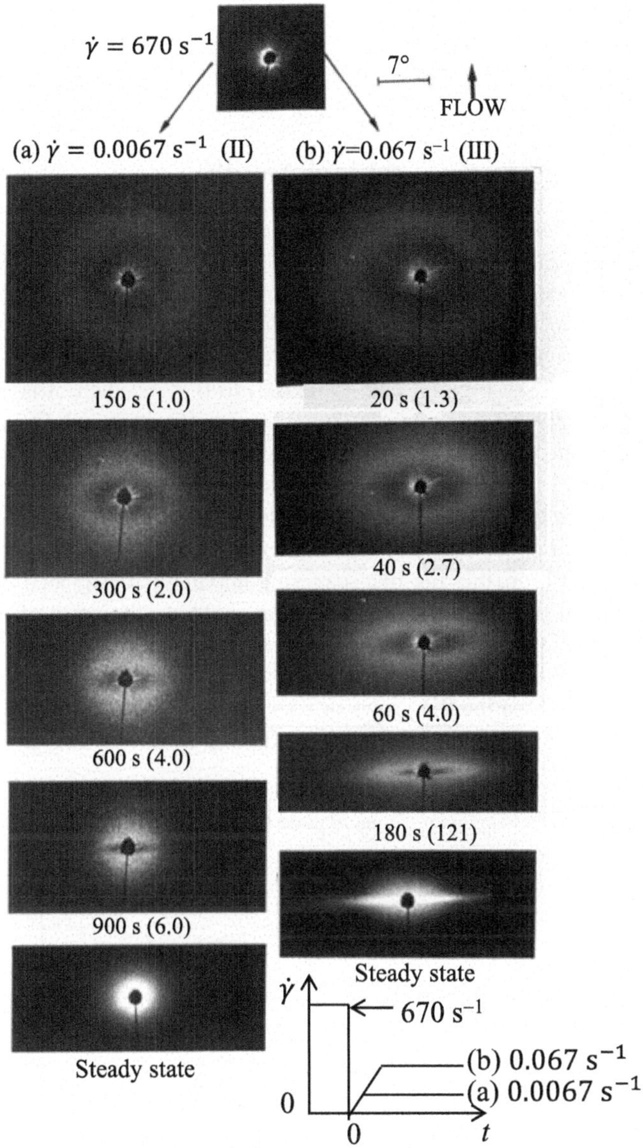

Fig. 4.29 Time evolution of the SALS patterns at $\dot{\gamma} = 0.0067\ s^{-1}$ in Regime II (part a) and $\dot{\gamma} = 0.067\ s^{-1}$ in Regime III (part b) after the shear drop from $\dot{\gamma} = 670\ s^{-1}$ in Regime V. The bottom right corner presents the shear protocol for the shear drop from $\dot{\gamma} = 670\ s^{-1}$ at steady state to $\dot{\gamma} = 0.0067\ s^{-1}$ (a) and $0.067\ s^{-1}$ (b). The flow direction is vertical. PS/PB (50:50)/DOP 3.0 wt. % with $T_c(0) = 54°C$ at $T = 43.2°C$. The number in parenthesis underneath each SALS pattern indicates the total strain imposed to the sample $\dot{\gamma}t$. Based on [46]

center part) for the PS/PB (50:50)/DOP 3.0 wt. % with $T_c(0) = 54.0\,°C$ at $\Delta T(0) = 10.8\,°C$ [46].

This solution with $\Delta T(0) = 10.8\,°C$ is analyzed to have the following characteristic shear rates:

$$\dot{\gamma}_{c,conc} = 7.1\ s^{-1}, \dot{\gamma}_c = 165\ s^{-1}.$$

Thus, $\dot{\gamma} = 0.0067\ s^{-1}$ in Regime II (part a) and $\dot{\gamma} = 0.067\ s^{-1}$ in Regime III (part b) belong to the weak shear regime. In other words, the shear flow hardly contributes to the critical temperature drop of the solution, i.e., $\Delta T_c(\dot{\gamma}) = T_c(0) - T_c(\dot{\gamma}) \cong 0$.

The shear drop to the given $\dot{\gamma}$ was conducted according to the shear history shown in the bottom of part (b): the shear rate $\dot{\gamma}$ is first drop rapidly from 670 s^{-1} ($> \dot{\gamma}_c \simeq 165\ s^{-1}$, see Fig. 4.4 and Table 4.2) to 0 s^{-1} : then, it is immediately increased to the given shear rate of $\dot{\gamma} = 0.0067\ s^{-1}$(part a) or 0.067 s^{-1} (part b) at the constant rate of the increased shear rate $d\dot{\gamma}/dt = 5.6 \times 10^{-3}\ s^{-2}$. The time required to reach $\dot{\gamma} = 0.0067$ and 0.067 s^{-1} is 1.2 and 12 s, respectively, each time of which is short enough compared with the time scale of the observation regarding the results shown in Figs. 4.29 and 4.30 at the respective shear rate.

After the shear drop, the ring-like scattering maximum, designated as "**spinodal ring**", emerges at the scattering vector $\mathbf{q} \equiv \mathbf{q}_{max}$ due to the onset of the spinodal decomposition (SD) under the shear flow: the ring is circular or nearly circular at $(t\dot{\gamma}) \lesssim 1$ as observed at $t = 150$ s in part (a) or $t = 20$ s in part (b), respectively. The structure developed via the SD grows or coarsens with time and is simultaneously deformed under the shear flow at the time scale of $(t\dot{\gamma}) > 1$ before reaching the

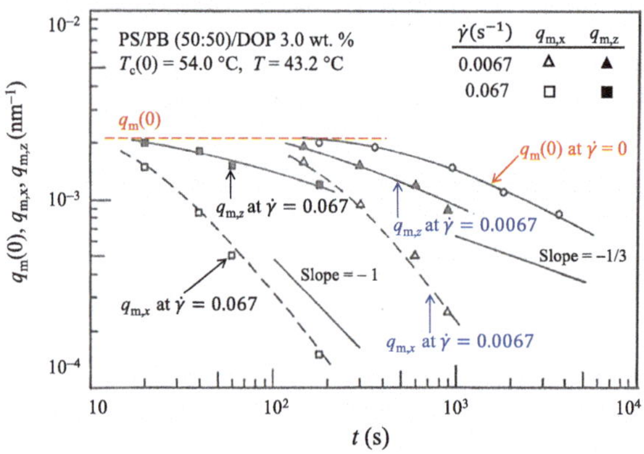

Fig. 4.30 Time evolution of the magnitude of the scattering vector $q_m(0)$ during the SD at $\dot{\gamma} = 0$ and at $T = 43.2\,°C$ after the temperature drop from the single-phase state versus time evolution of the magnitude of the scattering vectors $q_{m,x}$ and $q_{m,z}$ during the SD under the flow at $\dot{\gamma} = 0.0067\ s^{-1}$ (open and filled triangles, respectively) and 0.067 s^{-1} (open and filled squares, respectively) at $T = 43.2\,°C$ after the shear drop from the shear-induced single-phase state at $\dot{\gamma} = 670\ s^{-1}$. The red dashed line indicates the $q_m(0)$ in the early stage SD in the quiescent solution. PS/PB (50:50)/DOP 3.0 wt. % with $T_c(0) = 54\,°C$ at $T = 43.2\,°C$

steady-state structure at the given shear rate. The growth or the coarsening of the SD structure involves the shift of $\mathbf{q}_{max}$ of the spinodal ring toward small scattering vector $\mathbf{q}$, and the deformation of the SD structure involves deformation of the circular spinodal ring into the "elliptical spinodal ring", compressed and extended along and normal to the flow direction (FD), respectively, to eventually reach the steady-state SALS pattern at each $\dot{\gamma}$ as shown by the SALS pattern in the bottom of part a and b. The steady-state SALS pattern reveals the round droplets in Regime II ($\dot{\gamma} = 0.0067\ s^{-1}$ part a) and the elongated domains oriented along the flow direction in Regime III ($\dot{\gamma} = 0.067\ s^{-1}$ in part b), as schematically shown in Fig. 4.6a, b, respectively.

The results presented in Figure 4.29 reveal the interesting features in the early stage SD up to ~ 600 s at $\dot{\gamma} = 0.0067\ s^{-1}$(a) and up to $\sim$ 180 s at $0.067\ s^{-1}$ (b) as follows.

(1) In the SD process which occurs under the second-step shear flow at $\dot{\gamma} = 0.0067\ s^{-1}$ or $0.067\ s^{-1}$, the ordering or growth process of the structures evolved under SD is anisotropic with the different characteristic wave numbers (the magnitudes of the scattering vector) $q_{m,x}$ and $q_{m,z}$ of the dominant Fourier modes of the concentration fluctuations, at which the ring scattering intensity parallel and perpendicular to the FD become maximum, respectively;
(2) The growth rate of the SD structures as observed by the decrease of $q_{m,x}$ and $q_{m,z}$ with time is much faster in the direction parallel to the FD than perpendicular to it;
(3) The larger is the second-step shear rate, the faster are the growth rates in both directions.

Those features revealed in Fig. 4.29 are presented more quantitively in Fig. 4.30 [46]. The figure presents the time evolution of the magnitude of the scattering vector at the maximum scattered intensity of the spinodal ring, $q_m(0)$, at $\dot{\gamma} = 0$ and that parallel to the FD, $q_{m,x}$, and perpendicular to the FD, $q_{m,z}$, under the shear flow at $\dot{\gamma} = 0.0067$ and 0.067, respectively. In the figure $q_{m,x}$ and $q_{m,z}$ are plotted against the time after the onset of the second-step shear flow together with $q_m(0)$ in the quiescent solution ($q_m(0) = q_{m,x} = q_{m,z}$ at $\dot{\gamma} = 0$). The results reveal the SD in the time span from the early stage to the beginning of the intermediate stage under the very weak shear (the given each shear rate satisfying $\dot{\gamma} \ll \dot{\gamma}_{c,conc} = 7.1\ s^{-1}$, Table 4.2): $q_{m,x}(t)$ and $q_{m,z}(t)$ start to decrease from the time-independent constant value of $q_m(0) \cong 2 \times 10^{-3} nm^{-1}$ indicated by the red dashed line. The growth of the SD structures as manifested by the decrease of $q_{m,x}$ and $q_{m,z}$ with t from $q_m(0)$ starts to occur at an earlier time under the shear in comparison with the case of $\dot{\gamma} = 0$.

Figure 4.31 presents the characteristic wave numbers $q_{m,x}$ and $q_{m,z}$ plotted as a function of the reduced time $\dot{\gamma}t$ [46]. The reduced time has physical meanings of "total strain" imposed to the solution or a dimensionless time t reduced by the shear distortion time $1/\dot{\gamma}$ (149 and 14.9 s at $\dot{\gamma} = 0.0067\ s^{-1}$ and $0.067\ s^{-1}$, respectively) imposed to the concentration fluctuations of the solutions under shear flow. The effect of the shear on the SD processes become significant in the time scale of $t > 1/\dot{\gamma}$. In the time scale $t < 1/\dot{\gamma}$, the effect of the shear is very small, and the anisotropy of the

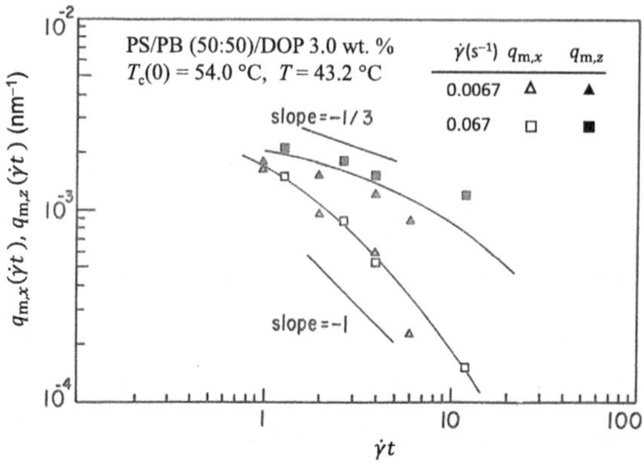

Fig. 4.31 Time evolution of the magnitudes of the scattering vectors $q_{m,x}(\tilde{t})$ and $q_{m,z}(\tilde{t})$ at the maximum SALS intensity $I_m(q_{m,x}, q_{m,z})$ along the flow direction (Ox axis) and the neutral direction (along O_z axis), respectively, as a function of the reduced time $\tilde{t} \equiv t\dot{\gamma}$ after the shear drop from $\dot{\gamma} = 670\ s^{-1}$ ($> \dot{\gamma}_c$) to $\dot{\gamma} = 0.0067$ (open and filled triangles, respectively) and 0.067 s^{-1}(open and filled squares, respectively)

growth as observed by $q_{m,x}$ versus t and $q_{m,z}$ versus t is very small and nearly equal to the growth observed by $q_m(0)$ versus t in the quiescent solution. It is interesting to note the two sets of data on the time dependence of the characteristic wave numbers ($q_{m,x}$ versus t and $q_{m,z}$ vs t) at the different shear rates appear to be superimposed to each other in the reduced plot. The anisotropy in the coarsening becomes significant at $t > 1/\dot{\gamma}^{-1}$: in the long-time limit, the time dependence of the characteristic wave numbers approaches to $q_{m,x} \sim t^{-1}$ and $q_{m,z} \sim t^{-1/3}$, respectively.

It is important to note that the shear flow does not affect the ordering of the dissipative structures driven by the drop of $T_c(\dot{\gamma})$ with increased $\dot{\gamma}$ in the very weak shear regime at $\dot{\gamma} = 0.0067$ and 0.067 s^{-1}, because $\Delta T(\dot{\gamma}) \cong \Delta T(0)$ or $\Delta T_c(\dot{\gamma}) \cong 0$ (at these shear rates ($\Delta T(\dot{\gamma})$ and $\Delta T_c(\dot{\gamma})$) being defined by Fig. 4.23 and Eq. 4.5.2, respectively). In the weak shear regime where $\Delta T(\dot{\gamma}) \cong \Delta T(0)$, the shear affects the ordering under the $\dot{\gamma}$-independent quench depth $\Delta T(0)$. Especially in the early stage, the ordering is dominantly controlled by the shear deformation rather than the ordering driven by the quench depth-dependent thermodynamic driving force. This is why the early stage dissipative structures evolved at the different $\dot{\gamma}$ values are scaled with the total strain $\dot{\gamma}t$ which is identical to the reduced $\dot{\gamma}$ with $1/t$ or the reduced time with $1/\dot{\gamma}$ (Fig. 4.31). In the later stage, however, the role of the thermodynamic driving force on the coarsening becomes important to compete with the role of the shear flow at the given shear rate $\dot{\gamma}$: the competition results in the self-organization of the $\dot{\gamma}$-dependent steady-state structures which cannot be scaled with $\dot{\gamma}t$, as evidenced at $\dot{\gamma}t > 5$ (see the last two data points for $q_{m,x}$ and $q_{m,z}$) in Fig. 4.31.

4.7.2 Shear Drop to the Strong Shear Regime

Figure 4.32 presents the time evolution of the SALS pattern after the shear drop to $\dot{\gamma} = 0.214\ \mathrm{s}^{-1}$ in the Regime III from the shear-induced single-phase solution at $\dot{\gamma} = 8.91\ \mathrm{s}^{-1}$ for the PS/PB (50:50)/DOP at 3.3 wt.% at $\Delta T(0) = 4\,°\mathrm{C}$. ($\dot{\gamma}_\mathrm{c} = 2.5\ \mathrm{s}^{-1}$, Table 4.2). This sheared solution has the characteristic shear rate $\dot{\gamma}_{\mathrm{c,conc}}$ for the concentration fluctuations, which satisfies the following relationship with the applied shear rate as given below

$$\dot{\gamma}_{\mathrm{c,conc}} = 0.08\ s^{-1} < \dot{\gamma} = 0.214\ \mathrm{s}^{-1},$$

so that the sheared solution belongs to the strong shear regime (see Table 4.3 in Sect 4.6). Since $\dot{\gamma}_{\mathrm{c,conc}} \sim R_\mathrm{m}(0)$, the maximum growth rate of the concentration fluctuations in the early stage SD (see Table 4.3 in Sect. 4.6), the given $\dot{\gamma} = (0.214\ \mathrm{s}^{-1})$ is faster than the characteristic growth rate $R_\mathrm{m}(0)$ ($= 0.08\ \mathrm{s}^{-1}$). Thus, the shear flow at $\dot{\gamma} = 0.214\ \mathrm{s}^{-1}$ imposed to the solution deforms and/or burst the characteristic concentration fluctuations grown in the sheared solution driven by the thermodynamic driving force. That is, the flow at the $\dot{\gamma}$ is the flow which belongs to the strong shear flow.

In contrast to the above solution used for Fig. 4.32, the solution used for Figs. 4.29, 4.30 and 4.31 has $\dot{\gamma}_{\mathrm{c,conc}} = 7.1\ \mathrm{s}^{-1}$ and the imposed flow rates of $\dot{\gamma} = 0.0006\ \mathrm{s}^{-1}$ and $\dot{\gamma} = 0.067\ \mathrm{s}^{-1}$, so that $\dot{\gamma}_{\mathrm{c,conc}} = 7.1\ \mathrm{s}^{-1} \gg \dot{\gamma} = 0.0067\ \mathrm{s}^{-1}$ and $0.067\mathrm{s}^{-1}$, and thereby the shear flow at the given shear rate imposed to the solution can hardly affect the growth of the characteristic concentration fluctuations intrinsic to the solution under the shear flow. That is, the given $\dot{\gamma}$ is the one which belongs to the weak shear flow.

Although the solution in Fig. 4.32 is under the strong shear regime, the self-organization of the solution apparently has the similar features to those of the 3.0 wt.% solution in Figs. 4.29, 4.30, and 4.31 in the weak shear regime: that is, the features (1) and (2) described earlier in conjunction with Fig. 4.29 in Sect. 4.7.1 are similar, except for such a difference that the anisotropy of the coarsening in this solution is validated to occur at much shorter time ($\dot{\gamma}^{-1} = 4.7$ s) than in the 3.0 wt.% solutions ($\dot{\gamma}^{-1} = 149$ s and 14.9 s at $\dot{\gamma} = 0.0067$ and $\gamma = 0.0067$, respectively). The reduced time scale $\dot{\gamma}t$ covered in this solution is 21.4 to 1926, whereas the reduced time scales in the 3.0 wt.% solutions are 1.0 to 6.0 and 1.3 to 21 at $\dot{\gamma} = 0.0067$ and $\dot{\gamma} = 0.067$, respectively. The time evolution of the SALS intensity profile in the direction normal to the flow direction $I(q_z, t)$ in the strong shear regime V is investigated below to clarify the difference in the self-organization, if any, in between the strong shear regime and the very weak shear regime.

Scattering analyses at the strong shear regime. When the shear rate is dropped from a given $\dot{\gamma}$ in the shear-induced single-phase state (in Regime V at the given shear rate $\dot{\gamma}$ above the critical shear rate $\dot{\gamma}_\mathrm{c}$) to the given shear rate $\dot{\gamma}$ satisfying $\dot{\gamma} < \dot{\gamma}_\mathrm{c}$ for the shear-induced phase transition from a single-phase solution to a two-phase solution, the solution is brought into the thermodynamic unstable state under

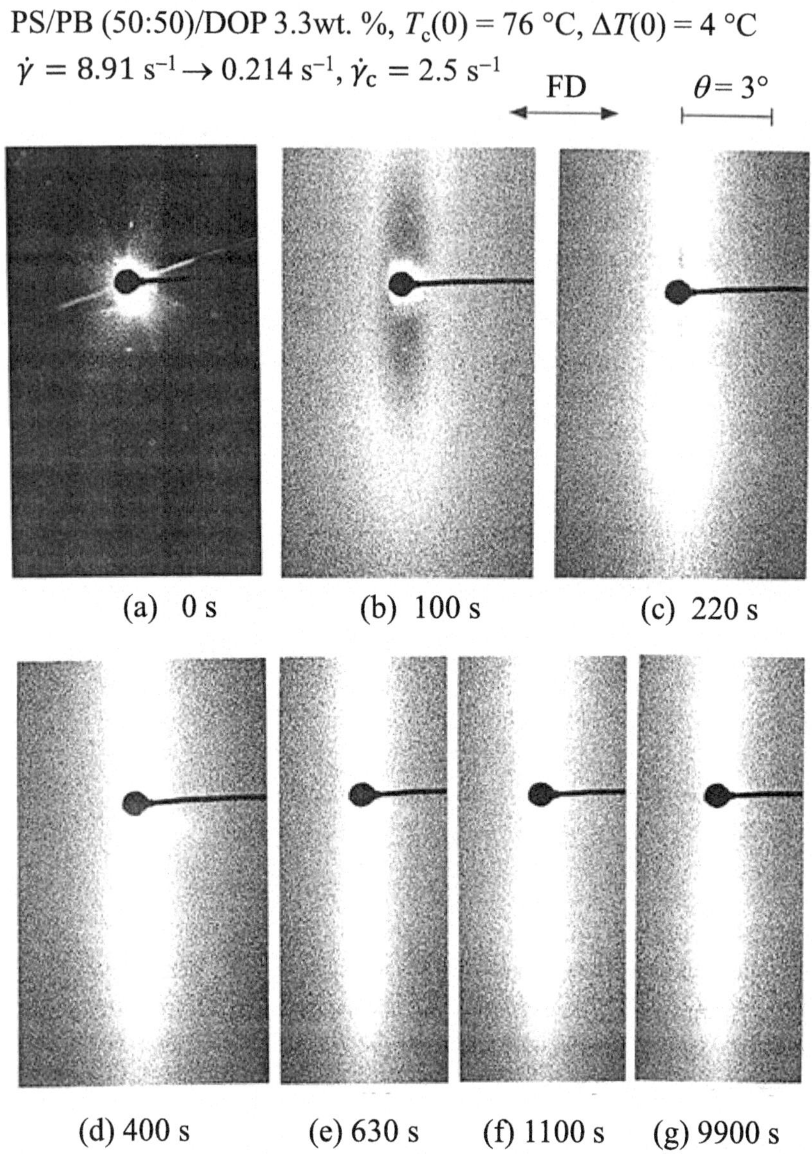

Fig. 4.32 Time evolution of the SALS pattern after the shear drop from the shear-induced single phase in the steady-state at $\dot{\gamma} = 8.91$ s^{-1} ($> \dot{\gamma}_c = 2.5$ s^{-1}) to $\dot{\gamma} = 0.214$ s^{-1}. The time described underneath of the each pattern indicates the time elapsed after the shear drop from $\dot{\gamma} = 8.91s^{-1}$ to $\dot{\gamma} = 0.214$ s^{-1}. The flow direction (FD) is horizontal. PS/PB (50:50)/DOP 3.3wt. %, $T_c(0) = 76$ °C, $\Delta T(0) = 4$ °C

the shear flow at the given $\dot{\gamma}$. This is because the thermodynamic force intrinsic to the solution outweighs the shear force imposed externally to the solution. The thermodynamic instability of the solution under the given shear flow gives rise to the phase separation via the spinodal decomposition (SD) for the solution under the flow at the given $\dot{\gamma}$. The SD grows spatial composition fluctuations with time. Figure 4.33 presents the growth of the dominant Fourier mode of the composition (or

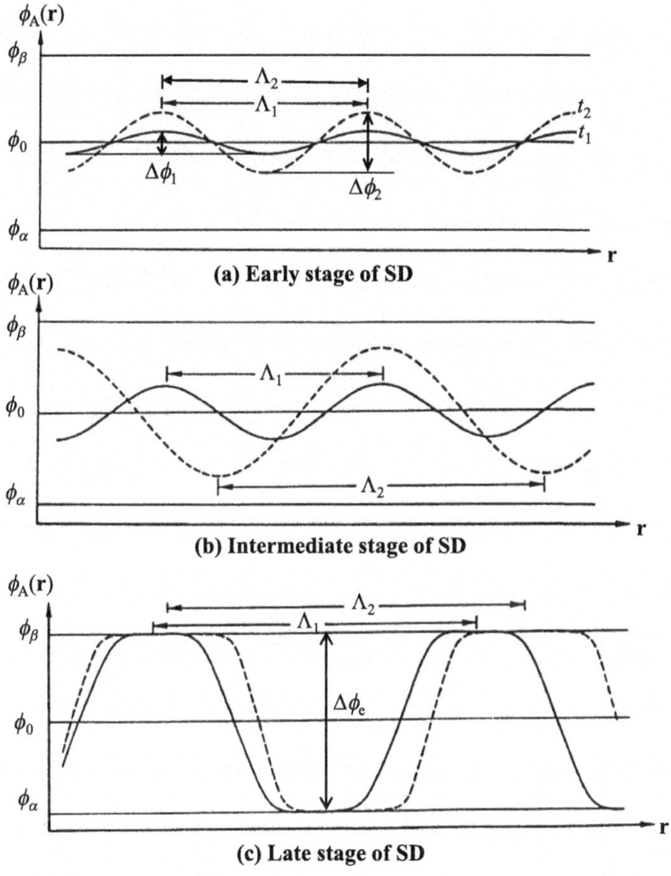

Fig. 4.33 Schematic presentation of the dominant Fourier mode of the spatial concentration fluctuations of one component A in a binary mixture of component A and B, $\phi_A(\mathbf{r})$, at the three different stages of the phase separation through SD (spinodal decomposition): **a** the early stage of SD, **b** the intermediate stage of SD, and **c** the late stage of SD. The time evolution of the fluctuations within each stage is illustrated by the solid line at time t_1 and the broken line at t_1 with $t_1 < t_2$. ϕ_α and ϕ_β are the equilibrium concentration of one component A in the coexisting two liquid phases. The wavelength Λ_1 and Λ_2 of the dominant Fourier modes of the fluctuations in part (a) corresponds to the time-independent wavelength $\Lambda_m = 2\pi/q_m(t)$ where q_m is the magnitude of the scattering vector q at the maximum scattered intensity $I_m = I(q)$ at $q = q_m$. The wavelengths Λ_1 and Λ_2 in part (b) and (c) correspond to the wavelengths of the dominant Fourier modes of the fluctuations $\Lambda(t) = 2\pi/q_m(t)$ at t_1 and t_2. From [69], reprinted by AIP Publishing with permission

the concentration) fluctuations of one of the component A in the binary mixture with the B component in the early stage (a), the intermediate stage (b), and the late-stage SD (c) in quiescent state [48, 69, 70].

In the early stage (part a), the growth of the fluctuations is weakly nonlinear so that it can be well approximated by the linearized theory by Cahn [68]. The fluctuation $\Delta\phi$ grows exponentially with time t from $\Delta\phi_1$ (solid line) to $\Delta\phi_2$ (broken line) with the time-independent periodicity $\Lambda(\Lambda_1 = \Lambda_2)$, giving rise to the time evolution of the scattered intensity $I(q, t)$ as described by Eqs. 4.6.1 and 4.6.2: the scattering maximum emerges with time at the magnitude of the scattering vector $q_{\rm m}(t)$ which is independent of the time and defined by $q_{\rm m}(0)$. In the later time, the intermediate stage of the SD (part b) and the late stage of the SD (part c), the characteristic length Λ increases from Λ_1 to Λ_2 with time as well as the increase of $\Delta\phi$ with time. Correspondingly, $q_{\rm m}(t) = 2\pi/\Lambda(t)$ decreases with the time t. The time evolution of the spatial composition fluctuations is theoretically described by the so-called time-dependent Ginzburg–Landau equation [50, 70, 71]

In the late stage (part c), the local compositions of the phase-separated domains reach locally equilibrium compositions: ϕ_β and ϕ_α for the component A in the A-rich and the A-poor domain in the A/B binary mixture, respectively. The characteristic length only increases from Λ_1 to Λ_2 with time, and thereby $q_{\rm m}(t)$ decreases and $I_{\rm m}(t)$ increases with the time t, respectively.

In the solution at $\dot{\gamma} = 0$, the equilibrium composition of A in the phase-separated domains at the late stage, defined by ϕ_α and ϕ_β in Fig. 4.33, depend only on T or $\Delta T(0) = T_{\rm c}(0) - T$, where $\Delta T(0)$ and $T_{\rm c}(0)$ are the quench depth and the transition temperature at $\dot{\gamma} = 0$, respectively. On the other hands, in the solution under the shear flow at the given shear rate satisfying $\dot{\gamma}(< \dot{\gamma}_{\rm c})$, the phase-separated domains attain the shear-rate dependent steady-state compositions $\phi_\alpha(\dot{\gamma})$ and $\phi_\beta(\dot{\gamma})$, as shown in Fig. 4.23b, instead of the corresponding equilibrium compositions ϕ_α and ϕ_β as shown in the Fig. 4.33c. In Fig. 4.23b, the equilibrium compositions ϕ_α and ϕ_β attained at $\dot{\gamma} = 0$ are redefined as $\phi_\alpha(0)$ and $\phi_\beta(0)$ to distinguish them from the compositions $\phi_\alpha(\dot{\gamma})$ and $\phi_\beta(\dot{\gamma})$ attained at the given $\dot{\gamma} \neq 0$ in the steady-state phase-separated structure. The steady state is achieved under a dynamic equilibrium between the growth of the phase-separated domains driven by the thermodynamic driving force and the burst of the overgrown domains by the shear-rate dependent external mechanical force imposed to the solution.

In the quiescent solution, the fluctuations are isotropic, so that Fig. 4.33 can be applied to any given directions. On the other hand, the solution under the shear, the fluctuations are anisotropic with respect to the flow direction (FD), so that Fig. 4.33 can be applied to the neutral direction perpendicular to the FD, i.e., along the Oz axis.

After the shear drop from the shear-induced single-phase state at the $\dot{\gamma} > \dot{\gamma}_{\rm c}$ to the given $\dot{\gamma}$ below the critical shear rate $\dot{\gamma}_{\rm c}$, the scattered intensity distribution changes with time as schematically illustrated in Fig. 4.34a from the blue curve at t_1 with the maximum intensity $I_{z,{\rm m}}(t_1; \dot{\gamma})$ at $q_z = q_{z,{\rm m}}(t_1; \dot{\gamma})$ to the black curve at t_2 with the maximum intensity $I_{z,{\rm m}}(t_2: \dot{\gamma})$ at $q = q_{z,{\rm m}}(t_2, \dot{\gamma})$. The maximum intensity increases from $I_{z,{\rm m}}(t_1; \dot{\gamma})$ to $I_{z,{\rm m}}(t_2; \dot{\gamma})$ with time at the given shear rate and the magnitude of

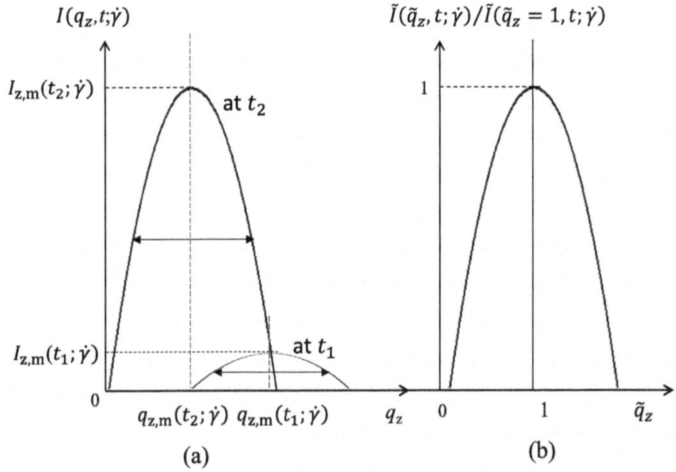

Fig. 4.34 Schematic illustration of the time evolution of the SALS intensity $I(q_{z,}t;\dot{\gamma})$ along the neutral direction at a given shear rate $\dot{\gamma}(<\dot{\gamma}_c)$ (part a) and the reduced scattered intensity $\tilde{I}(\tilde{q}_z,t,\dot{\gamma})/\tilde{I}(\tilde{q}_z=1,t,\dot{\gamma})$ versus the reduced magnitude of the scattering vector $\tilde{q}_z$ defined by $\tilde{q}_z\equiv q_z(t;\dot{\gamma})/q_{z,m}(t;\dot{\gamma})$ (part b). In the reduced scales, the time evolution of the scattered intensity profiles becomes the universal scaling function with t which characterizes the shape of self-similarly grown dissipative structures at the given $\dot{\gamma}$

the scattering vector at the maximum scattered intensity decreases from $q_{z,m}(t_1;\dot{\gamma})$ to $q_{z,m}(t_2;\dot{\gamma})$ with time at the given shear rate, due to the onset of the SD under the shear flow. The time evolution of the scattered intensity along the neutral direction (Oz direction) under the flow at the given shear rate is generally given by

$$I(q_z,t;\dot{\gamma})=K<\eta(t;\dot{\gamma})^2>\Lambda_z(t;\dot{\gamma})^3\,F(q_z,t;\dot{\gamma}), \tag{4.7.1}$$

where $<\eta(t;\dot{\gamma})^2>$ and $\Lambda_z(t;\dot{\gamma})$ are the mean square fluctuations of the scattering length density (SLD) and the characteristic length of the phase-separated structure under the flow at the given $\dot{\gamma}$ along the O_z, respectively. K is a proportionality constant, independent of q_z, t, and $\dot{\gamma}$. $F(q_z,t;\dot{\gamma})$ is the scaled structure factor universal with t and $\dot{\gamma}$, i.e., the structure factor which depends only on the shape of the scattering profile with respect to q_z, independent of t and $\dot{\gamma}$. The universal scaled structure factor reveals the universal shape of the dissipative structures self-organized in the given mixture under the shear flow, in the given length scale relevant to the q_z scale. The absolute intensity depends on the prefactor of $F(q_z,t,\dot{\gamma})$.

$\langle\eta(t;\dot{\gamma})^2\rangle$ is proportional to the mean square amplitude of the composition difference $\Delta\phi(t;\dot{\gamma})$ between the phase-separated domains (Figs. 4.23b and 4.33a), i.e.,

$$\langle\eta(t;\dot{\gamma})^2\rangle\sim\langle\Delta\phi(t;\dot{\gamma})^2\rangle. \tag{4.7.2}$$

The characteristic length of the phase-separated structure $\Lambda_z(t; \dot{\gamma})$ is reciprocally related to the magnitude of the scattering vector $q_{z,m}(t; \dot{\gamma})$ at the scattering maximum of $I(q_z, t; \dot{\gamma})$, as given by,

$$\Lambda_z(t; \dot{\gamma}) = 2\,\pi / q_{z,m}(t; \dot{\gamma}). \tag{4.7.3}$$

From Eqs. 4.7.1 to 4.7.3, one obtains that

$$I(q_z, t; \dot{\gamma}) = K_1\ \langle \Delta\phi(t; \dot{\gamma})^2 \rangle\ q_{z,m}(t; \dot{\gamma})^{-3}\ F(q_z, t; \dot{\gamma}), \tag{4.7.4}$$

where K_1 is the proportionality constant. The maximum scattered intensity $I_{z,m}(t; \dot{\gamma})$ is given from Eq. 4.7.4 by

$$I_{z,m}(t; \dot{\gamma}) = K_1\ \langle \Delta\phi(t; \dot{\gamma})^2 \rangle\ q_{z,m}(t; \dot{\gamma})^{-3}, \tag{4.7.5}$$

since

$$F\left(q_z = q_{z,m}, t; \dot{\gamma}\right) = 1. \tag{4.7.6}$$

Equation 4.7.5 predicts the followings in conjunction with Figs. 4.33 and 4.34a.

(1) In the early stage SD, the magnitude of the scattering vector $q_{z,m}(t; \dot{\gamma})$ at the maximum TRSALS intensity $I(q_z, t; \dot{\gamma})$ at the given $\dot{\gamma}$ is independent of t, i.e., in reference to Fig. 4.34a, $q_{z,m}(t; \dot{\gamma})$ at t_1 and that at t_2 at the given $\dot{\gamma}$ satisfies $q_{z,m}(t_1; \dot{\gamma}) = q_{z,m}(t_2; \dot{\gamma})$ at the any $\dot{\gamma}$; the maximum intensity $I_{z,m}(t; \dot{\gamma})$ at any $\dot{\gamma}$ increases with t only due to the exponential increase of the $\langle \Delta\phi(t; \dot{\gamma})^2 \rangle$ with t from $\Delta\phi_1$ to $\Delta\phi_2$ as illustrated in Fig. 4.33a.

(2) In the intermediate stage SD, $I_{z,m}(t; \dot{\gamma})$ increases with t due to the increase of the $\Delta\phi(t; \dot{\gamma})^2$ at the any $\dot{\gamma}$ and $q_{z,m}(t; \dot{\gamma})$ decreases with t due to the increase of $\Lambda_z\ (t; \dot{\gamma})$ with t at the any $\dot{\gamma}$.

(3) In the late-stage SD,

$$\langle \Delta\phi(t; \dot{\gamma})^2 \rangle = \Delta\phi_e(\dot{\gamma})^2, \tag{4.7.7}$$

i.e., in reference to the Fig. 4.23b, $\Delta\phi(t; \dot{\gamma})$ reaches the equilibrium value of the $\Delta\phi_e(\dot{\gamma})$ at the any given $\dot{\gamma}$, so that the maximum TRSALS intensity $I_{z,m}(t; \dot{\gamma})$ increases only due to the decrease of the $q_{z,m}(t; \dot{\gamma})$ or increase of $\Lambda_z(t; \dot{\gamma})$ with t.

Equation (4.7.4) can be rewritten as Eq. 4.7.10 given below by using the reduced scattered intensity $\tilde{I}(q_z, t; \dot{\gamma})$ and the reduced magnitude of the scattering vector $\tilde{q}_z(t; \dot{\gamma})$ with respect to the $q_{z,m}$ defined by

$$\tilde{I}(q_z, t; \dot{\gamma}) \equiv q_{z,m}(t; \dot{\gamma})^3\ I(q_z, t; \dot{\gamma}), \tag{4.7.8}$$

$$\tilde{q}_z(t;\dot{\gamma}) \equiv q_z/q_{z,\mathrm{m}}(t;\dot{\gamma}), \tag{4.7.9}$$

$$\tilde{I}(\tilde{q}_z, t;\dot{\gamma}) = K_1 \langle \Delta\phi(t;\dot{\gamma})^2 \rangle F(\tilde{q}_z, t;\dot{\gamma}). \tag{4.7.10}$$

Then, the reduced maximum scattered intensity $\tilde{I}_{z,\mathrm{m}}(t;\dot{\gamma})$ is given by

$$\begin{aligned}\tilde{I}_{z,\mathrm{m}}(t;\dot{\gamma}) &\equiv \tilde{I}(\tilde{q}_z = 1;t;\dot{\gamma}) \\ &= K_1 \langle \Delta\phi(t;\dot{\gamma})^2 \rangle,\end{aligned} \tag{4.7.11}$$

since

$$F(\tilde{q}_z = 1;t;\dot{\gamma}) = 1. \tag{4.7.12}$$

Equations 4.7.10 and 4.7.11 lead to

$$\tilde{I}(\tilde{q}_z, t;\dot{\gamma})/\tilde{I}_{z,\mathrm{m}}(t;\dot{\gamma}) = F(\tilde{q}_z, t;\dot{\gamma}), \tag{4.7.13}$$

where $F(\tilde{q}_z, t;\dot{\gamma})$is the reduced (or scaled) structure factor universal with t and $\dot{\gamma}$: it predicts the shape of the scattering profile around the scattering peak position $q_{z,m}(t,\dot{\gamma})$ at the given $\dot{\gamma}(\neq 0)$, i.e., the profile around the reduced scattering vector $\tilde{q}_z = 1$ as presented in Fig. 4.34b. When $F(\tilde{q}_z;t,\dot{\gamma})$ is universal with$t$ at the given $\dot{\gamma}$, as illustrated in Fig. 4.34b, the phase-separated structure grows with **the dynamical self-similarity** with respect to the time evolution of the characteristic length $\Lambda(t,\dot{\gamma}) = 2\pi/q_{z,\mathrm{m}}(t,\dot{\gamma})$ at the given $\dot{\gamma}(\neq 0)$. In this case, $F(\tilde{q}_z;t,\dot{\gamma})$ predicts the shape of the structure factor universal with t and $\dot{\gamma}$.

The **dynamically self-similar growth** of the structure at the given shear rates predicts the time evolution of $q_{z,\mathrm{m}}(t,\dot{\gamma})$ and $I_{z,\mathrm{m}}(t,\dot{\gamma})$ to be given by the power law with as follows t:

$$q_{z,\mathrm{m}}(t;\dot{\gamma}) \sim t^{-\alpha_\mathrm{F}}, \tag{4.7.14}$$

$$I_{z,\mathrm{m}}(t;\dot{\gamma}) \sim t^{\beta_\mathrm{F}}. \tag{4.7.15}$$

where the subscript "F" in the power-law exponent α_F and β_F specifies the respective exponents under the shear-flow-field. In the late-stage SD, the power-law exponents have the relationship given by

$$\beta_\mathrm{F} = 3\alpha_\mathrm{F} \tag{4.7.16}$$

based on Eqs. 4.7.5 and 4.7.7.

Figure 4.35 presents the time evolution of the TRSALS intensity at the given $\dot{\gamma}(\neq 0)$, $I(q_z, t;\dot{\gamma})$, for the same solution employed in Fig. 4.32. However, the shear rate $\dot{\gamma} = 0.446$ employed in this solution is higher than the shear rate $\dot{\gamma} = 0.214\ \mathrm{s}^{-1}$

employed in the solution in the Fig. 4.32: $\dot{\gamma}/\dot{\gamma}_{c,conc} = 5.6$ and 2.7 for the solution in Figs. 4.35 and 4.32, respectively, i.e., the solution in Fig. 4.35 is under the stronger shear flow than that in Fig. 4.32. In the figure the TRSALS intensity at the given $\dot{\gamma}(\neq 0)$ $I(q_z, t; \dot{\gamma})$ is presented as a function of the time after the shear drop from 8.91 s^{-1} (Regime V) to 0.446 s^{-1} (Regime IV) for the PS/PB (50:50)/DOP 3.3 wt.% with $T_c(0) = 76\,°C$ and at $\Delta T(0) = 4\,°C$. After the shear drop, the scattering maximum with the maximum intensity $I_{z,m}(t; \dot{\gamma})$ emerges at the q_z defined by the $q_{z,m}(t; \dot{\gamma})$ as presented in Fig. 4.35a.

The results in Fig. 4.35 presents such a general trend that $I_{z,m}(t; \dot{\gamma})$ and $q_{z,m}(t; \dot{\gamma})$ increases and decreases with t, respectively, under the given constant shear rate of 0.446 s^{-1}. In part (a), the scattering maximum at $t = 40$ and $100s^{-1}$ exists at $q_{z,m} >$ 0.005 and $q_{z,m} \cong 0.027\text{nm}^{-1}$, respectively, beyond the q-scale covered in the figure. The results in part (a) indicates the coarsening of the anisotropic dissipative structures with time evolved in the intermediate stage of the SD, as shown in Fig. 4.33b, under the strong shear regime. On the other hands, the results shown in parts (b) and (c) indicate the coarsening in the late stage as shown in Fig. 4.33c and as clarified below.

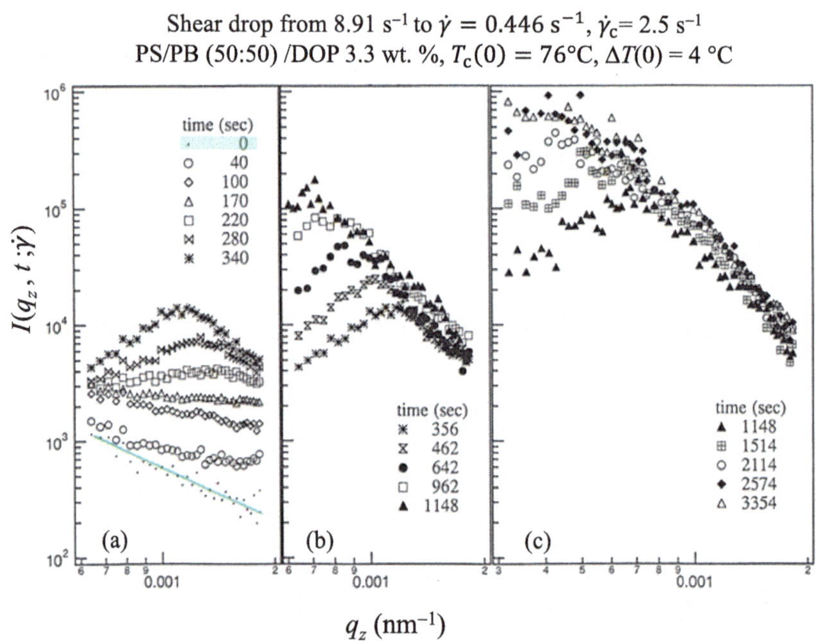

Fig. 4.35 Time evolution of the scattered intensity profile along ND, $I(q_z, t; \dot{\gamma})$, at the give $\dot{\gamma} = 0.446\ s^{-1}$ after the shear drop from the shear-induced single-phase at $\dot{\gamma} = 8.91\ s^{-1} (> \dot{\gamma}_c = 2.5\ s^{-1})$ for the PS/PB (50:50)/DOP 3.3 wt. % with $T_c(0) = 76\,°C$ at $T = 72\,°C$. Part (a): the early to intermediate stage SD and Part (b) and (c): the late- stage SD

It is important to note that the time evolution of the scattering profile in the late stage shown in the Fig. 4.35b,c is eventually **pinned to reach the steady-state** scattering profile controlled by the given shear rate, i.e., the increase of $I_{z,\mathrm{m}}(t;\dot{\gamma})$ and decrease of $q_{z,\mathrm{m}}(t;\dot{\gamma})$ with t are pinned at $t \gtrsim 2754$ s. The steady state is accomplished by the dynamic balance between the shear force and thermodynamic force where the former drives the burst of the overgrown phase-separated structures and the latter drives the growth of the over-burst phase-separated structures. In the quiescent solution where the former force is absent, the structure continues to grow with time to reach the macroscopic two-phase structure at the thermal equilibrium as shown in the left edge in Fig. 4.6 and Fig. 4.7 in Sect. 4.3.2.

Figure 4.36 presents the scattered intensity $I(\tilde{q}_z, t; \dot{\gamma})$ reduced by the maximum scattered intensity $I_{z,\mathrm{m}}(t;\dot{\gamma})$ at the given $\dot{\gamma} = 0.446\ \mathrm{s}^{-1}$ and at the different times after the given shear drop versus the magnitude of the scattering vector q_z reduced by the magnitude of the scattering vector $q_{z,\mathrm{m}}(t;\dot{\gamma})$ at the maximum scattered intensity $I_{z,\mathrm{m}}(t;\dot{\gamma})$ based on the time evolution of $I(q_z, t; \dot{\gamma})$ in the time span of $t = 356$ to 2574 s prior to reach the steady state, as shown in parts (b) and (c) in Fig. 4.35. The result shown in Fig. 4.36 reveals that the time evolution of the scattered intensity profile $I(q_z, t; \dot{\gamma})$ versus q_z at the given $\dot{\gamma}$ which are reduced by the maximum scattered intensity $I_{z,\mathrm{m}}(t;\dot{\gamma})$ and the magnitude of the scattering vector $q_{z,\mathrm{m}}(t;\dot{\gamma})$ at the maximum scattered intensity is universal with$t$ for the given $\dot{\gamma}$. Since one obtains

$$I(\tilde{q}_z, t; \dot{\gamma})/I_{z,\mathrm{m}}(t;\dot{\gamma}) = F(\tilde{q}_z, t; \dot{\gamma}) \tag{4.7.17}$$

from Eqs. 4.7.4, 4.7.5, and 4.7.9, the scaled structure factor $F(\tilde{q}_z, t; \dot{\gamma})$ with reduced $\tilde{q}_z$ is universal with t. This structure factor predicts the shape of the structure which self-similarly grows over the given time span at the given shear rate.

Figure 4.37 presents the time evolution of $q_{z,\mathrm{m}}(t;\dot{\gamma})$ and $I_{z,\mathrm{m}}(t;\dot{\gamma})$ determined from the scattering maximum in $I(q_z, t; \dot{\gamma})$ measured at $\dot{\gamma}$ =

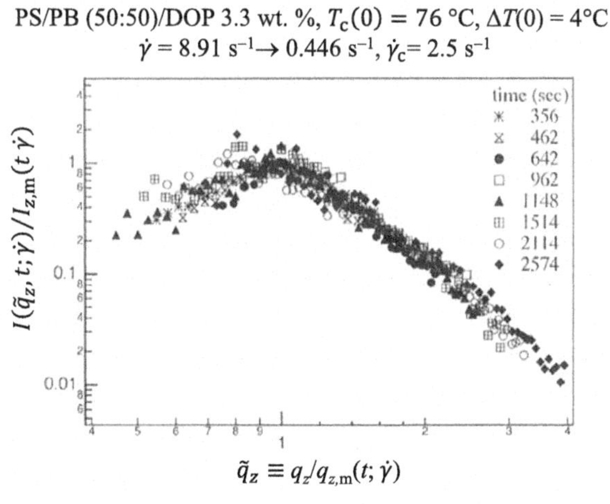

Fig. 4.36 Scaled structure factor constructed from the time evolution of $I(q_z, t; \dot{\gamma})$ at $\dot{\gamma} = 0.446\ \mathrm{s}^{-1} (< \dot{\gamma}_c = 2.5\ \mathrm{s}^{-1})$ shown in Fig. 4.35 with respect to the reduced scattered intensity $I(q_z, t; \dot{\gamma})/I_{z,\mathrm{m}}(t;\dot{\gamma})$ and the reduced magnitude of the scattering vector $q_z/q_{z,\mathrm{m}}(t;\dot{\gamma})$. PS/PB(50:50) /DOP 3.3 wt. % with $T_c(0) = 76\,°\mathrm{C}$ at $T = 72\,°\mathrm{C}$

0.446, 0.624, and 0.891 s^{-1} ($< \dot{\gamma}_c = 2.5$ s^{-1}) after the shear drop of the single-phase solution at $\dot{\gamma} = 8.91$ s^{-1}. The solution employed the same solution as employed in Fig. 4.35. The results reveal that $q_{z,m}(t;\dot{\gamma})$ (part a) and $I_{z,m}(t;\dot{\gamma})$ (part b) decreases and increases with t, respectively, approximately according to the power law given by Eqs. 4.7.14 and 4.7.15, respectively, at all the shear rates covered, implying a dynamically self-similar growth of the dissipative structures along the direction normal to the shear flow at each $\dot{\gamma}$.

It is striking to note that both $q_{z,m}(t;\dot{\gamma})$ and $I_{z,m}(t;\dot{\gamma})$ shift toward longer time scale with increased $\dot{\gamma}$ as highlighted by the green arrows in Fig. 4.37: the larger the shear rate is, the slower are the change of $q_{z,m}(t;\dot{\gamma})$ and $I_{z,m}(t;\dot{\gamma})$, and thereby, the slower is the growth of the dissipative structures, as the green arrows pointing toward right demonstrate. This shear effect on the growth of the dissipative structures in the strong shear regime is opposite to that in the very weak shear regime shown in Fig. 4.30. If $q_{z,m}(t;\dot{\gamma})$ and $I_{z,m}(t;\dot{\gamma})$ shown in Fig. 4.37 are plotted with the reduced time, $\tau \equiv \dot{\gamma}t$, the disparity in $q_{z,m}(\tau)$ versus τ and that in $I_{z,m}(\tau)$ versus τ at the different values of $\dot{\gamma}$ are further enlarged, in contrast to the case shown in Fig. 4.31 for the solutions in the very weak shear regime.

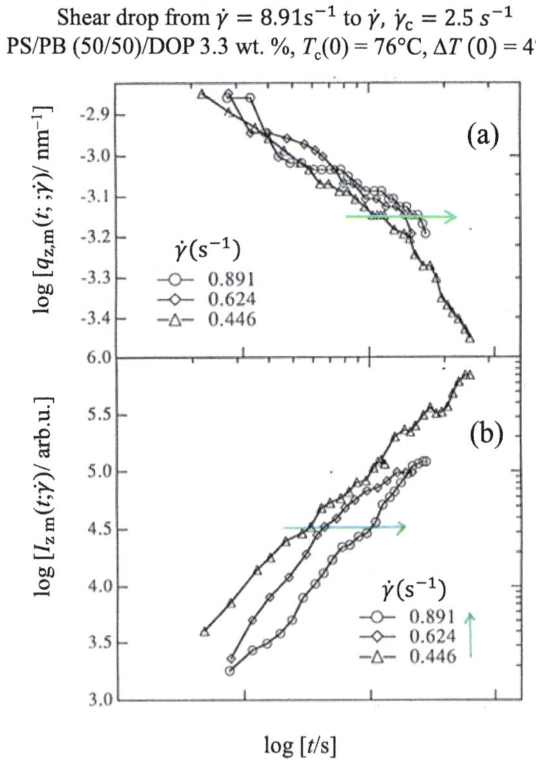

Fig. 4.37 Time evolution of $q_{z,m}(t;\dot{\gamma})$ (part a) and $I_{z,m}(t;\dot{\gamma})$ (part b) after the shear drop from $\dot{\gamma} = 8.91$ s^{-1} to $\dot{\gamma} =$ 0.891, 0.624, and 0.446 s^{-1} for the PS/PB (50:50)/DOP 3.3 wt. % with $T_c(0) = 76$°C at $T = 72$°C. $\dot{\gamma}_c = 2.5$ s^{-1}

4.7.3 Scaling Law on the Effect of Shear in the Strong Shear Regime

To investigate the origin of the discrepancy in the self-organization behavior in between the strong shear regime and the very weak shear regime, the following scaling analyses are conducted regarding the effect of the shear flow on the time evolution of $q_{z,\mathrm{m}}(t;\dot{\gamma})$ and $I_{z,\mathrm{m}}(t;\dot{\gamma})$ shown in Fig. 4.37. The scaling analyses are based on the deep insight into the effect of the increased shear rate, which brings about the increased critical temperature drop $\Delta T_{\mathrm{c}}(\dot{\gamma})$ as elucidated in Fig. 4.23b, which in turn brings about the decreased quench depth $\Delta T(\dot{\gamma})(=\Delta T(0)-\Delta T_{c}(\dot{\gamma}))$ with the increased shear rate $\dot{\gamma}$, and as elucidated also by the basic law on the effects of the $\dot{\gamma}$ on $\Delta T(\dot{\gamma})$ given by Eq. 4.7.27 below in this section. We define the reduced wave number at the maximum intensity $\tilde{q}_{\mathrm{m},z}(t;\dot{\gamma})$ and the reduced maximum intensity $\tilde{I}_{\mathrm{m},z}(t;\dot{\gamma})$ in the late-stage SD, and the reduced time $\tau(\dot{\gamma})$ as follows:

$$\tilde{q}_{z,\mathrm{m}}(t;\dot{\gamma}) \equiv q_{z,\mathrm{m}}(t;\dot{\gamma})/q_{z,\mathrm{m}}(0;\dot{\gamma}), \tag{4.7.18}$$

$$\tilde{I}_{z,\mathrm{m}}(t;\dot{\gamma}) \equiv \frac{I_{z,\mathrm{m}}(t;\dot{\gamma})}{\Delta\phi_{\mathrm{e}}(\dot{\gamma})^2\,\xi_z(\dot{\gamma})^3}, \tag{4.7.19}$$

$$\tau(\dot{\gamma}) \equiv t/t_{\mathrm{c}}(\dot{\gamma}), \tag{4.7.20}$$

where $q_{z,\mathrm{m}}(0;\dot{\gamma})$ and $\xi_z(\dot{\gamma})$ are respectively the characteristic wave number and the characteristic length of the structures developed along the z axis (neutral direction) in the early stage SD under the shear flow at the given shear rate $\dot{\gamma}$, and they are interrelated by

$$q_{z,\mathrm{m}}(0;\dot{\gamma}) \sim \xi_z(\dot{\gamma})^{-1}. \tag{4.7.21}$$

$\Delta\phi_{\mathrm{e}}(\dot{\gamma})$ is the composition difference between the two phases at the given $\dot{\gamma}$ and temperature T in the dynamic equilibrium (or steady-state) (Fig. 4.23b). The time $t_{\mathrm{c}}(\dot{\gamma})$ in Eq.4.7.20 is the characteristic time of the solution at the given $\dot{\gamma}$, which is given as follows according to the diffusion law for the concentration fluctuations,

$$t_{\mathrm{c}}(\dot{\gamma}) = \xi_z(\dot{\gamma})^2/D_{\mathrm{app}}(\dot{\gamma}), \tag{4.7.22}$$

where $D_{\mathrm{app}}(\dot{\gamma})$ is the collective diffusivity of the thermal concentration fluctuations at the given $\dot{\gamma}$.

In the context of the mean-field (MF) approximation, the following relations can be applied to the solutions under the shear flow at the given $\dot{\gamma}$,

$$\xi_z(\dot{\gamma}) \sim q_{z,\mathrm{m}}(0;\dot{\gamma})^{-1} \sim \Delta T(\dot{\gamma})^{-1/2}, \tag{4.7.23}$$

$$\Delta\phi_e(\dot{\gamma}) \sim \Delta T(\dot{\gamma})^{1/2}, \tag{4.7.24}$$

$$D_{app}(\dot{\gamma}) \sim \Delta T(\dot{\gamma}), \tag{4.7.25}$$

where $\Delta T(\dot{\gamma})$ is defined by

$$\Delta T(\dot{\gamma}) \equiv T_c(\dot{\gamma}) - T = \Delta T(0) - \Delta T_c(\dot{\gamma}), \tag{4.7.26}$$

in the case when $T \leq T_c(\dot{\gamma})$ or $\dot{\gamma} < \dot{\gamma}_c$ in the shear-induced two-phase state (see Fig. 4.23b), and the $\dot{\gamma}$-dependence of $\Delta T(\dot{\gamma})$ is given in this case by

$$\Delta T(\dot{\gamma}) = \Delta T(0) \left[1 - (\dot{\gamma}/\dot{\gamma}_c)^{1/2}\right] \tag{4.7.27}$$

instead of Eq. 4.4.13 which corresponds to the case of the shear-induced single-phase solution at $\dot{\gamma} > \dot{\gamma}_c$, as in the case shown in Fig. 4.23a. Equation 4.7.23 to Eq. 4.7.25 at $\dot{\gamma} = 0$ correspond to those in the MF approximation in the quiescent solutions.

From Eqs. 4.7.18–4.7.20 and 4.7.22–4.7.25, the reduced parameters are given by

$$\tilde{q}_{z,m}(t;\dot{\gamma}) \sim q_{z,m}(t;\dot{\gamma})\ \Delta T(\dot{\gamma})^{-1/2}, \tag{4.7.28}$$

$$\tilde{I}_{z,m}(t;\dot{\gamma}) \sim I_{z,m}(t;\dot{\gamma})\ \Delta T(\dot{\gamma})^{1/2}, \tag{4.7.29}$$

$$\tau(\dot{\gamma}) \sim t\ \Delta T(\dot{\gamma})^2. \tag{4.7.30}$$

Equations 4.7.5 and 4.7.7 lead to the equation given below,

$$I_{z,m}(t;\dot{\gamma}) \sim \Delta\phi_e(\dot{\gamma})^2\ q_{z,m}(t;\dot{\gamma})^{-3}, \tag{4.7.31}$$

in the late-stage SD. Equations 4.7.19, 4.7.23, and 4.7.31 lead to the equation given below,

$$\tilde{I}_{z,m}(t;\dot{\gamma}) \sim \tilde{q}_{z,m}(t;\dot{\gamma})^{-3}, \tag{4.7.32}$$

in the late-stage SD.

4.7.4 Results of the Scaling Analyses: Shear Controls Space–Time Self-Organization of Dissipative Structures

Figure 4.38 presents the results of the scaling analyses, $\tilde{q}_{z,m}(\tau;\dot{\gamma})$ versus τ (part a) and $\tilde{I}_{z,m}(\tau;\dot{\gamma})$ versus τ (part b), regarding the effects of the shear flow on the time evolution of the characteristic wave number $q_m(t;\dot{\gamma})$ and the maximum scattered

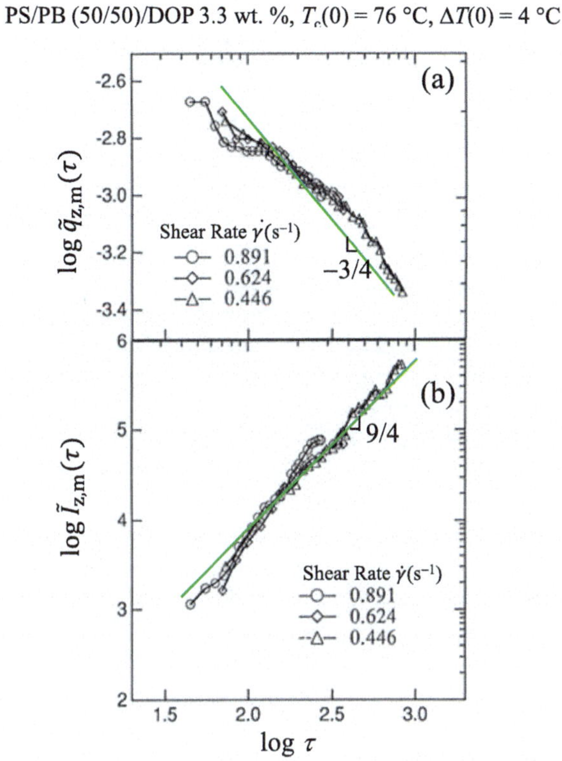

Fig. 4.38 Magnitude of the reduced scattering vector at the maximum intensity $\tilde{q}_{z,m}(\tau)$ (part a) and the reduced maximum intensity $\tilde{I}_{z,m}(\tau)$ plotted against the reduced time τ (part b). The time evolution of $q_{z,m}(t; \dot{\gamma})$ and $I_{z,m}(t; \dot{\gamma})$ at the different shear rates becomes universal with $\dot{\gamma}$ in the reduced plot. The green line indicates the power law of $q_{z,m}(\tau) \sim \tau^{-\alpha_F}$ with $\alpha_F = 3/4$ (a) and $I_{z,m}(\tau) \sim \tau^{\beta_F}$ with $\beta_F = 9/4$ (b). PS/PB (50:50)/DOP 3.3 wt. % with $T_c(0) = 76\,°C$ at $T = 72\,°C$. $\dot{\gamma}_c = 2.5\ s^{-1}$

intensity $I_m(t; \dot{\gamma})$ of the dominant Fourier modes of the growing dissipative structures with respect to the real-time scale t, as shown in Fig. 4.37. The results reveal the following conclusions:

(1) $q_m(t; \dot{\gamma})$ versus t and $I_m(t; \dot{\gamma})$ versus t at the different shear rates in the strong shear regime fall onto the respective master curve universal with the shear rate, when they are plotted in the reduced scales of $\tilde{q}_m(\tau; \dot{\gamma})$ versus τ (part a) and $\tilde{I}_m(\tau; \dot{\gamma})$ versus τ (part b).

(2) $\tilde{q}_m(\tau; \dot{\gamma})$ versus τ and $\tilde{I}_m(\tau; \dot{\gamma})$ versus τ are given by the following power laws,

$$\tilde{q}_m(\tau; \dot{\gamma}) \sim \tau^{-\alpha_F}, \alpha_F \cong 3/4 \tag{4.7.33}$$

and

$$\tilde{I}_m(\tau; \dot{\gamma}) \sim \tau^{\beta_F}, \beta_F \cong 9/4. \tag{4.7.34}$$

(3) The power-law exponents α_R and β_F are interrelated by

$$\beta_F = 3\alpha_F. \tag{4.7.35}$$

Here, the subscript F in Eqs. 4.7.33 to 4.7.35 specifies the exponent of τ for the reduced wave number and the reduced maximum scattered intensity under the shear "flow".

The conclusion (1) reveals that the $\dot{\gamma}$-dependence of the time evolution of the dissipative structures in the strong shear regime is consistently scaled with the $\dot{\gamma}$-dependence of the quench depth $\Delta T(\dot{\gamma})$; the drop of the quench $\Delta T(\dot{\gamma})$ with increased $\dot{\gamma}$ is indispensable regarding the self-organization of the dissipative structures under the shear flow at the any given shear rates. That is, the larger the $\dot{\gamma}$ is, the larger is the drop of $T_c(\dot{\gamma})$, and thereby the smaller is the quench depth $\Delta T(\dot{\gamma})$ (Fig. 4.23b), which predicts the slower growth of the dissipative structure with increased $\dot{\gamma}$ as evidenced by the shift of $q_{z,m}(t;\dot{\gamma})$ and $I_{z,m}(t;\dot{\gamma})$ toward the longer time scale, as highlighted by the green arrow in Fig. 4.37. On the other hands, in the weak shear regime where $\dot{\gamma} \ll \dot{\gamma}_c$, $T_c(\dot{\gamma}) \approx T_c(0)$ and $\Delta T(\dot{\gamma}) \approx \Delta T_c(0)$, and thereby the drop of $\Delta T(\dot{\gamma})$ is essentially absent. Thus, the shear does not affect the thermodynamic driving force for the growth of the dissipative structure but only affects the shear deformation of the dissipative structures developed at the given T. On the other hand, in the strong shear regime, **the shear controls the space–time organization of the dissipative structures through $\Delta T(\dot{\gamma})$** as given by Eq. 4.7.27, together with Eqs. 4.7.28–4.7.30.

The conclusion (2) discloses the dynamic self-similarity of the growth of the dissipative structures in the self-organization process under the shear flow. In conjunction of the conclusion 2, it is also important to note that α_F and β_F do not reach the limiting value of 1 and 3, respectively, under the flow at given quench depths $\Delta T(0)$ and shear rates $\dot{\gamma}$, in contrast to the case of phase-separated domains of the liquid mixtures in quiescent state at given quench depths $\Delta T(0)$ which reach the limiting value 1 and 3 for α_F and β_F, respectively, in the long time limit where the hydrodynamic interactions play a significant role on the coarsening process. The reason why the limiting values are not attained under the flow may be because the shear flow controls the steady-state dissipative structures determined by $\Delta T(0)$ and $\dot{\gamma}$: overgrown domains driven by the $\Delta T(0)$-dependent thermodynamic driving force are bust by the $\dot{\gamma}$-dependent viscous drag, and the over-burst small domains by the viscous drag are grown by the thermodynamic driving force. That is, the flow controls the limiting values.

The conclusion (3) elucidates the self-organization process in the late-stage SD under the shear flow occurs with the dynamic self-similarity before reaching the shear-rate-controlled steady state where the growing concentration difference between the two phases with t, $\Delta\phi(t;\dot{\gamma})$, reaches the dynamic equilibrium value $\Delta\phi_e(\dot{\gamma})$ (see Fig. 4.23b).

4.8 Tracing Back of Growth History of Phase-Separated Structures

4.8.1 Equivalence Principle between $\xi_{d,z}(\dot{\gamma})$ and $\Lambda_m(t)$

Here, $\xi_{d,z}(\dot{\gamma})$ is the characteristic length of the dissipative structure evolved along the neutral direction (along Oz axis) at a shear rate $\dot{\gamma}$ below $\dot{\gamma}_c$ at the steady-state, whereas $\Lambda_m(t)$ is the characteristic length of the domain structure at the given t after the cessation of the shear from shear-induced single phase (Regime V).

One can study the time evolution of phase-separated domains in the quiescent state (at $\dot{\gamma} = 0$) after the cessation of the steady-state shear flow in Regime V (the shear-induced single-phase state). The time-resolved light scattering patterns reveal the so-called **spinodal ring** which has the intensity maximum at $q = q_m(t)$, independent of the azimuthal angle (the circular spinodal ring). Figure 4.39 presents the time evolution of the characteristic length $\Lambda_m(t) \equiv 2\pi/q_m(t)$ and the shear-rate dependence of $\xi_{d,z}(\dot{\gamma})$ for the PS/PB (50:50)/DOP 3.0 wt. % at $\Delta T(0) = 6\,°\text{C}$ and $8\,°\text{C}$, respectively [9]. In this figure, the values of $\Lambda_m(t)$ are given in the right-hand side of the ordinate scale, while those of time t can be found at the top abscissa. The t-scale was deliberately chosen in such a manner that the value increases from right to left. The plot of $\log \Lambda_m(t)$ versus $\log t$ shows the coarsening of the domain structure as revealed by the increase of $\Lambda_m(t)$ with t given by the power law of $\Lambda_m(t) \sim t^{p_1}$ such that the power-law exponent p_1 approximately changes in the order of 1/3 or 1/4 → 1/2 → 1/4 with the increased time [9, 18]. On the other hand, $\xi_{d,z}(\dot{\gamma})$ decreases with the increased $\dot{\gamma}$ according to the power law of $\xi_{d,z}(\dot{\gamma}) \sim \dot{\gamma}^{-p_2}$ such that the slope of $\log \xi_{d,z}$ versus $\log \dot{\gamma}$ approximately changes in the order of $-1/4 \rightarrow -1/2 \rightarrow -1/4$ or $-1/3$ with the increased $\dot{\gamma}$.

The earlier part of the curve $\Lambda_m(t)$ versus t showing the change in the slope from 1/3 to 1/2 or 1/4 to 1/2 with the increased t was elucidated to correspond to the coarsening process of the bicontinuous phase-separated domains rich in the polymers composed of PB/PB (50: 50) in the matrix phase poor in the polymers also composed of PS/PB (50: 50). Here, the phase separation is driven by the spinodal decomposition, whereas the later part of the curve showing the change in the slope from 1/2 to 1/4 with increased t was elucidated to correspond to the percolation-to-cluster (PC) transition (PCT) [71, 72] which brings about the slowing down of the coarsening rate after the transition. Note that the PCT is the reverse transition of the cluster-to-percolation (CP) transition (CPT) brought about by the increased shear rate $\dot{\gamma}$ as discussed in conjunction with Figs. 4.5 and 4.6. After the completion of the PCT, we expect that the clusters of droplets grow with $t^{1/3}$ in the long-time hydrodynamic limit, although this long-time limit was not covered in this experiment.

Here, we note an intriguing possibility that the increased shear rate enables us to trace back the history of the phase separation process in the quiescent solutions as observed by $\Lambda_m(t)$. This is because the shear rate determines the time $(1/\dot{\gamma})$ available for the phase-separated domains to grow effectively free from the shear flow at the given shear rate $\dot{\gamma}$. When the domains are observed in Oxz (or q_x-q_z) plane, one

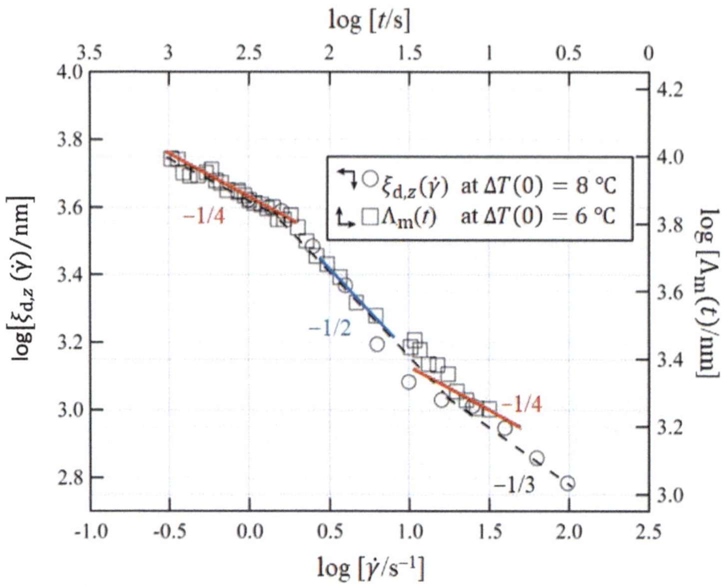

Fig. 4.39 Log $\xi_{d,z}(\dot{\gamma})$ vs $\log \dot{\gamma}$ (left ordinate and bottom abscissa) for the PS/PB (50:50)/DOP 3.3 wt. % at $\Delta T(0) = 8\,°C$ and log $\Lambda_m(t)$ versus log t (right ordinate and top abscissa) where $\log t \equiv$ (constant) $- \log \dot{\gamma}$ for the PS/PB (50:50)/DOP 3.0 wt. % at $\Delta T(0) = 6\,°C$ obtained at $\dot{\gamma} = 0$ after the cessation of the shear from the shear-induced single-phase state. The (constant) was arbitrary chosen so that the two curves have superposed each other. Based on Hashimoto, T. 2008 In *Soft Matter Characterization*; Borsali, R., Pecora, R., Eds., Springer. New York; Chapter 8, pp378-462 [9]

should consider the domain size along the neutral axis (along Oz-axis or q_z axis), because the deformation effect on the domain size is less significant along the neutral axis than along the flow axis (Ox-axis or q_x axis). Thus, one may be able to compare the experimental results on $\Lambda_m(t)$ versus t obtained for the quiescent solution at the quench depth $\Delta T(0) = 6°C$ and concentration 3.0 wt. % with the experimental results on $\xi_{d,z}(\dot{\gamma})$ versus $\dot{\gamma}$ obtained for the nearly same solution, i.e., the same solution with slightly different quench depth $\Delta T(0) = 8$ °C and concentration 3.3 wt. %, under the steady-state shear by taking into account the fact that $\dot{\gamma}$ should correspond to t^{-1}.

Figure 4.39 demonstrates validity of the above conjuncture where the results on $\log \xi_{d,z}(\dot{\gamma})$ versus $\log \dot{\gamma}$ is plotted together with the results on $\log \Lambda_m(t)$ versus $\log t$ on the same double logarithmic scale. The scales in the left ordinate and the bottom abscissa refer to $\log \xi_{d,z}(\dot{\gamma})$ and $\log \dot{\gamma}$, respectively [9]. It should be noted that the top abscissa corresponds to $\log t =$ (constant) $- \log \dot{\gamma}$ where the constant was chosen such that the curve on $\log \Lambda_m(t)$ versus $\log t$ and the curve on $\log \xi_{d,z}(\dot{\gamma})$ versus $\log \dot{\gamma}$ are superposed each other. The fact that the $\Delta T(0)$ employed for the two experiments are slightly different should not alter the self-assembling process itself but slightly alter the time and spatial scales of the self-assembling process in the two experiments. Thus, we shift the two curves both vertically and horizontally in the double logarithmic scale. Upon shifting, the two results are demonstrated to well superpose each other, revealing the **equivalence principle** of $\xi_{d,z}(\dot{\gamma})$ and $\Lambda_m(t)$ under the concept of $\dot{\gamma} \sim t^{-1}$. Thus, in principle, one can "trace back" a growth history (time evolution pathway) of phase-separated domain structures in the liquid mixtures from the quiescent single-phase mixtures by applying the increased shear rate to the given phase-separated liquid mixtures, of which the growth history is to be explored.

Naturally, one cannot trace back the history without the applied shear, but one can accomplish this with applied shear in the context of the concept regarding the reversible change of the steady-state domain structures with shear rates.

4.8.2 Equivalence in Evolution of SALS Patterns in the Following Two Processes: Increased Time after Cessation of Shear from the Given $\dot{\gamma}$ and Increased Shear up to a Given $\dot{\gamma}$

Figure 4.40 presents the time evolution of the SALS pattern after the cessation of the steady-state shear flow at $\dot{\gamma} = 0.67\ \text{s}^{-1}$ in Regime III for the PS/PB (50:50)/DOP 3.0 wt. % with $T_c(0) = 54.0\ °\text{C}$ at $\Delta T(0) = 10.8\ °\text{C}$. The results reveal the followings.

(1) The strong equatorial SALS streak arising from the elongated long domains oriented along the flow direction loses its intensity with time after the cessation as presented by the pattern shown from part (b) to (e).
(2) Simultaneously with the event (1), the circular diffuse SALS pattern appears, and its intensity increases with time as presented by the patterns illuminated from part (c) to (f).
(3) The events (1) and (2) reveal that the elongated long domains stabilized at the given shear rate at the steady-state shear flow in Regime III, defined as $\dot{\gamma}_{II}$ (see Table 4.2), are destabilized and burst into round droplets after the cessation of the steady-state shear flow: the cessation involves the percolation-to-cluster transition (PCT) with t.
(4) These changes with t after the cessation of the shear are opposite to those observed in Fig. 4.3 for the same 3.0 wt. % solution or Fig. 4.5 for the 3.3 wt. % solution with increased $\dot{\gamma}$; the increase of $\dot{\gamma}$ causes the cluster-to-percolation transition (CPT) as summarized in Fig. 4.6, whereas the increase of the time after the cessation of the shear causes the percolation-to-cluster transition (PCT).
(5) Thus, with regard to the evolution of the SALS patterns and the relevant dissipative structures, we can validate the following conclusion. The increased time, e.g., t_2, after the cessation of the shear from the given steady-state shear rate, e.g., $\dot{\gamma}_1$ (e.g., 0.67 s^{-1}, lower than critical shear rate $\dot{\gamma}_c$) gives rise to the effect on the SALS pattern changing from the steady-state SALS pattern at $\dot{\gamma}_1$ (Fig. 4.40a)

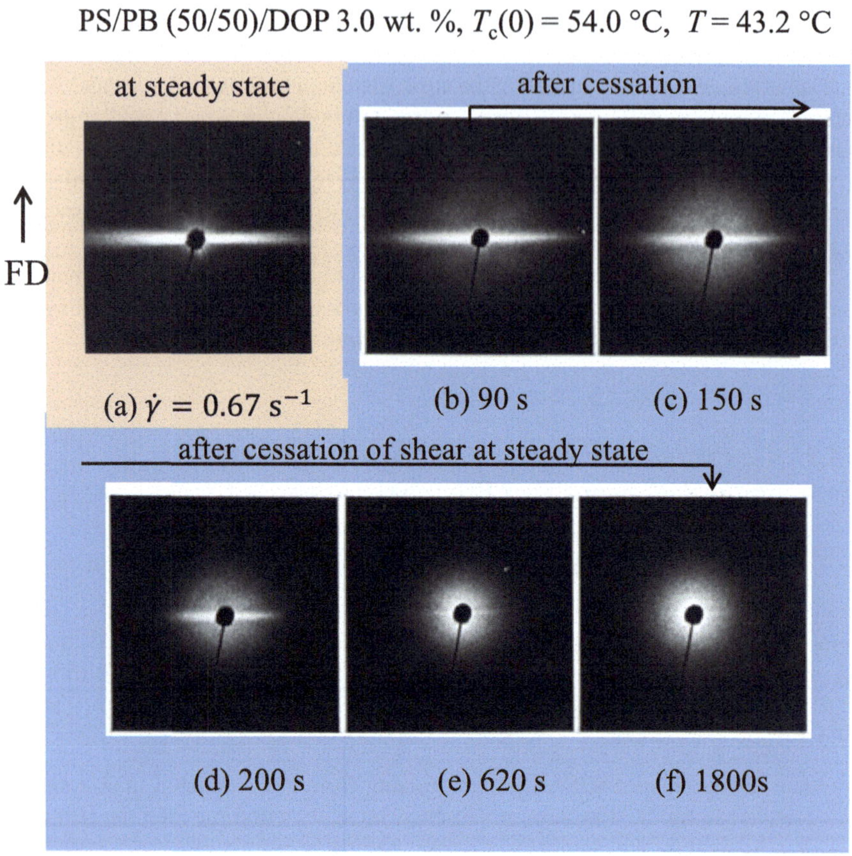

Fig. 4.40 Steady-state SALS pattern at $\dot{\gamma} = 0.67\ \text{s}^{-1}$ in Regime III (part a) and the time evolution of the SALS pattern after the cessation of the steady-state shear flow at $\dot{\gamma} = 0.67\ \text{s}^{-1}$ (part b to f). PS/PB (50:50)/DOP 3.0 wt. % with $T_c(0) = 54.0\,°\text{C}$ at $T = 43.2\,°\text{C}$

to the pattern at the given time t_2 after the cessation (as shown by the change of the SALS pattern from Fig. 4.40a to one of Fig. 4.40b to 4.40f, depending on the time elapsed after the cessation). This effect is found to give rise to the effect opposite to the effect of the increased shear rate from $\dot{\gamma}_2$ to the given shear rate $\dot{\gamma}_1$ at the steady-state on the SALS pattern (as shown by the change of the SALS pattern from one of Fig. 4.40b–4.40f into Fig. 4.40a). This evidence predicts the principle of the reversibility of the evolution of the SALS pattern in the following two processes: the increased shear rate and the increased time after the cessation of the shear rate. It predicts and further generalizes the **equivalence principle** of $\xi_{d,z}(\dot{\gamma})$ and $\Lambda_m(t)$ under the principle of $\dot{\gamma} \sim t^{-1}$ in the case of $\dot{\gamma} < \dot{\gamma}_c$. Note that the equivalence principle validated in Fig. 4.39 is valid in the case of $\dot{\gamma} > \dot{\gamma}_c$. Thus, the principle is valid generally at any given $\dot{\gamma}$ below and above $\dot{\gamma}_c$.

4.8.3 *Reversible Changes of the Steady-State Structures with $\dot{\gamma}$*

The reversible changes of the steady-state SALS patterns and the corresponding dissipative structures with the shear at the given shear rate $\dot{\gamma}_0$ are already suggested in Section 4.3.4. Figure 4.41 summarizes the representative results demonstrating the reversible changes of the steady-state SALS patterns at the given shear rate. The reversibility is confirmed by the identity of the SALS pattern (1) and (2) in Regime II under the shear protocol indicated in part (a) and the SALS pattern (3) and (4) in Regime III under the shear protocol indicated in part (b). These patterns are obtained for the given solution PS/PB (50:50)/DOP 3.0 wt.% at $T_c(0) = 54.0\,°\text{C}$ and $\Delta T(0) = 10.8\,°\text{C}$.

Identity of the steady-state SALS pattern (1) and (2). We look into Fig.4.41 and focus mainly on the "SALS pattern (1)" and the "SALS pattern (2)" in the shear rate protocol (a). The two SALS patterns (1) and (2) are obtained at the same $\dot{\gamma}$ in the Regime II and at the steady state. The steady-state SALS pattern (1) is obtained by the increased shear-rate protocol (a) from $\dot{\gamma} = 0$ to the given $\dot{\gamma}$ in the Regime II. On the other hand, the steady-state SALS pattern (2) is obtained under the shear protocol involving the further increase of $\dot{\gamma}$ from the given $\dot{\gamma}$ in the Regime II to the given $\dot{\gamma}$ in the Regime V, followed by the shear-rate drop to the given $\dot{\gamma}$ in the Regime II. Here, it is important to note a piece of the experimental that the steady

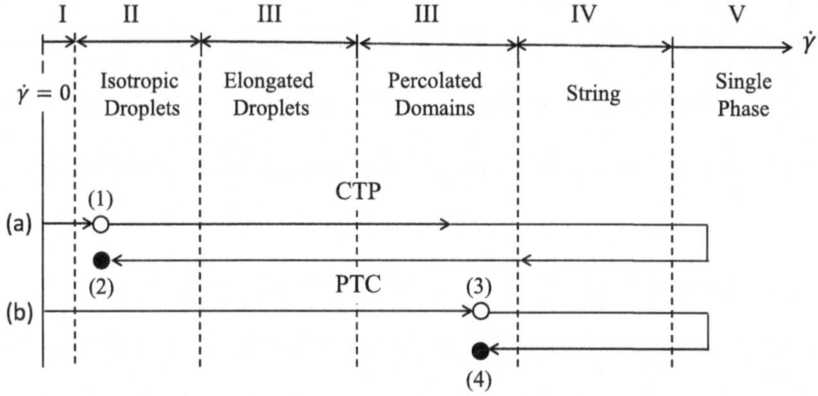

(1) Fig. 4.3 (b), (2) Fig. 4.29 (a) at steady state, (3) Fig. 4.3 (c), (4) Fig.4.29 (b) at steady state,

Fig. 4.41 Reversible change of the steady-state SALS pattern with the increased and decreased $\dot{\gamma}$ as revealed by the identity between the steady-state SALS patterns (1) and (2) under the shear protocol specified in part (a) and those (3) and (4) under the shear protocol specified in part (b). The patterns (1) to (4) are shown in the respective figures specified in the legends at the bottom of the figures state SALS pattern (1) is identical to the steady state SALS pattern (2), despite the difference in the shear history.

Identity of the steady-state SALS pattern (3) and (4). Again, we look into Fig. 4.41 and now focus mainly on the "SALS pattern (3)" and the "SALS pattern (4)" in the shear rate protocol (b). The two SALS patterns (3) and (4) are obtained at the same $\dot{\gamma}$ in Regime III and at the steady state. The steady-state SALS pattern (3) is obtained by the increased shear-rate protocol (b) from $\dot{\gamma} = 0$ to the given $\dot{\gamma}$ in Regime III. On the other hand, the steady-state SALS pattern (4) is obtained under the shear protocol involving the further increase of $\dot{\gamma}$ from the given $\dot{\gamma}$ in Regime III to the given $\dot{\gamma}$ in Regime V, followed by the shear-rate drop to the given $\dot{\gamma}$ in Regime III. Here again, it is important to note a piece of the experimental that the steady state SALS pattern (3) is identical to the steady state SALS pattern (4), despite the difference in the shear history.

4.9 Off-Critical Mixtures: Composition Dependence of Dissipative Structures and Flow-Induced Phase Transition

4.9.1 *Experiments*

In this section, we describe the flow-induced dissipative structures and phase transition as a function of the composition of the PS/PB in the PS/PB/DOP solution. The composition covered is the PS/PB = 20/80 (18.8/81.2), 30/70 (28.4/71.6), 40/60 (38.1/61.9), 50/50 (48.0/52.0), 60/40 (58.1/41.9), 70/30 (68.3/31.7), and 80/20 (78.7/21.3) by wt.%/wt.% (vol.%/vol.%). The binodal temperature for the each solution with the given composition at $\dot{\gamma} = 0$ is defined as $T_c(0)$. The binodal temperature $T_c(0)$ is assessed as follows,

$$T_c(0) \equiv \lim_{r \to 0} T_{cl}(r),\ r \equiv dT/dt, \tag{4.9.1}$$

where $T_{cl}(r)$ is the cloud point for the given solution at $\dot{\gamma} = 0$ which is assessed by measuring the transmitted intensity for the solution as a function of the temperature at the constant heating rate r. $T_{cl}(r)$ is the temperature at which the transmittance starts to increase at the given r. $T_c(0)$ for the given solution is determined by the extrapolation of the $T_{cl}(r)$ values measured as a function of r to the zero-heating rate. $T_c(0)$ for each solution described above is summarized in Table 4.1 in Sect. 4.2.

The steady-state SALS pattern at a given set of $\dot{\gamma}$, T, and composition is obtained by using the shear protocol presented in Fig. 4.2a for the stepwise shear up (part (a) for the increased $\dot{\gamma}$) and in Fig. 4.2b for the stepwise shear drop (part (b) for decreased $\dot{\gamma}$). Prior to measure the SALS pattern, the solution at a given T below $T_c(0)$ is first brought into a single-phase state by increasing $\dot{\gamma}$ from $\dot{\gamma} = 0$ to $\dot{\gamma} > \dot{\gamma}_c$

and subsequently kept under the shear rate at least for 20 min. in the shear-induced single-phase state. This single-phase solution is brought to $\dot{\gamma} = 0$ and kept for 48 h to attain the initial state common to all the experiments, i.e., the macroscopically phase-separated solution in Regime I, for all the experiments with the solutions at the varied temperatures and compositions. The intensity measurements of the 2D TRSALS patterns, $I(q_y = 0, q_x, q_z; \dot{\gamma})$, and the integrated SALS intensity along the z-axis (the neutral direction), $\mathcal{J}_z(\dot{\gamma})$, are measured as a function of $\dot{\gamma}$ at 20 min. after the stepwise shear-jump from $\dot{\gamma} - \Delta\dot{\gamma}$ to $\dot{\gamma}$ or shear-drop from $\dot{\gamma} + \Delta\dot{\gamma}$ to $\dot{\gamma}$. The measurements for the stepwise increase or decrease of $\dot{\gamma}$ are continued across the critical shear rate $\dot{\gamma}_c$ for the shear-induced two-phase to single-phase transition or single-phase to two-phase transition, respectively.

4.9.2 Flow-Induced Dissipative Structures

Figure 4.42 presents the typical steady-state 2DSALS patterns representative for the off-critical mixtures; the patterns are shown for the PS/PB (20:80)/DOP 3.3wt. %

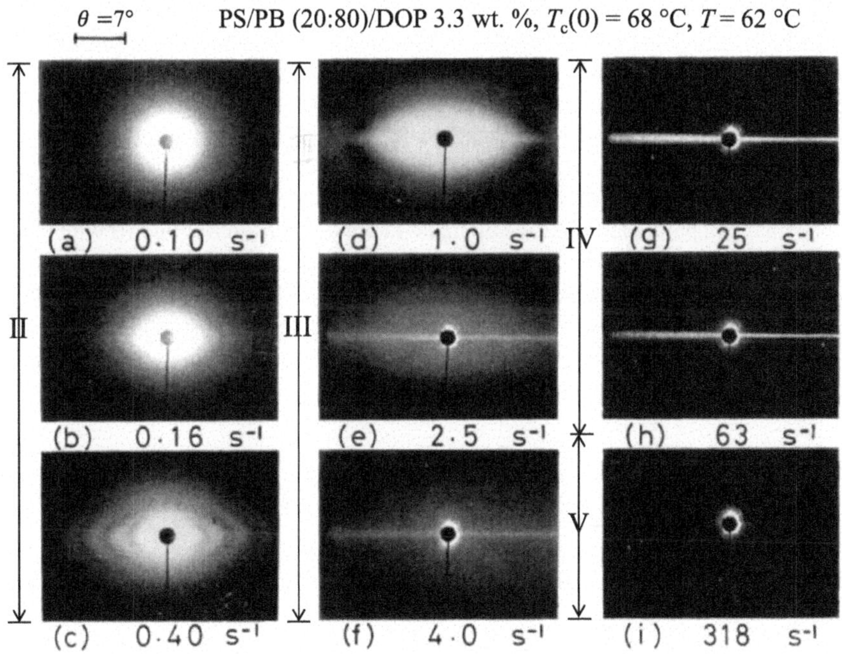

Fig. 4.42 SALS patterns for the typical off-critical mixture observed at the steady-state under the increased shear rate from (a) 0.10 s^{-1} to (i) 318 s^{-1} starting from the macrophase-separated solution at the thermal equilibrium (at $\dot{\gamma} = 0\ s^{-1}$). The flow direction is vertical. PS/PB (20:80)/DOP 3.3 wt. %, $T_c(0) = 68\,°C$ at $T = 62\,°C$. Based on [6]

with $T_c(0) = 68\,°C$ at $T = 62\,°C$ under the increased shear rate [6] according to the shear protocol shown in the Fig. 4.2a. The change of the patterns with the increased shear rate is quite similar to that shown in the Figs. 4.3 and 4.5 for the nearly critical solutions. The change of the patterns is also classified into the five regimes (I to V) as in the case of the near-critical mixtures:

Regime I (the macroscopically phase-separated two-phase structure, the corresponding SALS pattern being not included in the figure),

Regime II (the patterns in parts a to c),

Regime III (the patterns in parts d to f),

Regime IV (the patterns in parts g and h), and

Regime V (the pattern in part i).

The steady-state dissipative structures evolved in Regime II to IV have the same characteristics as those for the nearly critical mixtures as summarized in the Fig. 4.6. The SALS pattern (c) exhibits the multiple-order scattering maxima, especially clear along the equator, which originates from the form-factor maxima arising from the uniformity of the size of the ellipsoidal droplets along the neutral direction.

Figure 4.43 presents typical POM images representative for the off-critical mixtures. The images are taken at the steady-state shear at the shear rate as indicated in the figure legend under the increased shear-rate protocol shown in Fig. 4.2a; the POM images are shown for the PS/PB (80:20)/DOP 3.3 wt. % with $T_c(0) = 58\,°C$ at $T = 44\,°C$ [4]. The POM images shown are taken for the PS/PB with the composition opposite to the composition for which the SALS patterns shown in Fig. 4.42 are taken. Nevertheless, the variation of the characteristic features of the dissipative structures with the increased $\dot{\gamma}$ is the same, independent of the compositions. The POM images in part (a) imply ellipsoidal droplets with their major axes oriented along the flow direction, which represents the dissipative structures in the Regime II. The image in part (b) implies the elongated long domains in Regime III with their axes oriented along the flow direction which are self-organized through the cluster-to-percolation transition along the flow direction (FD). The image (c) represents the string structures in Regime IV which are highly oriented, persistent, and uniform along the FD and are stabilized due to the shear-driven suppression of the interface fluctuations.

4.9.3 *Uniformity of Droplet Size in Regime II*

The spheroidal droplets developed at the given shear rate in Regime II have the size characterized by $R_x \equiv v_a R$ and $R_z \equiv R$ along the flow direction and the neutral direction, respectively, with $v_a R_x / R_z$ being the aspect ratio of the ellipsoids. Upon decreasing the shear rate, the ellipsoidal droplets are found to grow into larger round droplets; further growth of the droplets are found to result eventually in the macroscopic phase separation with the PS-rich phase(or PB-poor phase) and PS-poor phase (or PB-rich phase) across the interface parallel to Oxy plane, respectively, in order to reduce the total interface area and hence the interfacial free energy, driven

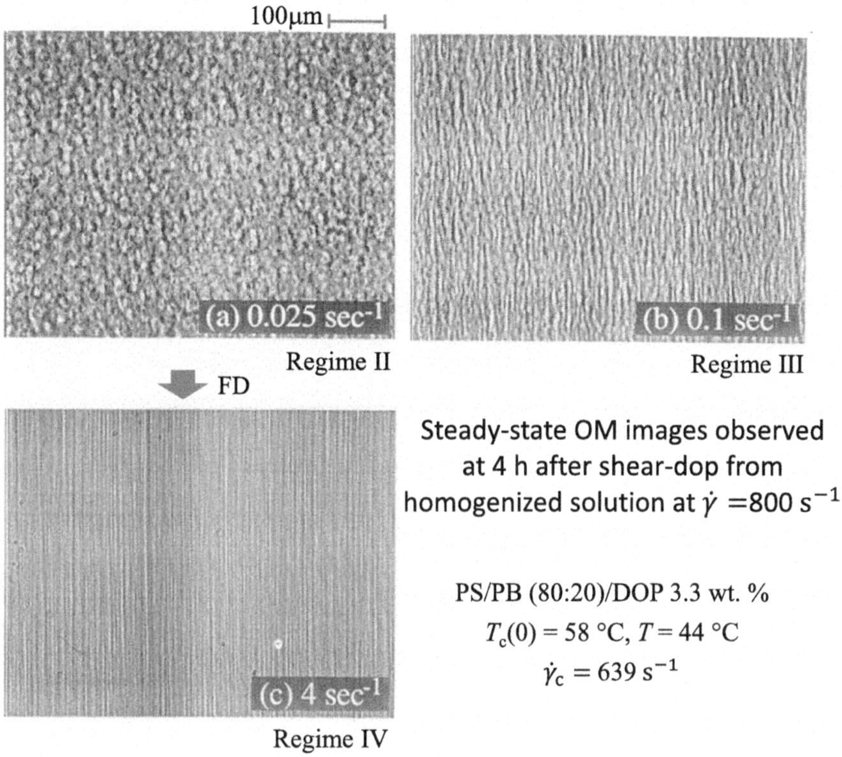

Fig. 4.43 Self-organized steady-state dissipative structures with the increased shear rate protocol. The OM images are taken with the increased shear rate starting from the macrophase-separated solution at the thermal equilibrium ($\dot{\gamma} = 0\ \text{s}^{-1}$). Part **a**: $\dot{\gamma} = 0.025\ \text{s}^{-1}$ in Regime II, part **b**: 0.1 s^{-1} in Regime III, part (c): 4 s^{-1} in Regime IV. The flow direction is vertical. PS/PB (80:20)/DOP 3.3 wt. %, $T_c(0) = 58\,°\text{C}$ at $T = 44\,°\text{C}$. $\dot{\gamma}_c = 639\ \text{s}^{-1}$. Based on [4]

by the thermodynamic driving force of the given system. The thermodynamic force acting on the droplets is proportional to $\gamma_{int}R$ where γ_{int} and R are the interfacial tension and the radius of the droplets normal to the Ox axis, respectively. However, the overgrown droplets will be broken into the smaller ones due to the viscous drag proportional to $\dot{\gamma}\eta R^2$ where η is the viscosity of the droplets and matrix solution, which is assumed to be identical and to be the one given by the Newtonian flow for simplicity. The droplets will repeat the processes of the coalescence-growth and the break-up and restoration before reaching the average size and the size distribution at the steady state. The balance of the two opposing forces described above predicts the average size R,

$$R \sim \gamma_{\text{int}}/(\dot{\gamma}\eta). \tag{4.9.2}$$

Figure 4.44 presents the SALS intensity distributions $I(q_x)$ versus q_x (open triangles) and the SALS intensity distributions $I(q_z)$ versus q_z (open circles), along the flow direction and the neutral direction, respectively, in the double logarithmic scale at the steady state and at the two representative shear rates [73]. The data were obtained by the rheo-optical apparatus described in Figs. 3.1 to 3.4. It is striking to note that the SALS intensity distributions $I(q_z)$ at $\dot{\gamma} = 0.143\ \text{s}^{-1}$, presented in part (a) and (b), oscillates with q_z, showing at least 13 scattering maxima and minima over the q range covered in the rheo-optical apparatus. Note that part (b) is an enlargement of the part of the profile in part (a) which is encompassed by the square to highlight the profile at the large q_x and q_z range. On the other hand, the SALS intensity distributions $I(q_x)$ do not show such a remarkable oscillation as that in the SALS intensity distributions $I(q_z)$, although there are some oscillations. The oscillation in $I(q_z)$ sifts toward the higher q values at the higher shear rate as shown in part (c), obtained at $\dot{\gamma} = 0.466\ \text{s}^{-1}$. The oscillation in the $I(q_x)$ at this shear rate is more distinct than that at the lower shear rate; one can recognize the three maxima or shoulders along q_x. The trajectory of the scattering intensity maxima with q_x or q_z obeys the Porod's law [15], q_x^{-4} or q_z^{-4}, respectively.

The scattering maxima are found to correspond to those from the form factor of the spheroids. The solid lines indicate the calculated scattering function $I_{\text{theor}}(q_z R_z)$ given below by Eq. 4.9.3 for the single spheroid with the fixed radius of $R_z = 12.0\ \mu\text{m}$ (part a and part b) and 3.9 μm (part c) and with the axis of the revolution oriented

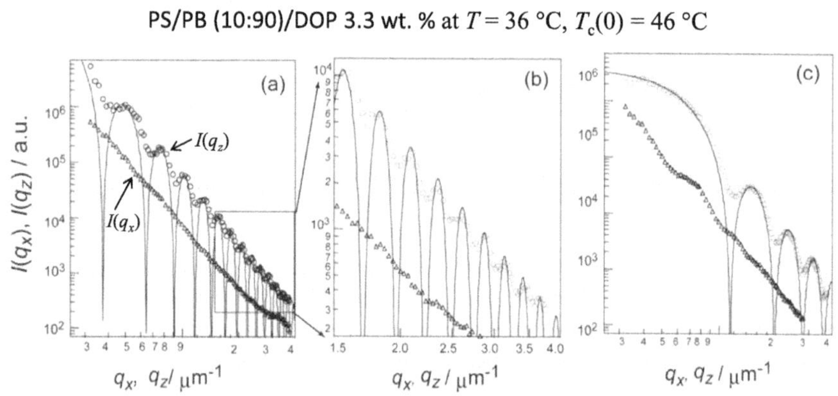

Fig. 4.44 Scattered intensity distribution $I(q_x)$ and $I(q_z)$ parallel (open triangles) and perpendicular (open circles) to the flow direction, respectively, measured at various shear rates for the PS/PB (10/90)/DOP 3.3 wt. % with $T_c(0) = 46\,°\text{C}$ at $T = 36\,°\text{C}$. Part (a) and (b): $\dot{\gamma} = 0.143\ \text{s}^{-1}$ and part (c): $0.446\ \text{s}^{-1}$. Part (b) is the expansion of the rectangular region of part (a) to clearly show the higher order maxima of $I(q_z)$ up to the 10th order at least. Each solid line represents a theoretical scattering function $I_{\text{theor}}(q_z)$ along q_z given by Eq. 4.9.10 for the single spheroid with its axis of revolution oriented parallel to the flow direction. The best fit between $I(q_z)$ and $I_{\text{theor}}(q_z)$ predicts that the half-length of the spheroid, R_z, is 12.0 μm for Part (a) and (b), and 3.9 μm for part (c), respectively. Based on Hashimoto, T. 2008 In *Soft Matter Characterization*; Borsali, R., Pecora, R., Eds., Springer. New York; Chapter 8, pp. 378–462 [9]

parallel to the flow direction (q_x-axis). $I_{\text{theor}}(q_z)$ under the given experimental setup shown in Fig. 4.45 is given below by Eq. 4.9.3 [13]

$$I_{\text{theor}}(\mathbf{q}) = K\ \Phi(U^*)^2, \tag{4.9.3}$$

$$K = \left(I_{\text{i}}/R_{\text{D}}^2\right)(p_{\text{s}} - p_{\text{m}})^2\ V_{\text{spheroid}}^2, \tag{4.9.4}$$

$$p_{\text{s}} = (2\ \pi/\lambda_{\text{s}})\ \alpha_{\text{s}},\ p_{\text{m}} = (2\ \pi/\lambda_{\text{m}})\ \alpha_{\text{m}},\ V_{\text{spheroid}} = (4\ \pi/3)v_{\text{a}}R^3, \tag{4.9.5}$$

$$\Phi(U) \equiv \frac{3}{U^3}\ (\sin U - U\ \cos U), \tag{4.9.6}$$

$$U^* \equiv qR\left[1 + \left(v_{\text{a}}^2 - 1\right)\cos^2\mu\right]^{1/2} \tag{4.9.7}$$

$$= \left[q_x^2\ R_x^2 + q_z^2\ R_z^2\right]^{1/2}, \tag{4.9.8}$$

$$q = (4\ \pi/\lambda_{\text{s}})\ \sin(\theta/2), \mathbf{q} \equiv (2\ \pi/\lambda_{\text{s}})\ (\mathbf{s}' - \mathbf{s}_0). \tag{4.9.9}$$

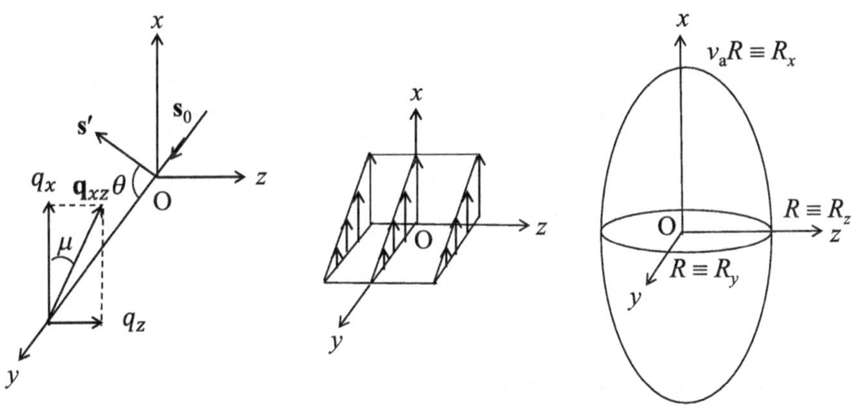

Fig. 4.45 (Part a) Cartesian coordinate Oxyz specifying the SALS experiments under the shear flow. The Ox, Oz, and Oy axis are set parallel to the flow and neutral direction, and the propagation direction of the incident beam, respectively. $\mathbf{s}_0$ and $\mathbf{s}'$ are unit vectors along the propagation direction of the incident and the scattered beam, respectively. The scattering vector $\mathbf{q} \equiv (2\pi/\lambda_{\text{s}})(\mathbf{s}' - \mathbf{s}_0)$ has its component $\mathbf{q}_{xz}$ in Oxz plane with the magnitude of the scattering vector q_x and q_z along Ox and Oz axis, respectively. (Part b) The velocity gradient of the shear flow exists parallel to Oxy plane. (Part c) The major axis and minor axes of the spheroid have the half-length $v_aR \equiv R_x$, $R \equiv R_y$, and $R \equiv R_z$ along Ox, Oy, and Oz axis, respectively. The ellipsoid rotates around Oz axis under the shear flow shown in part (b)

Here, the intensity I_i is the incident beam intensity; R_D is the sample-to-detector distance; p_s and p_m are the scattering length density of the spheroid and its surrounding matrix, respectively, defined by Eq. 4.9.5; λ_s and λ_m are the wavelength of the incident light beam within the spheroid and its matrix, respectively; α_s and α_m are the polarizability density in the corresponding medium, respectively; $V_{spheroid}$ is the volume of the spheroid; q is the magnitude of the scattering vector $\mathbf{q}$ given by the second line of Eq. 4.9.9 with s′ and s_0 being unit vectors along the propagation direction of the scattered beam and incident beam, respectively. $\mathbf{q}$ has the components q_x, q_y, and q_z along Ox, Oy, and Oz axis, respectively. Note that Eqs. 4.9.7 and 4.9.8 are valid under the small-angle approximation, $\cos^2(\theta/2) \cong 1$.

The theoretical scattered intensity along the neutral direction defined as $I_{theor}(q_z Rz)$ is given by

$$\begin{aligned} I_{theor}(q_z R_z) &\equiv I_{theor}(q_x = 0, q_y = 0, q_z R_z) \\ &= K\ \Phi(q_z R_z)^2 \end{aligned} \tag{4.9.10}$$

Similarly, the theoretical scattered intensity along the flow direction, $I_{theor}(q_x)$, is given by

$$I_{theor}(q_x R_x) \equiv I_{theor}\left(q_x R_x, q_y = 0, q_z = 0\right) \tag{4.9.11}$$

$$= K\ \Phi(q_x R_x)^2. \tag{4.9.12}$$

The function $\Phi(U)^2$ has the n-th order maximum at $U_{max,n}$ as given by

$$U_{max,n} = 5.763, 9.095, 12.32, 15.51, 18.69, 21.85, 25.01, \ldots \text{for } n = 1, 2, 3, 4, 5, 6, 7, \ldots \tag{4.9.13}$$

respectively. Thus, the size $R_{x,n}$ and the size $R_{z,n}$ can be evaluated by

$$R_{x,n} = U_{max,n}/q_{x,max,n},\ R_{z,n} = U_{max,n}/q_{z,max,n} \tag{4.9.14}$$

where $R_{x,n}$ and $R_{z,n}$ are the radius of the spheroid along Ox-axis and Oz-axis determined from the respective nth order maximum, respectively; $q_{x,max,n}$ and $q_{z,max,n}$ are the q_x and q_z at the nth order maximum of the experimental scattering profile $I(q_x)$ and $I(q_z)$, respectively. The average radius R_x or R_z may be estimated by averaging the radii $R_{x,n}$ or $R_{z,n}$ evaluated for different n's, respectively, based on Eqs. 4.9.13 and 4.9.14.

One can conduct the best-fit between the measured scattering profile $I(q_z)$ and the calculated scattering function $\langle I_{theor}(q_z)\rangle$ which incorporates the size distribution of R_z, with the distribution function given by

$$P(R_z) = (2\ \pi\sigma_z)^{1/2} \exp\left[(R_z - \langle R_z\rangle)^2/2\ \sigma_z^2\right]. \tag{4.9.15}$$

$\langle I_{\text{theor}}(q_z)\rangle$ is given by

$$\langle I_{\text{theor}}(q_z)\rangle = \int_0^\infty P(R_z)\, I_{\text{theor}}(q_z; R_z)\, dR_z / \int_0^\infty P(R_z) dR_z, \tag{4.9.16}$$

where $I_{\text{theor}}(q_z; R_z)$ is given by Eqs. 4.9.6 and 4.9.10. The best fit between experimental scattering profile $I(q_z)$ shown in Fig. 4.44 (open circles) and $\langle I_{\text{theor}}(q_z)\rangle$ given by Eq. 4.9.16 gives $\langle R_z\rangle = 11.95\mu$m and $\sigma_z/\langle R_z\rangle = 0.025$ at $\dot{\gamma} = 0.143$ s^{-1}, and $\langle R_z\rangle = 3.95$ μm and $\sigma_z/\langle R_z\rangle = 0.04$ at $\dot{\gamma} = 0.446$ s^{-1} [73], although the best-fitted calculated profiles $\langle I_{\text{theor}}(q_z)\rangle$ based on Eq. 4.9.16 are not shown in Fig. 4.44.

It is quite striking to note the following pieces of evidence.

(1) The size distribution of R_z becomes very narrow under shear flow.
(2) Compared with R_z, the size distribution of R_x is much broader.
(3) The narrow distribution of R_z and R_x are found to be a unique characteristic inherent to the off-critical mixtures: the droplets in the nearly critical mixtures hardly exhibit such multiple-order form factor maximas as presented in Figs. 4.3b and 4.8 in Sect. 4.3.

Evidence (2) may be the instability of the size R_x inherent in the flow-induced deformation of the droplets and their diffusion-coalescence and burst. It may be partially due also to the rotation of the ellipsoids around the neutral axis, which causes to enlarge the distribution of the effective size R_x along q_x even in the case when R_x has a fixed value (see the experimental condition to measure $I(q_x; \dot{\gamma})$ as presented in Fig. 4.45. Evidence (3) may be due to the smaller volume fraction of the ellipsoids in the off-critical mixtures than in the near-critical mixtures. In the near-critical mixtures, the volume fraction of the particles is as large as that of the matrix phase. Thus, a larger number of the particles may be involved in the process of their diffusion-coalescence and burst, which controls the size distribution of the particles in the dynamic equilibrium, compared with the case of the off-critical mixtures with a smaller volume fraction of the particles. This may cause the particle size distribution larger in the near-critical mixtures than in the off-critical mixture.

Figure 4.46 presents the time evolution of $\langle R_z\rangle$, the average radius of the ellipsoids along the neutral direction, and σ_z, the standard deviation of R_z from the average radius $\langle R_z\rangle$ which are measured as a function of time after the shear drop from $\dot{\gamma} = 89.1$ s^{-1} in Regime V to 0.0891 s^{-1} in Regime II for PS/PB (80:20)/DOP 3.3 wt. % with $T_c(0) = 58\,°$C at $T = 50\,°$C [9]. $\langle R_z\rangle$ and σ_z, which were evaluated by the best fit of the experimental $I(q_z)$ and theoretical $<I_{theor}(q_z)>$, are revealed to increase and decrease, respectively, toward their steady-state values at long time of ~ 8 h after the shear drop for this dilute droplet system. The slow energy dissipation process may involve the burst and diffusion-coalescence processes of the droplets repeatedly over and over to achieve the steady-state dissipative structure with the very narrow size distribution as clarified by $\sigma_z/\langle R_z\rangle \cong 0.0225$ at the steady state.

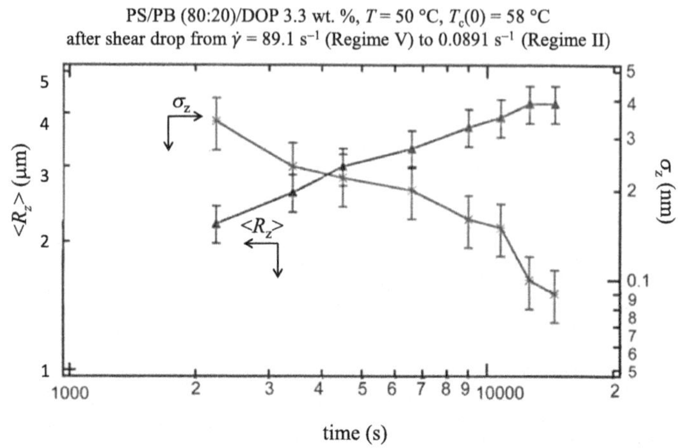

Fig. 4.46 Time evolution of $\langle R_z \rangle$ and σ_z after the shear drop of $\dot{\gamma}$ from 89.1 s^{-1} in Regime V to 0.0891 s^{-1} in Regime II for the PS/PB (80:20)/DOP 3.3 wt. % at $T = 50\,°\text{C}$, $T_c(0) = 58\,°\text{C}$. Based on Hashimoto, T. 2008 In *Soft Matter Characterization*; Borsali, R., Pecora, R., Eds., Springer. New York; Chapter 8, pp378-462 [9]

The steady-state values of R_z, R_x, σ_z, and σ_x are evaluated as a function of $\dot{\gamma}$ under the increased shear rate protocol for the same solution and at the same temperature as shown in Fig. 4.44. Figure 4.47 presents the average size of the spheroidal droplets $\langle R_z \rangle$ and $\langle R_x \rangle$ at the steady state as a function of $\dot{\gamma}$ [73]. The results reveal that both $\langle R_z \rangle$ and $\langle R_x \rangle$ fulfill the Taylor's law for the single droplet size under the shear flow at the any given shear rates corresponding to Regime II, given by Eq. 4.9.2,

$$\langle R_x(\dot{\gamma}) \rangle \sim \dot{\gamma}^{-1},\ \langle R_z(\dot{\gamma}) \rangle \sim \dot{\gamma}^{-1} (\text{for spheroids in Regime II}) \tag{4.9.17}$$

and the aspect ratio $\nu_a = R_x/R_z$ is found to be about 2, independent of $\dot{\gamma}$ [73].

It should be noted that both $I(q_z)$ and $I(q_x)$ as a function of q_z and q_x, respectively, have the asymptotic behavior of the Porod law,

$$I(q_x) \sim R_x^{-4}\, q_x^{-4}, I(q_z) \sim R_z^{-4}\, q_z^{-4}, \tag{4.9.18}$$

as shown in Fig. 4.44, indicating that the spheroids have a smooth interface with a negligible interfacial thickness $t_I : t_I/\langle R_z \rangle = t_I/R_x \cong 0$ over the q-range covered in this experiment. It should be also noted that the Porod law enables to evaluate the aspect ratio ν_a from the intensity ratio $q_x^4 I(q_x)/q_z^4 I(q_z)$,

$$\nu_a = \left\{ \frac{q_x^4\, I(q_x)}{q_z^4\, I(q_z)} \right\}^{-1/4}. \tag{4.9.19}$$

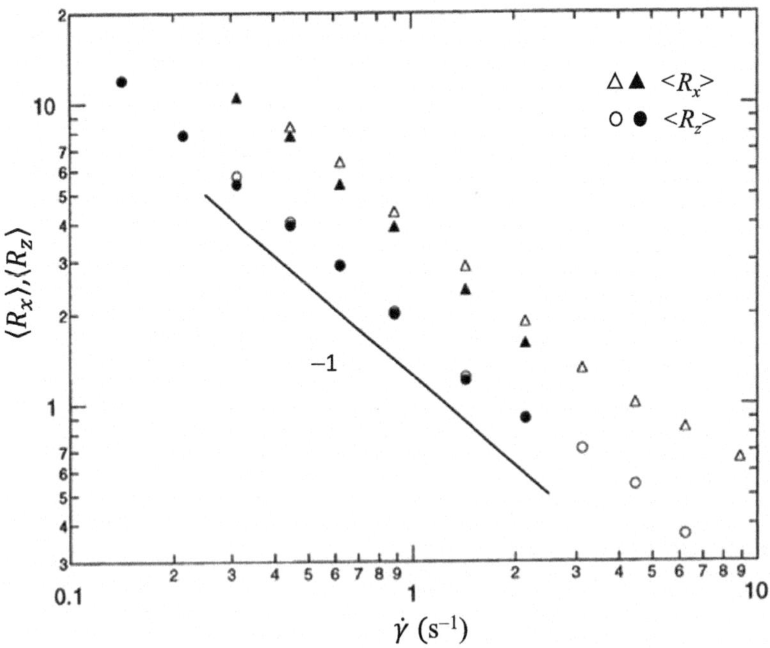

Fig. 4.47 $<R_x>$ and $<R_z>$ against $\dot{\gamma}$ evaluated at steady state under the increased shear rate in the Regime II for the off-critical mixture PS/PB (10:90)/DOP 3.3 wt. % at $T_c(0) = 46\,°C$ at $T = 36\,°C$. The filled triangles and the filled circles present the average radii of the droplets $<R_x>$ and $<R_z>$ parallel to FD and ND, respectively. The average radius $\langle R_x \rangle$ or $\langle R_z \rangle$ are estimated by averaging the radii $R_{x,n}$ or $R_{z,n}$ evaluated for the different order n's for the series of the higher-order form-factor maxima, respectively, based on the Eqs. 4.9.13 and 4.9.14. The unfilled triangles and circles present $<R_x>$ and $<R_z>$ estimated by conducting the Guinier plot, i.e., ln $I(q_i; \dot{\gamma})$ versus q_i^2 with $i = x$ orz. The straight line of slope -1 is shown to indicate $\langle R_i \rangle \sim \dot{\gamma}^{-1}$ ($i = x$ or z). Based on [73], reprinted by AIP Publishing with permission

Figure 4.48 presents the relative standard deviation $\sigma_z/\langle R_z \rangle$ and $\sigma_x/\langle R_x \rangle$ as a function of $\dot{\gamma}$ [73]. The result reveals that $\sigma_z/\langle R_z \rangle$ and $\sigma_x/\langle R_x \rangle$ concurrently increases and decreases to reach the respective constant value, 0.04 and 0.12, respectively.

4.9.4 Flow-Induced Phase Transition in Off-Critical Mixtures

In this section, we describe the flow-induced phase transition as a function of the composition X of the PS component in the polymer mixtures PS/PB in DOP for the PS/PB (X:1 − X)/DOP 3.3 wt. % where X/(1 − X) is varied from 20/80, 30/70, 40/60, 50/50, 60/40, 70/30 to 80/20 wt. %/wt. %. The phase transition is studied with

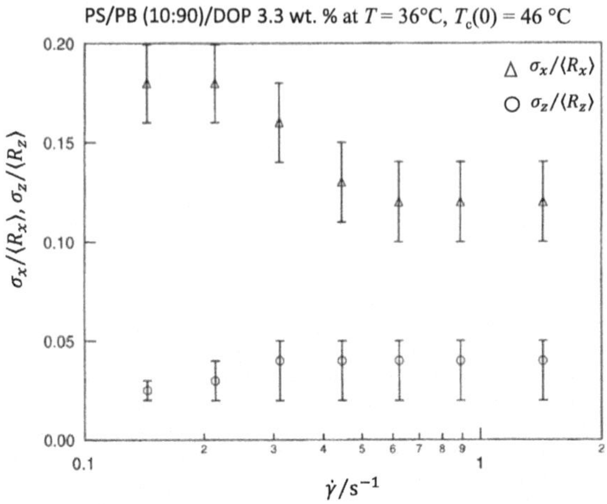

Fig. 4.48 Relative standard deviation of the droplet radius $\sigma_x/\langle R_x\rangle$ and $\sigma_z/\langle R_z\rangle$ plotted against $\dot{\gamma}$ at the steady-state shear with the increased shear rate protocol. The values $\sigma_x/\langle R_x\rangle$ and $\sigma_z/\langle R_z\rangle$ are estimated by taking the average values of $\sigma_x/<R_x>$ and $\sigma_z/<R_z>$ evaluated by the best-fit of the experimental scattering profiles $I(q_z)$ and $I(q_x)$, such as shown in Fig. 4.44 together with those at other shear rates, and $\langle I_{\text{theor}}(q_z)\rangle$ given by Eqs. (4.9.15) and (4.9.16) and $\langle I_{\text{theor}}(q_x)\rangle$ defined similarly to $\langle I_{\text{theor}}(q_z)\rangle$, respectively. PS/PB (10:90)/DOP 3.3 wt. % with $T_c(0) = 46\,°C$ at $T = 36\,°C$. Based on [73]], reprinted by AIP Publishing with permission

the increased $\dot{\gamma}$ for the two-phase to the single-phase transition and the decreased $\dot{\gamma}$ for the single-phase to the two-phase transition.

A. Hysteresis in phase transition under flow. Figure 4.49 presents the integrated SALS intensity along the neutral direction $\mathcal{J}_z(\dot{\gamma})$ as a function of $\dot{\gamma}$ with the increased shear rate and the decreased shear rate for the PS/PB (50:50)/DOP 3.3 wt.% at $T = 68\,°C$ and $\Delta T(0) = 8\,°C$ for the representative of the near-critical composition (part a) and for the PS/PB (80:20)/DOP 3.3 wt. % at $T = 51°C$ and $\Delta T_c(0) = 7\,°C$ for the representative of the off-critical mixture (part b) (see Table 4.1) [74]. With the increased $\dot{\gamma}$, $\mathcal{J}_z(\dot{\gamma})$ sharply decreases to reach the critical shear rate $\dot{\gamma}_{c,i}$ at which the shear-induced two-phase to single-phase transition occurs and above which $\mathcal{J}_z(\dot{\gamma})$ only slightly decreases with $\dot{\gamma}$. On the other hand, with the decreased $\dot{\gamma}$ in the shear-induced single-phase state, $\mathcal{J}_z(\dot{\gamma})$ slightly increases to reach $\dot{\gamma}_{c,d}$ at which the shear-induced single-phase to two-phase transition occurs and below which $\mathcal{J}_z(\dot{\gamma})$ sharply increases with the decreased $\dot{\gamma}$. These trends are common for both near-critical and off-critical solution. However, the striking differences can be found in these two solutions as follows.

(1) In the near-critical solution,

$$\delta\dot{\gamma}_c \equiv \dot{\gamma}_{c,i} - \dot{\gamma}_{c,d} \cong 0, \tag{4.9.20a}$$

Fig. 4.49 Hysteresis effect on the shear-rate dependence of $\mathcal{J}_z(\dot{\gamma}; T)$ for PS/PB (50:50)/DOP 3.3 wt. % with $T_c(0) = 76°C$ at $T = 68°C$ (part a) and for PS/PB (80:20)/DOP 3.3 wt. % with $T_c(0) = 58°C$ at $T = 51°C$ (part b). Based on [74]], reprinted by AIP Publishing with permission

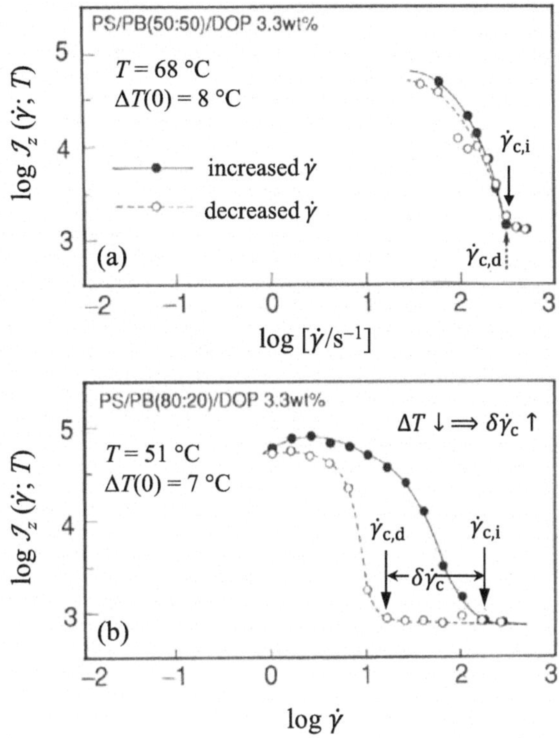

thereby $\mathcal{J}_z(\dot{\gamma})$ reversibly changes upon the increased $\dot{\gamma}$ and the decreased $\dot{\gamma}$.

(2) In the off-critical solution,

$$\delta\dot{\gamma}_c > 0, \tag{4.9.20b}$$

and the variation of $\mathcal{J}_z(\dot{\gamma})$ with $\dot{\gamma}$ has a large hysteresis.

$\dot{\gamma}_{c,i}$ and $\dot{\gamma}_{c,d}$ can be measured as a function of T or $\Delta T(0)$ and composition of the PS/PB. The results on the critical shear rates at the given temperatures and compositions, $\dot{\gamma}_{c,i}(T)$ and $\dot{\gamma}_{c,d}(T)$, thus obtained can be converted to the critical temperatures at the given shear rates and compositions, $T_{c,i}(\dot{\gamma})$ and $T_{c,d}(\dot{\gamma})$, or the quench depths, $\Delta T_{c,i}(\dot{\gamma})$ and $\Delta T_{c,d}(\dot{\gamma})$, respectively, at the given shear rates and compositions, according to the concept as schematically elucidated in Fig. 4.24a, b in Sect. 4.4. Here, the quench depths, $\Delta T_{c,i}(\dot{\gamma})$ and $\Delta T_{c,d}(\dot{\gamma})$, are defined by

$$\Delta T_{c,i}(\dot{\gamma}) \equiv T_c(0) - T_{c,i}(\dot{\gamma}), \tag{4.9.21a}$$

$$\Delta T_{c,d}(\dot{\gamma}) \equiv T_c(0) - T_{c,d}(\dot{\gamma}). \tag{4.9.21b}$$

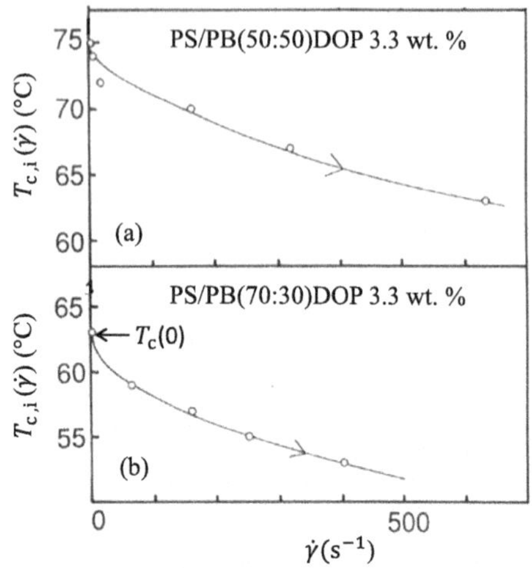

Fig. 4.50 The composition dependence of $T_{c,i}(\dot{\gamma})$ for the PS/PB/DOP 3.3 wt. % with the composition of PS/PB = 50/50 (**a**), and 70/30 (**b**)

B. The near-critical mixtures. Figure 4.50 presents some representative data on $T_{c,i}(\dot{\gamma})$ versus $\dot{\gamma}$ for the PS/PB/ DOP 3.3 wt. % solutions with respect to the two compositions, PS/PB(50/50) and (70/30) [45]. The critical temperature at a given composition is elucidated to decrease with increased $\dot{\gamma}$. The data also clarify that the drop of critical temperature $\Delta T_{c,i}(\dot{\gamma}) = T_c(0) - T_{c,i}(\dot{\gamma})$ increases with $\dot{\gamma}$. Essentially the same trend is observed on $T_{c,d}(\dot{\gamma})$ versus $\dot{\gamma}$ for each of the same solutions, although the data are not shown here.

Figure 4.51 presents the drop of the critical temperature, $\Delta T_c(\dot{\gamma})$, reduced with the critical temperature at $\dot{\gamma} = 0$, $T_c(0)$, as a function of $\dot{\gamma}$ for the near-critical mixtures with the different compositions as given in the figure legend [45]. The experiments for all the given compositions reveal that the values $\Delta T_{c,i}(\dot{\gamma})$ obtained in the increased $\dot{\gamma}$ are identical to the values $\Delta T_{c,d}(\dot{\gamma})$ obtained in the decreased $\dot{\gamma}$ from the shear-induced single-phase solutions, although the corresponding data on $\Delta T_{c,d}(\dot{\gamma})/T_c(0)$ versus $\dot{\gamma}$ are not presented here. Thereby, the following low as given below is experimentally validated,

$$\Delta T_{c,i}(\dot{\gamma}) \cong \Delta T_{c,d}(\dot{\gamma}) \equiv \Delta T_c(\dot{\gamma}) \text{ for all the compositions from the 30/70 to the 70/30,} \quad (4.9.22)$$

so that the critical temperature drops in Fig. 4.51 are simply defined by the $\Delta T_c(\dot{\gamma})$ without using the subscript "i" or "d". It is revealed that the reduced critical temperature is universal with the composition and the shear protocol, as demonstrated by the straight line in Fig. 4.51 which predicts the experimental law as given by

$$\Delta T_c(\dot{\gamma})/T_c(0) = (1.3 \pm 0.1) \times 10^{-3} \dot{\gamma}^{0.50 \pm 0.02}, \quad (4.9.23)$$

for the 3.3 wt.% solutions with the compositions between 30/70 and 70/30 measured under the increased $\dot{\gamma}$ and the decreased $\dot{\gamma}$.

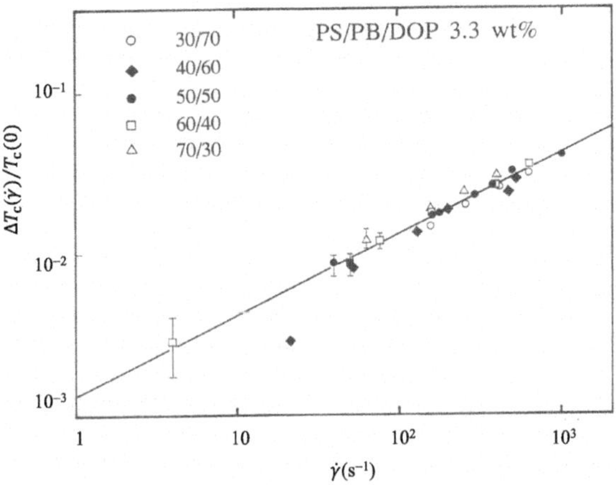

Fig. 4.51 The composition dependence of the reduced critical temperature drop $\Delta T_c(\dot{\gamma})/T_c(0)$ under the increased shear-rate protocol for the PS/PB/DOP 3.3 wt. % with the composition of the PS/PB = 30/70, 40/60, 50/50, 60/40, and 70/30. The straight line is given by Eq. (4.9.23). Essentially identical result is obtained under the decreased shear-rate protocol. Based on [45], reprinted by AIP Publishing with permission

The result for the 3.3 wt.% solutions given by Eq. 4.9.23 and Fig. 4.51 is consistent with Eq. 4.5.3 for PS/PB(50: 50)/DOP 3.3 wt. % and also with Eq. 4.5.2 obtained for the PS/PB (50:50) 3.0 wt.% with respect to the power law on the shear-rate dependence of the critical temperature drop. This power law is further confirmed by the mean-field theory (Eq. 4.5.2) in Sect. 4.5.

$$\Delta T_c(\dot{\gamma})/T_c(0) \sim \dot{\gamma}^{1/2}$$

The small difference of the prefactors for the 3.0 and 3.3 wt. % solutions, as presented in Eqs. 4.5.2 and 4.5.3, respectively, can be accounted for by the small difference in the characteristic time τ_p for the concentration fluctuations of these two solutions having slightly different concentrations. That is, with the decreased concentration c, $\tau_p^{1/2}$ increases and thereby the prefactor increases with the decreased concentrtion. This is because

$$\tau_p = \xi_c^2/D_c, \xi_c \sim c^{-1}, D_c \sim c^{-3/2} \tag{4.9.24}$$

for the semidilute Θ solutions, where ξ_c is the correlation length for the concentration fluctuations, D_c is the self-diffusivity of the PS and PB employed in this work, which are validated to be essentially identical, and c is the concentration of polymers.

C. The off-critical mixtures. The SALS experiments regarding the drops of the reduced critical temperatures for the off-critical mixtures PS/PB (20:80)/DOP 3.3 wt. % and (80:20)/DOP 3.3 wt. %, as given by Eqs. 4.9.25, 4.9.26, 4.9.27 and 4.9.28 below, respectively, reveal the results different from the result for the near-critical mixtures given by Eq. 4.9.23 as follows.

For the 20/80 mixture,

$$\Delta T_{c,d}(\dot{\gamma})/T_c(0) = (2.0 \pm 0.3) \times 10^{-3}\, \dot{\gamma}^{0.5\pm0.1}, \tag{4.9.25}$$

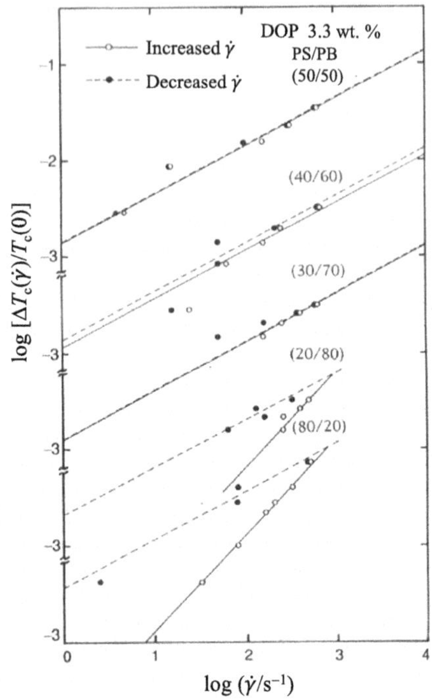

Fig. 4.52 The hysteresis effect on the reduced critical temperature drop $\Delta T_c(\dot{\gamma})/T_c(0)$ versus $\dot{\gamma}$ with the increased and decreased $\dot{\gamma}$ for the PS/PB/DOP 3.3 wt.% with the composition of the PS/PB = 50/50, 40/60, 30/70, 20/80, and 80/20. Based on [45], reprinted by AIP Publishing with permission

$$\Delta T_{c,i}(\dot{\gamma})/T_c(0) = (6.6 \pm 3.0) \times 10^{-5}\ \dot{\gamma}^{1.0 \pm 0.1}. \tag{4.9.26}$$

For the 80/20 mixture,

$$\Delta T_{c,d}(\dot{\gamma})/T_c(0) = (3.7 \pm 0.5) \times 10^{-3}\ \dot{\gamma}^{0.5 \pm 0.1}, \tag{4.9.27}$$

$$\Delta T_{c,i}(\dot{\gamma})/T_c(0) = (1.3 \pm 0.1) \times 10^{-4}\ \dot{\gamma}^{1.0 \pm 0.1}. \tag{4.9.28}$$

Figure 4.52 presents $\Delta T_c(\dot{\gamma})/T_c(0)$ versus $\dot{\gamma}$ in the double logarithmic scale for the representative near-critical mixtures with the composition (50/50), (40/60), and (30/70) and for the off-critical mixtures with the composition (20/80) and (80/20) in the increased $\dot{\gamma}$ (the solid lines with the unfilled circles) and the decreased $\dot{\gamma}$ (the broken lines with the filled circles) [45]. All the data on $\Delta T_c(\dot{\gamma})/T_c(0)$ versus $\dot{\gamma}$ are presented by the power law of $\dot{\gamma}^n$ with the exponent n being either 1/2 or 1. Upon the decreased $\dot{\gamma}$, the power law for the off-critical mixtures is identical to that for the near-critical mixtures, both mixture having $n = 1/2$, whereas upon the increased $\dot{\gamma}$, the power law for the off-critical mixtures is different from that for the near-critical mixtures:the exponent having $n = 1$ and $1/2$ for the off-critical mixtures and for the near-critical mixtures, respectively. Both off-critical mixtures, PS/PB (20/80) and (80/20), have the common characteristic that they have different power laws $\dot{\gamma}^1$ and

$\dot{\gamma}^{1/2}$ on the increased $\dot{\gamma}$ and the decreased $\dot{\gamma}$, respectively, which reveals existence of the hysteresis on the flow-induced phase transition.

It should be noted that the power law for the increased shear rate is valid in the relatively high shear-rate range $\dot{\gamma} \gtrsim 30\ \mathrm{s}^{-1}$, whereas it is less accurate at $\dot{\gamma} < 30\ \mathrm{s}^{-1}$. The reason for this will be described below as the note on Eq. 4.9.38 .

D. The summary regarding the hysteresis in the phase transition of the off-critical mixtures. Figure 4.53 schematically presents the summary of the hysteresis effect on the phase transition for the off-critical mixtures. The hysteresis found in the $\mathcal{J}_z(\dot{\gamma}; T)$ in the increased and decreased $\dot{\gamma}$ at the given T or $\Delta T(0)$ (part a) gives rise to the hysteresis in $\dot{\gamma}_c(T)$ with the decreased and increased T or $\Delta T(0)$ (part b). That is, upon the increased $\Delta T(0)$ as shown by the arrow in part (b), $\dot{\gamma}_{c,i}$ increases according to Eq. 4.9.29a: upon the decreased $\Delta T(0)$ as shown by the arrow in part (b), $\dot{\gamma}_{c,d}(T)$ decreases according to Eq. 4.9.29b,

$$\dot{\gamma}_{c,i}(T) \sim \Delta T(0), \tag{4.9.29a}$$

$$\dot{\gamma}_{c,d}(T) \sim \Delta T(0)^2. \tag{4.9.29b}$$

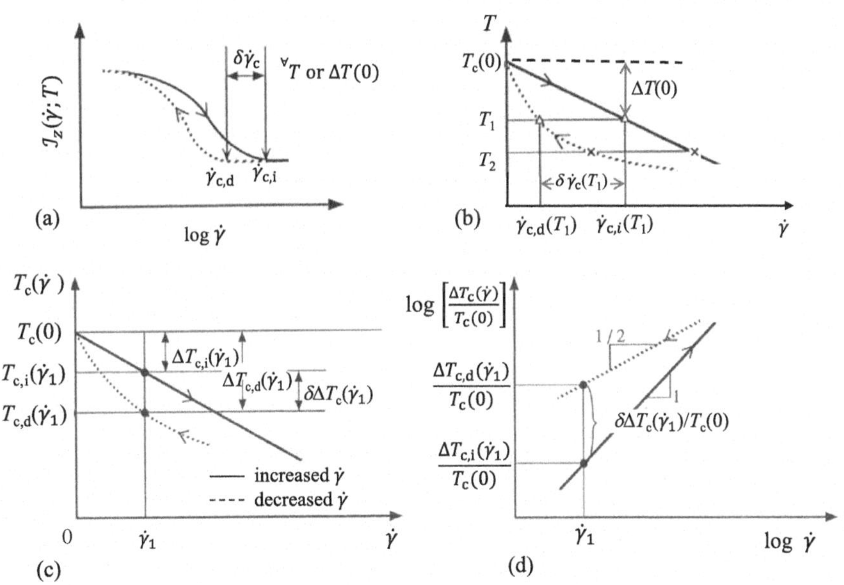

Fig. 4.53 The summary of the hysteresis effect with the increased and decreased $\dot{\gamma}$ on $\mathcal{J}_z(\dot{\gamma}; T)$ versus $\dot{\gamma}$ at a given T or $\Delta\ T(0)$ (part a), $\dot{\gamma}_c(T)$ versus T (part b), $T_c(\dot{\gamma})$ versus $\dot{\gamma}$ (part c), and $\Delta T_c(\dot{\gamma})/T_c(0)$ versus $\dot{\gamma}$ (part d), where the respective quantity with the increased and decreased shear-rate protocol are indicated by the solid line with the arrow directed toward right and the dotted line with the arrow directed toward left, respectively

This difference in the temperature-dependencies of the critical shear rates with the increased and decreased shear rates as specified by Eqs. 4.9.29a and 4.9.29b, respectively, brings about the relationship between the two critical shear rates as given below by Eq. 4.9.30.

$$\dot{\gamma}_{\mathrm{c,i}}(T) > \dot{\gamma}_{\mathrm{c,d}}(T), \tag{4.9.30}$$

which accounts for the observed hysteresis as described by $\delta\dot{\gamma}_{\mathrm{c}}(T)$ at the given T as shown in part (a) and (b). The relation between $\dot{\gamma}_{\mathrm{c}}(T)$ and T or $\Delta T(0)$ as given by Eq. 4.9.29 and shown in Fig. 4.53b can be converted into the relation between $T_{\mathrm{c}}(\dot{\gamma})$ or $\Delta T_{\mathrm{c}}(\dot{\gamma})$ versus $\dot{\gamma}$ as shown in part (c); the relation gives the following equations,

$$\Delta T_{\mathrm{c,i}}(\dot{\gamma}) \sim \dot{\gamma}^{1}, \text{ and } \Delta T_{\mathrm{c,d}}(\dot{\gamma}) \sim \dot{\gamma}^{1/2}. \tag{4.9.31}$$

This difference as specified by the first line of the Eq. 4.9.31 and the second line of Eq. 4.9.31 gives,

$$\Delta T_{\mathrm{c,d}}(\dot{\gamma}) > \Delta T_{\mathrm{c,i}}(\dot{\gamma}), \tag{4.9.32}$$

which accounts for the observed hysteresis $\delta\Delta T_{\mathrm{c}}(\dot{\gamma})$ presented by Fig. 4.53c or the observed hysteresis $\delta\Delta T_{\mathrm{c}}(\dot{\gamma})/T_{\mathrm{c}}(0)$ at the given $\dot{\gamma}$ presented by Fig.4.53d.

Here, it should be noted again that the relation between $\gamma_{\mathrm{c,i}}(T)$ and T shown in part b and Eq. 4.9.29a at the high temperatures T close to $T_{\mathrm{c}}(0)$ is experimentally less accurate. The same note as given above is also applied to the relation between $\Delta T_{\mathrm{c,i}}(\dot{\gamma})$ and $\dot{\gamma}$ as shown in part c and d as well as in the first line of Eq. 4.9.31 at the low shear-rate range.

E. Shear-rate dependent phase diagram. Figures 4.54 and 4.55 respectively present the steady-state phase diagrams of the PS/PB DOP 3.3 wt.% solution under the shear flow at $\dot{\gamma} = 160$ and $1000\ \mathrm{s}^{-1}$ upon the increased and decreased shear rate in reference to the equilibrium phase diagram for the quiescent solution at $\dot{\gamma} = 0$ [45]. The phase diagram under the shear is elucidated based on the results shown in Figs. 4.51 and 4.52. Under the high shear rate of $\dot{\gamma} = 1000\ \mathrm{s}^{-1}$, the phase diagram for the 2-phase to 1-phase transition is almost identical to that for the 1-phase to 2-phase transition, i.e., $\Delta T_{\mathrm{c,i}}(\dot{\gamma}) \cong T_{\mathrm{c,d}}(\dot{\gamma})$ at all the compositions covered, which is well expected from the results shown in Figs. 4.51 and 4.52. At $\dot{\gamma} = 160\ \mathrm{s}^{-1}$, the phase diagram for the 2-phase to 1-phase transition is almost identical to that for the 1-phase to 2-phase transition for the near-critical mixtures with ϕ_{PS} between $\phi_{\alpha}\left(\dot{\gamma} = 160\ \mathrm{s}^{-1}\right)$ and $\phi_{\beta}\left(\dot{\gamma} = 160\ \mathrm{s}^{-1}\right)$. However, they are different for the off-critical mixtures, i.e., $T_{\mathrm{c,i}}(\dot{\gamma}) > T_{\mathrm{c,d}}(\dot{\gamma})$.

F. The interpretation of the scaling laws for off-critical mixtures: $\Delta T_c(\dot{\gamma}) \sim \dot{\gamma}^n$ and $\dot{\gamma}_c(T) \sim \Delta T_c(0)^{1/n}$. Let us discuss the physical picture for the scaling law or the power the law for the off-critical mixtures with $n = 1$ given by Eqs. 4.9.26 and 4.9.28 from the two-phase to single-phase transition, defined as "2 → 1 phase transition", with the increased shear rate and that with $n = 1/2$ given by Eqs. 4.9.25 and 4.9.27 from the single-phase to the two-phase transition, defined as "1 → 2 phase

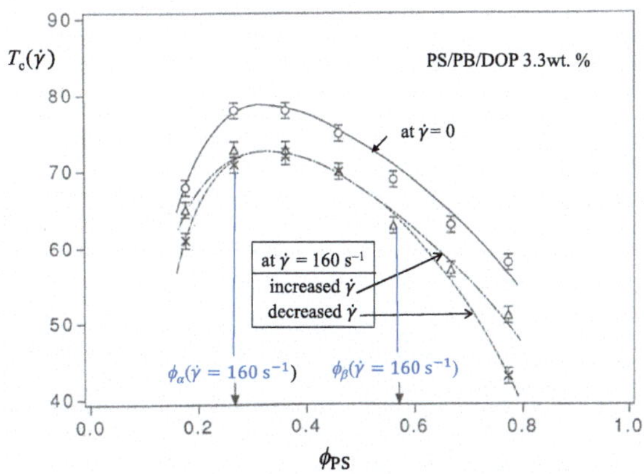

Fig. 4.54 Equilibrium phase diagram at $\dot{\gamma} = 0$ and the steady-state phase diagram at $\dot{\gamma} = 160\ s^{-1}$ under the increased and decreased shear-rate protocol for the PS/PB/DOP 3.3 wt. %. ϕ_{PS} is the volume fraction of PS in the mixture of PS and PB

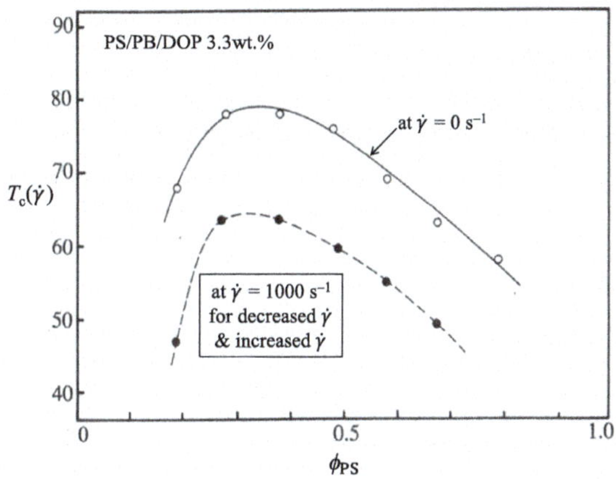

Fig. 4.55 Equilibrium phase diagram at $\dot{\gamma} = 0$ (solid line) and the steady-state phase diagram at $\dot{\gamma} = 1000\ s^{-1}$ (broken line) for the PS/PB/DOP 3.3 wt. %. The hysteresis effect on the phase diagram with the increased and decreased $\dot{\gamma}$ is not observed at $\dot{\gamma} = 1000\ s^{-1}$. Based on [45], reprinted by AIP Publishing with permission

transition", with the decreased shear rate, respectively. To facilitate the discussion for the off-critical mixtures, we first discuss the scaling law for the near-critical mixtures given by Eq. 4.9.23.

The near-critical mixtures. The binodal temperature at $\dot{\gamma} = 0$, defined as $T_c(0)$, is close to the spinodal temperature at $\dot{\gamma} = 0$, defined as $T_s(0)$, for the near-critical mixtures. Thus, both the critical shear rate for the $1 \rightarrow 2$ transition $\dot{\gamma}_{c,d}(T)$ and the $2 \rightarrow 1$ transition $\dot{\gamma}_{c,i}(T)$ at the given $T < T_c(0)$ are determined by the criterion that the shear rate in the increased and decreased shear protocol, respectively, reach the growth rate $R(q_m(0;T)) \equiv R_m(T)$ as given below by Eq. 4.9.33. Here, $R_m(T)$ is the dominant Fourier modes of the concentration fluctuations with the wave number or the magnitude of the scattering vector $q = q_m(0;T)$ at the given T in the early stage spinodal decomposition (SD),

$$\dot{\gamma}_{c,d}(T) = \dot{\gamma}_{c,i}(T) = R(q = q_m(0;T)) \equiv R_m(T), \tag{4.9.33}$$

Here, $R(q = q_\mathrm{m}(0; T))$ is given by $R(q)$ at $q = q_\mathrm{m}(0; T)$ in Eq. 4.6.2 together with Eq. 4.6.7 in Sect. 4.6,

$$R(q = q_\mathrm{m}(0; T)) = (1/2)\, D_\mathrm{app}(T)\, q_\mathrm{m}(0; T)^2. \tag{4.9.34}$$

In the case where $\dot{\gamma} < \dot{\gamma}_\mathrm{c,d}(T) = \dot{\gamma}_\mathrm{c,i}(T) = R_\mathrm{m}(T)$, the concentration fluctuations grow over the time span corresponding to $1/\dot{\gamma}$ into the dissipative structure which is the phase-separated domain structure attained in the steady-state at the given $\dot{\gamma}$ and $T < T_\mathrm{c}(0)$. On the other hand, when $\dot{\gamma} > \dot{\gamma}_\mathrm{c,d}(T) = \dot{\gamma}_\mathrm{c,i}(T) = R_\mathrm{m}(T)$, the concentration fluctuations, which can grow according to the SD at $\dot{\gamma} = 0$, do not have the time to grow under the shear at the given $\dot{\gamma}$ and thereby remain to those in the shear-induced single-phase state, according to the **equivalence principle of $\dot{\gamma}^{-1}$ and t or $\dot{\gamma}$ and t^{-1} under the shear flow.**

In the context of the mean-field approximation under the flow,

$$D_\mathrm{app}(T) \sim \Delta T(\dot{\gamma}),\ q_\mathrm{m}(0; T)^2 \sim \Delta T(\dot{\gamma}) \tag{4.9.35}$$

with $\Delta T(\dot{\gamma}) \equiv T_\mathrm{c}(\dot{\gamma}) - T$ as defined in Fig. 4.23b in Sect. 4.4 and by Eq. 4.3.34 in Sect. 4.3.7. Eq. 4.9.33 to 4.9.35 lead to

$$\dot{\gamma}_\mathrm{c,d}(T) = \dot{\gamma}_\mathrm{c,i}(T) \sim \Delta T(\dot{\gamma})^2,\ \Delta T_\mathrm{c,d}(\dot{\gamma}) = \Delta T_\mathrm{c,i}(\dot{\gamma}) \sim \dot{\gamma}^{1/2}, \tag{4.9.36}$$

which account for the experimental results given by Eq. 4.9.23 and shown by Fig. 4.51,

The off-critical mixtures. We postulate that $\dot{\gamma}_\mathrm{c,d}(T)$ for the off-critical mixtures obey the same law as given by the Eq. 4.9.33, i.e.,

$$\dot{\gamma}_\mathrm{c,d}(T) = R(q = q_\mathrm{m}(0; T)), \tag{4.9.37}$$

and thereby, substituting Eqs. 4.9.34 and 4.9.35 into Eq. 4.9.37, $\dot{\gamma}_\mathrm{c,d}(T)$ and $\Delta T_\mathrm{c,d}(\dot{\gamma})$ are given by

$$\dot{\gamma}_\mathrm{c,d}(T) \sim \Delta T(\dot{\gamma})^2, \tag{4.9.38}$$

$$\Delta T_\mathrm{c,d}(\dot{\gamma}) \sim \dot{\gamma}^{1/2}. \tag{4.9.39}$$

The postulate given by Eq. 4.9.37 may be validated, because the nucleation rate in polymers is very much suppressed; the critical free energy for the nucleation is $\propto N^{1/2}$ [75], large for polymers with the large N (the degree of the polymerization).

Upon increasing $\dot{\gamma}$ toward $\dot{\gamma}_\mathrm{c,i}$ at a given T, we should account for the string-like structure with its characteristic size normal to the flow direction $\xi_{\mathrm{d},z}$ varying with the $\dot{\gamma}$ in Regime IV before reaching the shear-induced single phase in Regime V. We postulate that the $\dot{\gamma}$-dependence of $\xi_{\mathrm{d},z}$ for the off-critical mixtures obeys the same law as that for the string-like structures for the near-critical critical mixtures in

Regime IV, i.e.,

$$\xi_{\mathrm{d},z} \sim \dot{\gamma}^{-1/4}, \tag{4.9.40}$$

as shown by the red straight line in Fig. 4.15 in Sect. 4.3.7 and in Fig. 4.39 in Sect. 4.8.1. Based on the principle of the equivalence between $\dot{\gamma}^{-1}$ and t, $\xi_{\mathrm{d},z}$ grows with t according to

$$\xi_{\mathrm{d},z}(t) \sim t^{1/4}. \tag{4.9.41}$$

Equation 4.9.41 is consistent with the growth law on $\Lambda_{\mathrm{m}}(t) \sim t^{1/4}$ shown in Fig. 4.39. This is also consistent with the growth law of the domains with the characteristic size $R_{\mathrm{d}}(t)$ in the absence of the shear flow predicted by Ohta [76] based on the interface diffusion of the two components in the mixtures. In the nonhydrodynamic regime, the prediction by Ohta is given by

$$R_{\mathrm{d}}(t) = \left[\Lambda(0)\ \gamma_{\mathrm{int}}\ \xi_{int}\right]^{1/4} t^{1/4}, \tag{4.9.42}$$

where $\Lambda(0)$ is the Onsager transport coefficient in the limit of $q \to 0$ given by,

$$\Lambda(0) = N\ \phi\ (1-\phi)\ D_{\mathrm{c}} \sim \Delta T(\dot{\gamma})^{0} \tag{4.9.43}$$

with ϕ being the composition of the one component in the mixtures. γ_{int} is the interfacial tension given by

$$\gamma_{\mathrm{int}} \sim \left[\Delta T(\dot{\gamma})\right]^{3/2}, \tag{4.9.44}$$

in the weak segregation. ξ_{int} is the interface width at T close to $T_{\mathrm{c}}(\dot{\gamma})$ given by

$$\xi_{\mathrm{int}} \sim \xi_{\mathrm{T}} \sim \left[\Delta T(\dot{\gamma})\right]^{-1/2}. \tag{4.9.45}$$

The growth rate of the domain $\Gamma_{\mathrm{d}}(t;T)$ is given by

$$\Gamma_{\mathrm{d}}(t;T) = \frac{d}{dt}R_{\mathrm{d}}(t;T) \sim \left[\Lambda(0)\ \gamma_{\mathrm{int}}\ \xi_{int}\right]^{1/4} t^{-3/4}, \tag{4.9.46}$$

and thereby

$$\Gamma_{\mathrm{d}}(\dot{\gamma};T) \sim \left[\Lambda(0)\ \gamma_{\mathrm{int}}\ \xi_{int}\right]^{1/4} \dot{\gamma}^{3/4}. \tag{4.9.47}$$

Equation (4.9.47) is obtained based on the equivalence principle between t and $\dot{\gamma}^{-1}$. The critical shear rate under the increased shear-protocol $\dot{\gamma}_{\mathrm{c,i}}(T)$ is given by

$$\dot{\gamma}_{\mathrm{c,i}}(T) = \Gamma_{\mathrm{d}}(\dot{\gamma};T) \text{ at } \dot{\gamma} = \dot{\gamma}_{\mathrm{c,i}}(T). \tag{4.9.48}$$

Equations 4.9.43–4.9.48 lead to

$$\dot{\gamma}_{\mathrm{c,i}} \sim \Lambda(0)\ \gamma_{\mathrm{int}}\ \xi_{\mathrm{int}} \sim \left[\Delta T(\dot{\gamma})\right]^{1}. \tag{4.9.49}$$

Thus, one can predict the experimental law,

$$\Delta T_{\mathrm{c,i}}(\dot{\gamma}) \sim \dot{\gamma} \text{ for the off-critical mixtures.} \tag{4.9.50}$$

Note on Eq. 4.9.39. $\Delta T_{\mathrm{c,d}}(\dot{\gamma})$ in Eq. 4.9.39 is defined as $\Delta T_{\mathrm{c,d}}(\dot{\gamma}) \equiv T_{\mathrm{c}}(0) - T_{\mathrm{c,d}}(\dot{\gamma})$ as shown in Fig. 4.53c where $T_{\mathrm{c}}(0)$ is the binodal temperature at the given ϕ and at $\dot{\gamma} = 0$, and $T_{\mathrm{c,d}}(\dot{\gamma})$ is the critical temperature for the $1 \rightarrow 2$ phase transition at the given ϕ and $\dot{\gamma}$. The prediction given by Eq. 4.9.39 should be rigorously valid for $\Delta T_{\mathrm{s,d}}(\dot{\gamma})$ instead of $\Delta T_{\mathrm{c,d}}(\dot{\gamma})$, i.e.,

$$\Delta T_{\mathrm{s,d}}(\dot{\gamma}) = C_{\mathrm{d}}\dot{\gamma}^{1/2}, \tag{4.9.51}$$

where $\Delta T_{\mathrm{s,d}}(\dot{\gamma})$ is defined by

$$\Delta T_{\mathrm{s,d}}(\dot{\gamma}) \equiv T_{\mathrm{s}}(0) - T_{\mathrm{c,d}}(\dot{\gamma}). \tag{4.9.52}$$

as indicated in Fig. 4.56. $T_{\mathrm{s}}(0)$ is the spinodal temperature at the given ϕ and at $\dot{\gamma} = 0$; C_{d} is the proportionality constant, independent of $\dot{\gamma}$, Thus, the experimentally measured $\Delta T_{\mathrm{c,d}}(\dot{\gamma})$ is related to $\Delta T_{\mathrm{s,d}}(\dot{\gamma})$ as follows:

$$\Delta T_{\mathrm{c,d}}(\dot{\gamma}) = \Delta T_{\mathrm{c,s}}(0) + \Delta T_{\mathrm{s,d}}(\dot{\gamma}) = \Delta T_{\mathrm{c,s}}(0) + C_{\mathrm{d}}\dot{\gamma}^{1/2}, \tag{4.9.53}$$

where $\Delta T_{\mathrm{c,s}}(0) \equiv T_{\mathrm{c}}(0) - T_{\mathrm{s}}(0)$. Thus, in the case where $\Delta T_{\mathrm{c,s}}(0)$ is small and/or $\dot{\gamma}$ is high so that $\Delta T_{\mathrm{c,s}}(0) \ll C_{\mathrm{d}}\dot{\gamma}^{1/2}$, Eq. 4.9.53 reduces to Eq. 4.9.39. On the other hand, in the case when $\dot{\gamma}$ is low, the first term in the second line of the rhs of Eq. 4.9.53 becomes significant, so that Eq. 4.9.39 is not valid. Thus, the agreement between the theoretical and experimented $\Delta T_{\mathrm{c}}(\dot{\gamma})/T_{\mathrm{c}}(0)$ becomes less reliable upon decreasing $\dot{\gamma}$.

4.10 Further Remarks

The author would like to point out a series of the interesting works by Mewis and Van Puyvelde and their co-workers on the binary systems at the low shear rates (corresponding to Regime III in Fig. 4.6). They elucidated that the SALS can discriminate between the different modes of the relaxation for the droplets: the retraction, the end-pinching and the break up by the Rayleigh instabilities [27]. The latter pattern evolution was modeled by means of the diffraction approach [28]. The proposed model together with the Tomotika theory is demonstrated to provide the elegant

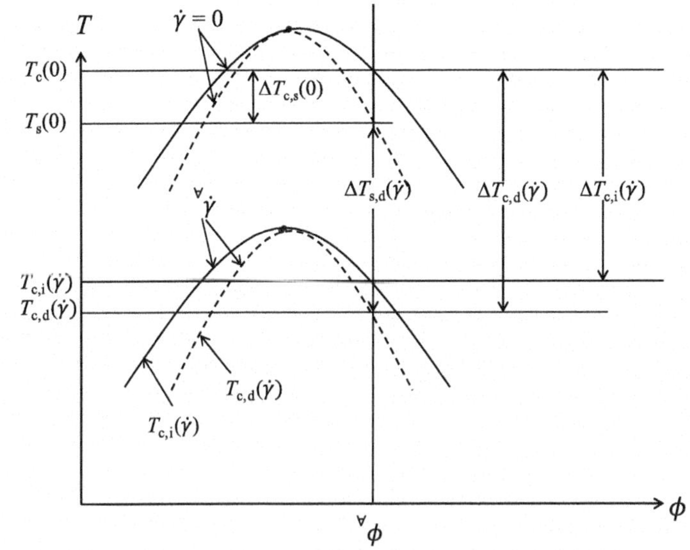

Fig. 4.56 Schematic presentation of the equilibrium phase diagram at $\dot{\gamma} = 0$ where the solid and broken line are the binodal and spinodal line, respectively, and the steady-state phase diagram at the given shear rate $\dot{\gamma}$ where the solid and broken line indicate the phase diagrams with the increased shear rate and decreased shear rate, respectively. The following quantities are defined at the given composition ϕ: $\Delta T_{c,s}(0) \equiv T_c(0) - T_s(0)$, $\Delta T_{s,d}(\dot{\gamma}) \equiv T_s(0) - T_{c,d}(\dot{\gamma})$, $\Delta T_{c,d}(\dot{\gamma}) \equiv T_c(0) - T_{c,d}(\dot{\gamma})$, and $\Delta T_{c,i}(\dot{\gamma}) \equiv T_c(0) - T_{c,i}(\dot{\gamma})$, where $T_c(0)$ and $T_s(0)$ are the binodal and spinodal temperature at the given composition ϕ at $\dot{\gamma} = 0$.

method of estimating the interfacial tension as low as 10^{-6} N/m for the immiscible blends. They demonstrated that SALS can be used to unambiguously determine the break-up time of droplets during the shear flow [77]. The systematic measurements of this break-up time led to the scaling law [78].

References

1. Onuki, A. and Hashimoto, T. (1989) *Macromolecules*, **22**, 879–890.
2. de Gennes, P.G. (1979) *Scaling Concept in Polymer Physics*, Cornell University, Ithaca.
3. Takebe, T., Sawaoka, R., and Hashimoto, T. (1989) *J. Chem. Phys.*, **91**, 4369–4379.
4. Hashimoto, T., Matsuzaka, K., Moses, E. and Onuki, A. (1995) *Phys. Rev. Lett.*, **71**, 126–129.
5. Hashimoto, T., Takebe, T., and Suehiro, S. (1988) *J. Chem. Phys.*, **88**, 5874–5881.
6. Hashimoto, T., Takebe, T., and Fujioka, K. (1990) In *Dynamics and Patterns in Complex Fluids*, Onuki, A. and Kawasaki, K. (eds), Springer, Berlin, p. 86–99.
7. Kume, T., Asakawa, K., Moses, E., Matsuzaka, K., and Hashimoto, T. (1995) *Acta Polym.*, **46**, 79–85.
8. Hashimoto, T. (2005) *Bull. Chem. Sco., Jpn.*, **78**, 1–39.
9. Hashimoto, T. 2008 In *Soft Matter Characterization*, Borsali, R., Pecora, R., (ed), Springer. New York; Chapter 8, pp.378–462.

10. Kume, T. and Hashimoto, T. (1995) In *Flow-Induced Structure in Polymers, ACS Symp. Series*, 597, 35–47.
11. Ornstein, L.S. and Zernicke, Z. (1914) *Proc. Akad. Sci. (Amsterdam)*, **17**, 783; ibid (1918) Phys.Z., **19**, 134; ibid (1926). *Phys. Z.*, **217**, 761.
12. Takebe, T., Fujioka, K., Sawaoka, R., Hashimoto, T. (1990) *J. Chem. Phys.*, **93**, 5271–5280.
13. Hashimoto, T. (2022) *Principles and Applications of X-ray, Light and Neutron Scattering*, Springer Nature, Singapore.
14. Debye, P. and Bueche, A. M. (1949) *J. Appl. Phys.*, **20**, 518.
15. Porod. G. (Brumberger, H. ed.) (1967) *Small-angle X-ray Scattering* (Chapter 1, pp. 1–17), N.Y., Gordon and Breach Science Publisher.
16. Onuki, A., Yamazaki, K., and Kawasaki, K. (1981) *Ann. Phys.*, **131**, 217.
17. Onuki, A. (2002) *Phase Transition Dynamics*, Cambridge Univ. Press., Cambridge, UK.
18. Hashimoto, T., Takebe, T., and Asakawa, K. (1993) *Physica A*, **194**, 338–351.
19. Hobbie, H. K., Kim, S., and Han, C. C. (1996) *Phys. Rev. E*, **54**, R5909.
20. Jeon, H.S., Nakatani, A.I., Hobbie, E.K., and Han, C.C. (2001) *Langmuir*, **17**, 3087.
21. Yu, J.-W., Douglass, J.F., Hobbie, E.K., Kim, S., and Han, C.C. (1997) *Phys. Rev. Lett.*, **78**, 2664.
22. Kim. S., Hobbie, E.K., Yu, J.-W., and Han, C.C. (1997) *Macromolecules*, **30**, 8245.
23. Tomotika, S. (1935) *Proc. Roy. Soc. London*, **150**, 322
24. Nakatani, A.I. and Han, C.C. (1998) In *Structure and Properties of Multiphase Polymeric Materials*, Araki, T., Trang-Cong, Q., and Shibayama, M. (eds), M. Dekker, N.Y., Chapter 7.
25. Vinckier, I., Mewis, J., and Moldenaers, P. (1997) *J. Rheol. Acta*, **36**, 513.
26. Vinckier, I., Moldenaers, P., and Mewis, J. (1997) *J. Rheol.*, **41**, 705.
27. Van Puyvelde, P., Yang, H., Mewis, J., and Moldenaers, P. (1998) *J. Colloid Interf. Sci.*, **200**, 86.
28. Mewis, J. Yang, H., Van Puyvelde, P., Moldenaers, P., and Walker, L.M. (1998) *Chem. Eng. Sci.*, **53**, 2231.
29. Yan, H., Zhang, H.J., Moldenaers, P., and Mewis, J. (1998) *J. Polym., Sci.*, **39**, 5731.
30. Minnale, M., Moldenaers, P., and Mewis, J. (1999) *J. Rheol.*, **43**, 815.
31. Vincker, I., Mewis, J., and Moldenaers, P. (1999) *Rheol. Acta*, **38**, 65 and 198.
32. Nakatani, A. I., Kim, H., Takahashi, Y., Matsushita, Y., Takano, A., Bauer, B., and Han, C.C. (1990) *J. Chem. Phys.*, **93**, 795.
33. Hair, D.W., Hobbie, E.K., Nakatani, A.I., and Han, C.C. (1992) *J. Chem. Phys.*, **96**, 9133.
34. Hair, D.W., Hobbie, E.K., Douglass, J.F., and Han, C.C. (1992) *Phys. Rev. Lett.*, **68**, 2476.
35. Hobbie, E.K., Nakatani, A.I., Yajima, H., Douglass, J.F., and Han, C.C. (1996) *Phys. Rev. E. Rapid Commun.*, **53**, R4322.
36. Hindawi, I.A., Higgins, J.S., Galambos, A.F., and Weiss, R.A. (1990) *Macromolecules*, **23**, 670.
37. Chen, Z.J., Wu, R.-J., Shaw, M.T., Weiss, R.A., Ferandez, M.L., and Higgins, J.S. (1995) *Polm. Eng. Sci.*, **35**, 92.
38. Debye, P., Anderson, H. R., and Brumberger, A. M. (1957) *J. Appl. Phys.*, **28**, 679.
39. Mani, S., Malone, M.F., Winter, H.H., Halary, J.L., and Monnerie, L. (1991) *Macromolecules*, **24**, 5451.
40. Mani, S., Malone, M.F., and Winter, H.H. (1992) *Macromolecules*, **25**, 5671.
41. Hindawi, I.A, Higgins, J.S., and Weiss, R.A. (1992) *Polymer*, **33**, 2522.
42. Gerard, H. and Higgins, J.S. (1991) *Phys. Chem. Chem. Phys.*, **1**, 3059.
43. Gerard, H., Cabral, J.T., and Higgins, J.S. (2003) *Phil. Trans. Soc. Lond. A*, **361**, 767.
44. Chopra, D., Vlassopoulos, D., and Hatzikiriakos, S.G. (1998) *J. Rheol.*, **42**, 1227.
45. Fujioka, T., Takebe, T., and Hashimoto, T. (1993) *J. Chem. Phys.*, **98**, 717–728.
46. Takebe, T. and Hashimoto T. (1988) *Polymer Commun.*, **29**, 261–263.
47. Beysens, D. and Gbadamassi, M. (1979) *J. Phys. Lett.*, **40**, L565.
48. Hashimoto, T. (1993) In *Materials Science and Technology*, Cahn, R.W., Kramer, E.J., and Haasen, P. (eds), *Vol.12, Structure and Properties of Polymers*, Thomas, E.L. (vol ed), VCH, Weinheim, Chapt. 6, pp. 251–300.

49. Hashimoto, T. ((2004) *J. Polym. Sci.: Part B: Polym. Phys.*, **42**, 3027–3062
50. Ma, S.K. (1976) *Modern Theory of Critical Phenomena,* Benjamin-Cummings, MA.
51. Gunton, J.D., San Miguel M., and Sahni, P.S. (1983) In *Phase Transitions and Critical Phenomena, Vol.8,* Domb, C. and Lebowitz, J. (ed), Academic Press, N.Y., p.269.
52. Onuki A. and Kawasaki K. (1979) *Ann. Phys.* (NY), **121**, 456.
53. Onuki, A. and Kawasaki, K. (1978) *Progr. Theor. Phys. Supp.*, **64,** 436, 1978.
54. Kawasaki K. (1970) *Ann. Phys* (NY), **61**, 1
55. Onuki, A. (1986) *Physica A*, **140**, 204.
56. Hashimoto, T., Izumitani, T., and Takenaka, M. (1989) *Macromolecules* **22**, 2293–2302.
57. Hashimoto, T. and Izumitani, T. (1993) *Macromolecules*, **26**, 3631–3638.
58. Ribbe, A. and Hashimoto, T. (2000) *Macromolecules* **33**, 7827–7834.
59. Rangel-Nafaile, C., Metzner, A.B., and Wissbrun, K.F. (1984) *Macromolecules*, **17**, 1187.
60. Wolf, B.A. (1984) *Macromolecules*, **17**, 615.
61. Horst, R. and Wolf, B.A. (1992) *Macromolecules*, **25**, 5291.
62. Horst, R. and Wolf, B.A. (1993) *Macromolecules*, **26**, 5676.
63. Fernandez, M.L., Higgins, J.S., Horst, R., and Wolf, B.A. (1995) *Polymer*, **36**, 149.
64. Soontaranun, W., Higgins, J.S., and Papathanaisou, T.D. (1996) *Fluid Phase Equilib.*, 121, 273.
65. Fukuda, T., Fujimoto, K., Tsuji, Y., and Miyamoto, T. (1996) *Macromolecules*, **29,** 3300.
66. Fukuda, T., Fujimoto, K., Murakami, M., Kawabata, K. and Miyamoto, T. (1996) *Macromol. Symp.*, **101**, 147.
67. Chopra, D., Haynes, C., Hatzikiriakos, S.G,. and D. Vlassopoulos (1999) *J. Non-Newtonian Fliud Mech.*, **82**, 367.
68. Cahn, J.W. (1965) *J. Chem. Phys.* **42**, 93.
69. Hashimoto, T., Itakura, M., and Hasegawa, H. (1986) *J. Chem. Phys.*, **85**, 6118–6128.
70. Hashimoto, T. (1988) *Phase Transitions*, **12**, 47–119..
71. Takeno, H., Hashimoto, T. (1997) *J. Chem. Phys.*, **107**, 1634–1644.
72. Takeno, H., Iwata, M., Takenaka, M., and Hashimoto, T. (2000) *Macromolecules*, **33**, 9657–9665.
73. Hashimoto, T., Matsuzaka, K., and Fujioka, K. (1998) *J. Chem. Phys.*, **108**, 6963–6972.
74. Asakawa, K. and Hashimoto, T., (1996) *J. Chem. Phys.*, **105**, 5216–5233.
75. Binder, K. (1983) *J. Chem. Phys.*, **79**, 6387.
76. Ohta, T. (1988) *J. Phys. C.*, **21**, L361.
77. Varmant, J., Van Puyvelde, P., Moldenaers, P., Mewis, J., and Fuller, G.G. (1998) *Langmuir*, **14**, 1612.
78. Van Puyvelde, P., Yang, H., and Mewis, J., and Moldenaers, P. (2000) *J. Rheology*, 44, 1401.

Chapter 5
Flow-Induced Demixing (Phase Separation)

Abstract This chapter will discuss self-organization of ultrahigh molecular weight atactic polystyrene (a-PS) in dioctyl phthalate (DOP) as a solvent under externally applied flow fields. The system is a dynamically asymmetric system, where a-PS and DOP are the slow and fast component, respectively. All the solutions studied form thermodynamically stable single-phase solutions in the absence of the applied flow and in the presence of the applied flow at the shear rate $\dot{\gamma}$ satisfying $0 \leq \dot{\gamma} < \Gamma_{MD}$, the relaxation rate of molecular deformation-orientation, at all temperatures covered in this study (Regime I). Upon *increasing the flow rate* at $\dot{\gamma} > \Gamma_{MD}$, the solutions are brought under the strong flow with respect to the molecular deformation-orientation, where the single-phase solutions become thermodynamically unstable under the flow, giving rise to a cascade evolution of a series of flow-rate dependent dissipative structures: from the dissipative structures in Regime II (where the flow-induced spinodal decomposition (SD) occurs to evolve the plane wave concentration fluctuations with the wave vectors oriented along the flow direction (FD)) to those in Regime IV at $\dot{\gamma} \geq \dot{\gamma}_a \gg \Gamma_{MD}$ (strings composed of phase-separated domains aligned along the string axis and FD) through those in Regime III (phase-separated domains). The strings have a weak and strong birefringence in the early and late stage of Regime IV, defined as Regime IV-1 and IV-2, respectively. In Regime IV-2, the domains in the strings are predicted to be interconnected by bundles of stretched chains oriented along the string axis and FD. The same series of dissipative structures are elucidated to be self-organized in cascade *with time after the step-up shear* to a constant $\dot{\gamma} \geq \dot{\gamma}_a$, the critical shear rate for the string formation, which is brought about through the dissipation of mechanical work imposed to the system upon *increased* $\dot{\gamma}$ *or increased time spent* after the step-up shear.

5.1 Introduction

In this chapter, we discuss the self-organization of ultrahigh molecular weight (UHMW) atactic polystyrene (a-PS) solutions in dioctyl phthalate (DOP) as a solvent under externally applied flow fields with varying flow rates. The system composed of UHMW a-PS and DOP is a dynamically asymmetric system with a large asymmetric

T. Hashimoto, *Polymer Phasing*, https://doi.org/10.1007/978-981-97-9648-9_5

parameter α_a, where a-PS and DOP are the slow and fast component, respectively. All the solutions studied form thermodynamically stable single-phase solutions in the absence of the applied flow at all temperatures covered in this study.

As discussed in Sect. 2.4, the stress-diffusion coupling or the viscoelastic effect on the thermal concentration fluctuations involved in dynamically asymmetric systems suppresses the relaxation rate of the concentration fluctuations $\Gamma(q)$ at $q > q_{ve}$, the characteristic wave number for the viscoelastic effect, relative to that in the dynamically symmetric systems, as discussed in conjunction with Fig. 2.15 in Sect. 2.5. This kind of the suppression is anticipated to cause the asymmetric systems to become increasingly susceptible to the weak shear with the increased asymmetry parameter α_a, in comparison with the dynamically symmetric systems with $\alpha_a = 1$.

In this Chapter and in the next Chapter also, we deal with the dynamically asymmetric systems. In references to Fig. 1.7 and Fig. 1.8 in the Sect. 1.4 in Chap. 1, we can generally increase the shear rate $\dot{\gamma}$ applied to the systems (as defined in the Fig. 1.7a) from the zero shear rate to the shear rate larger than Γ_{conc}, the characteristic relaxation rate of the systems for the concentration fluctuations (defined as CF) and Γ_{MD}, the characteristic relaxation rate of the systems for the molecular deformation and orientation (defined as MD), as indicated in Fig. 1.8d. Thus, the systems under consideration can be subjected first to the weak shear flow with respect to the CF at $\dot{\gamma} < \Gamma_{conc}$ as shown in the Fig. 1.7 (b-1) and 1.8e, subsequently to the strong flow with respect to the CF as shown in Fig. 1.7 (b-2) but yet the weak flow with respect to the MD at $\Gamma_{conc} < \dot{\gamma} < \Gamma_{MD}$ as shown in Fig. 1.7 (c-2), 1.8e and 1.8f and finally to the strong flow with respect to both the CF and the MD at $\dot{\gamma} > \Gamma_{MD}$ as shown in Fig. 1.7 (b-3) and Fig. 1.7 (c-3), and 1.8f, as presented in the three characteristic shear-rate Regimes designated as Regime $[\dot{\gamma}_I]$ to Regime $[\dot{\gamma}_{III}]$ in Fig. 1.7b and Fig. 1.8d.

In the dynamically symmetric systems discussed in Chap. 4, the systems are subjected only to Regime $[\dot{\gamma}_I]$ and Regime $[\dot{\gamma}_{II}]$, unable to reach Regime $[\dot{\gamma}_{III}]$, and thereby, the polymers remain undeformed conformations even under the strong shear flow: only the phase-separated domains and the concentration fluctuations are deformed. Under this condition, the flow forms such various characteristic dynamical structures as those specified in Regime I to Regime V as summarized in Fig. 4.6 (the classification with respect to the shear-induced structures).

In contrast to the dynamically symmetric systems, the a-PS/DOP solutions are typical dynamically asymmetric solutions. The a-PS/DOP solutions under the weak flow with respect to the molecular deformation-orientation form the stable single-phase solutions with respect to the concentration fluctuations in the shear-rate Regime $[\dot{\gamma}_I]$ or the suppressed concentration fluctuation by the flow in the shear-rate Regime $[\dot{\gamma}_{II}]$; the suppression in Regime $[\dot{\gamma}_{II}]$ occurs dominantly along the flow direction as discussed in Sect. 4.3.8 in conjunction with Fig. 4.16. Upon further increasing the flow rate, the a-PS/DOP solutions are brought into the shear-rate in Regime $[\dot{\gamma}_{III}]$ under the strong flow with respect to the molecular deformation-orientation, where the single-phase solutions under the flow become thermodynamically unstable, giving rise to a cascade evolution of flow-rate dependent dissipative structures, as will be presented below in this chapter.

5.2 Systems with Dynamic Asymmetry

The dynamic asymmetric systems employed in this chapter are listed in Table 5.1. M_w and M_n are the weight-average and number-average molecular weight, respectively, and $T_{cl}(0)$ is the cloud point of the respective polymer at the given concentration in the quiescent state. All the polymer solutions are homogeneous, single-phase solutions at temperature above the $T_{cl}(0)$ in quiescent state.

Figure 5.1 presents the cloud point curve for the PS548/DOP 6.0 wt.%. The points marked by the symbol × along the vertical line at 6 wt.% concentration indicate specific temperatures where the respective data are shown in the text. DOP is the Θ solvent for the solution of the a-PS548/DOP 6.0 wt.% at 22 °C.

Table 5.1 Dynamically asymmetric systems in this chapter

Sample code	$M_w \times 10^{-6}$	M_w/M_n	Concentration	T_{cl} (0) °C
PS548/DOP[a]	5.48	1.15	1.5, 3.0, 6.0	14.9, 16.3, 13.8[b]
PS384/DOP	3.84	1.06	6.0	10.0
PS288/DOP	2.88	1.09	5.0	14.0
dPS200/DOP[c]	2.00	-	8.0	< 2

[a] Atactic polystyrene (a-PS) with weight-average molecular weight 5.48 million in dioctyl phthalate (DOP), Θ solvent at 32 °C
[b] T_{cl} (0) (°C) = 14.9, 16.3, and 13.8 for 1.5, 3.0, and 6.0 wt.% solution of a-PS/DOP, respectively
[c] Deuterated a-PS

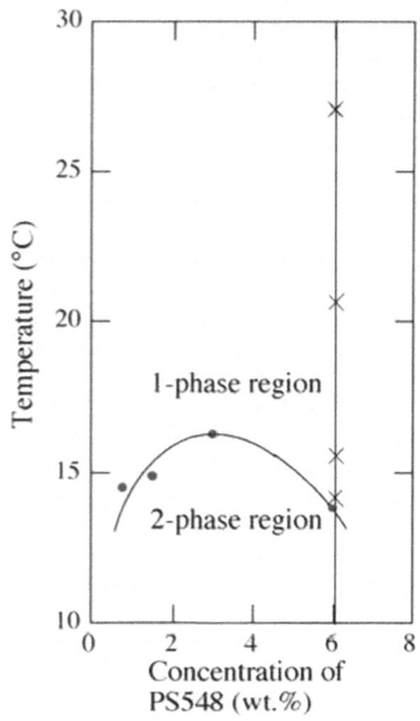

Fig. 5.1 Cloud point curve for the PS548/DOP 6.0 wt.%. The points marked by the symbol × indicate the temperatures where the shear flow is applied to the solution

5.3 Steady-State Dissipative Structures Evolved by Increased Flow Rate

When the semidilute solution in the single-phase state was subjected to the shear flow with the shear rate greater than the critical value, the solution was found to change from the transparent solution to the turbid solution [1]. It was also reported that this shear-induced turbidity brings a related change in the rheological behaviors to the solution [2–4]. The intriguing phenomenon, as elucidated from the transmitted light intensity and the rheological properties as a function of the shear rate $\dot{\gamma}$, suggests that the flow induces formation of the dissipative structure in the solution, reflecting either the shear-enhanced concentration fluctuations or the shear-induced phase separation.

The experimental studies described above were further advanced by a series of studies with the shear-rheo-optical methods as described in the Chaps. 3 and 4. The shear small-angle light scattering (shear-SALS) [5–13] and the shear small-angle neutron scattering [14, 15] elucidate the dissipative structures in the q space through the scattering structure factor for the systems under the shear flow, whereas the shear optical microscopy (shear-OM) elucidates the dissipative structures in the real space [13, 16–19].

Figure 5.2 presents the typical experimental results on the shear-SALS patterns and shear-OM images in Oxz plane for the PS548/DOP 6.0 wt.% at 27 °C, the typical system with the dynamic asymmetry having its cloud point at rest at 13.8 °C [16, 17, 20]. The experiments were carried out at 27°C, 13.2 °C above the cloud point (Table 5.1), where the solution at rest ($\dot{\gamma} = 0$) is in the single-phase state.

The solution at $\dot{\gamma} = 0\,\text{s}^{-1}$ is in the single-phase state, so that it does not exhibit appreciable SALS intensity at the small angles as shown in part (a) and no features in the OM image as shown in part (e). The bright halo around the direct beam stop (appearing as the dark circle) in part (a) is an artifact arising from the high q tail of the incident beam; the true scattering is very weak so that it is buried in the dark background in this pattern. Upon imposing the shear flow at the $\dot{\gamma} = 0.23\,\text{s}^{-1}$, the SALS pattern with the strong scattering intensity appears along the flow direction but the scattering along the neutral direction (the horizontal direction in the figure) remains almost the same as the scattering from the quiescent solution, so that the scattering exhibits the so-called **butterfly pattern** with the butterfly wings spreading along the flow direction and with the dark sector appearing along the neutral axis, as shown in the part (b). The butterfly pattern reveals the shear-enhanced concentration fluctuations which are anisotropic; the dominant Fourier modes of the concentration fluctuations have the wave vectors preferentially oriented along the flow direction as will be discussed below in detail in this section.

The OM images shown in part (f) to part (h) exhibit some anisotropic structures with the characteristic length of the order of 10 μm along the flow direction, which may reflect anisotropic dissipative structures developed by the shear-induced phase separation or concentration fluctuations. The fact that the structures observed under the OM images reflect true structures developed in the sheared solution was confirmed by the piece of evidence that the first Fourier transform (FFT) patterns of the OM images are essentially identical to the real scattering patterns. For example, the FFT patterns of the OM images in parts (e) and (f) are, respectively, presented in parts

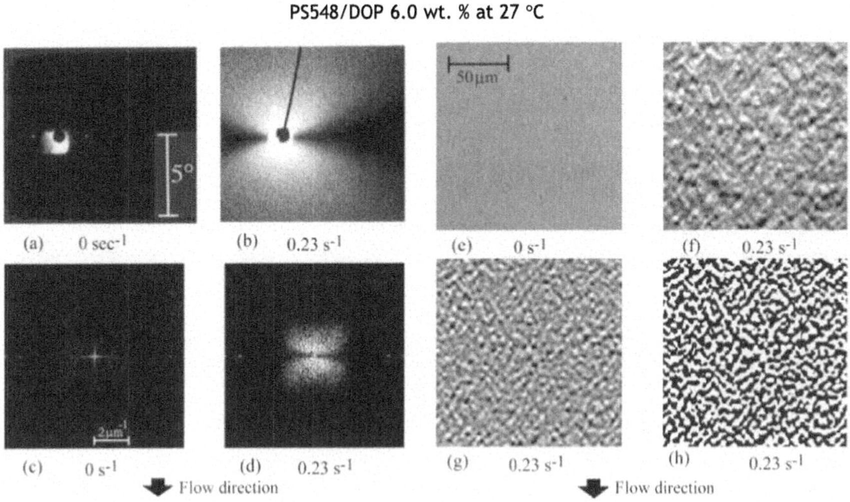

Fig. 5.2 Typical shear-SALS patterns in the quiescent state ($\dot{\gamma} = 0$) (**a**) and at the $\dot{\gamma} = 0.23\,\mathrm{s}^{-1}$ at steady-state (**b**). The SALS pattern (**c**) and (**d**) are obtained by the FFT of the transmission shear-OM image at the $\dot{\gamma} = 0$ (**e**) and at the $\dot{\gamma} = 0.23\,\mathrm{s}^{-1}$ (**f**), respectively. The OM image (**g**) is the filtered image of the raw OM image (**f**) obtained by using the bandpass filter that is tuned to the dominant Fourier modes of the system corresponding to the SALS pattern in (**b**). The OM image (**h**) is obtained after the binarization of the filtered image (**g**). The scale bars in part (**a**) and (**c**) indicate, respectively, the scattering angle in the solution (common to part **a** and **b**) and the magnitude of the scattering vector in the solution (common to the FFT pattern shown in part **c** and **d**). The scale bar in part (**e**) indicates the scale bar common to the OM images in part (**e**) to (**h**). The flow direction and neutral (vorticity) direction are vertical and horizontal, respectively. The SALS patterns (**a**) and (**b**) were taken with the shutter speed of 1/30 and 1/250 s, respectively, while the OM images were taken with the shutter speed of 10^{-4} s. Based on Hashimoto [20]

(c) and (d). One can judge that the FFT pattern (d) exhibits the characteristics of the butterfly SALS pattern shown in pattern (b) and that patterns (c) and (e) consistently represent the homogeneous solution.

The butterfly patterns were observed at various temperatures and concentrations [21] and for various solvents [21], such as cyclohexane and diethyl malonate, both of which are the Θ solvents for the a-PS at 35 °C, as well as dibutyl phthalate and tricresyl phosphate, both of which are the good solvents for the PS. They were also observed for the semidilute solutions of polyethylene in paraffin as a solvent (athermal solution) [19, 22, 23] and for the sheared colloidal suspensions [24]. They are thus quite general for sheared dynamically asymmetric systems. It should be noted that this shear-induced concentration fluctuations and/or phase separation (demixing) was also reported for the polymer mixtures of the PS/PVME by Winter and co-workers [25] and Higgins and co-workers [26] and for the other polymer mixtures [27–30].

Figure 5.3 presents the evolution of the steady-state SALS patterns in the increased shear-rate protocol from $\dot{\gamma} = 0$ to $14.3\,\mathrm{s}^{-1}$ for the PS548/DOP 6.0 wt.% at $T = 27\,°\mathrm{C} > T_{\mathrm{cl}}(0) = 13.8\,°\mathrm{C}$. At $\dot{\gamma} = 0$, the solution is in a single-phase state,

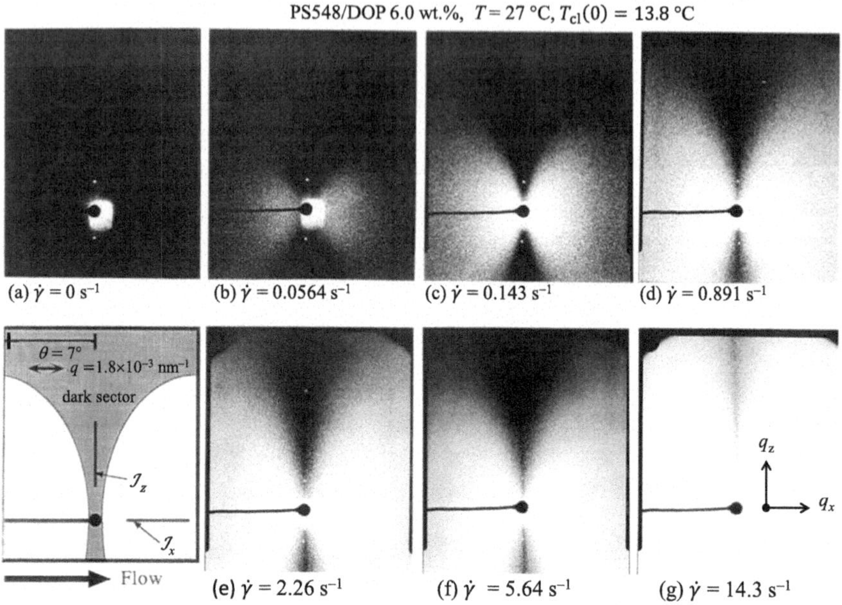

Fig. 5.3 The steady-state SALS patterns evolved in the increased shear-rate protocol for the PS548/DOP 6.0 wt.% at $T = 27\,°\mathrm{C}$, $T_{cl}(0) = 13.8\,°\mathrm{C}$

so that the solution does not exhibit the SALS pattern with the scattered intensity $I(q_x, q_z)$ excess to that from the thermal concentration fluctuations in the single-phase solution, defined as $I_{\mathrm{therm}}(q_x, q_z)$, with q_x and q_z being the magnitude of the scattering vectors along the flow and neutral direction, respectively (part a). Upon increasing $\dot{\gamma}$, the strong butterfly SALS pattern appears along q_x direction: The lateral spread of the butterfly pattern along q_z is very narrow (part b). Thus, the scattered intensity $I(q_x, q_z = 0)$ along q_x in the flow direction starts to increase from $I_{\mathrm{therm}}(q_x, q_z = 0)$ at the shear rate higher than the critical shear rate defined as $\dot{\gamma}_{c,x}$, whereas the scattered intensity $I(q_x = 0, q_z)$ along q_z in the neutral direction of the flow remains that from $I_{\mathrm{therm}}(q_x = 0, q_z)$. Upon further increasing $\dot{\gamma}$, the scattered intensity of the butterfly pattern and the lateral spread of the butterfly pattern further increase (part c to part g).

The change of the butterfly pattern with $\dot{\gamma}$ as described above develops the dark sectors emanated from the center of the butterfly pattern, toward the q_z direction normal to the flow direction, as illustrated by the gray zone in the schematic illustration of the butterfly pattern at the bottom-left corner of the figure. The dark sectors have the weak scattered intensity comparable with that from the single-phase solution at the given temperature, because the scattered intensity in the dark sectors arises from the thermal concentration fluctuations essentially free from the shear flow effect as evidenced by the SALS pattern shown in the part (a) in the Fig. 5.3 (see Fig. 4.16, Sect. 4.3.8 in detail). Moreover, the tips of the dark sector located at $(q_x = 0, q_z)$ are found to shift from the origin O in the q space at $(q_x = q_z = 0)$ toward the larger $|q_z|$

values with increased $\dot{\gamma}$ from $0.143\,\mathrm{s}^{-1}$ (part c) to $14.3\,\mathrm{s}^{-1}$ (part g), and the trend of which is especially clear in the change of the pattern from part (f) to (g). The shift of the dark sector tips with $\dot{\gamma}$ reveals that the scattered intensity $I(q_x = 0, q_z)$ along q_z direction starts to increase at the shear rate higher than the critical shear rate defined as $\dot{\gamma}_{c,z}$: $\dot{\gamma}_{c,z} > \dot{\gamma}_{c,x}$.

To investigate quantitatively the shear effects on the steady-state SALS patterns at the respective shear rates, the integrated intensity of the SALS butterfly pattern is measured along the q_x and the q_z directions, defined as $\mathcal{J}_x$ and $\mathcal{J}_z$, respectively, as shown in the schematic illustration at the bottom-left corner of Fig. 5.3,

$$\mathcal{J}_x(\dot{\gamma}) \equiv \int_{-\alpha}^{\alpha} dq_z \int_{\delta}^{q'} dq_x I(q_x, q_z; \dot{\gamma}), \tag{5.3.1}$$

$$\mathcal{J}_z(\dot{\gamma}) \equiv \int_{-\alpha}^{\alpha} dq_x \int_{\delta}^{q'} dq_z\, I(q_x, q_z; \dot{\gamma}). \tag{5.3.2}$$

$I(q_x, q_z : \dot{\gamma})$ is the scattered intensity distribution in the $q_x - q_z$ plane of the 2D detector at the given $\dot{\gamma}$. The limits of the integration $|\alpha|$, δ, and q' are the 3.4×10^{-4}, 6.2×10^{-4}, and the $2.7 \times 10^{-3}\,\mathrm{nm}^{-1}$, respectively.

Figure 5.4 presents the integrated SALS intensity $\mathcal{J}_j(\dot{\gamma})$ normalized with respect to the integrated SALS intensity at $\dot{\gamma} = 0$ as a function of $\dot{\gamma}$, $\mathcal{J}_j(\dot{\gamma})/\mathcal{J}_j(\dot{\gamma} = 0)$ with $j = x$ or z, for the a-PS548/DOP 6.0 wt.% solution [9]. The figure also presents the steady-state shear viscosity η and coefficient of the first normal stress difference ψ_1 as a function of $\dot{\gamma}$. The steady-state properties are measured at the time longer than $20/\dot{\gamma}$ seconds after the specified shear rate is attained in the increased shear-rate protocol from $\dot{\gamma} = 0$.

The scattering and rheological behavior as a function of $\dot{\gamma}$ can be classified into the four regimes, as shown in the top part of Fig. 5.4. In Regime I, the concentration fluctuations and hence the relevant integrated scattered intensity $\mathcal{J}_x(\dot{\gamma})$ and $\mathcal{J}_z(\dot{\gamma})$ are essentially identical to those in the quiescent solution in the single-phase state: the shear viscosity η and the coefficient of the first normal stress difference ψ_1 are nearly independent of $\dot{\gamma}$, exhibiting the Newtonian behavior. The Regime I exists in the shear-rate range at $\dot{\gamma} < \dot{\gamma}_{c,x}$, the critical shear rate above which the scattered intensity along the flow direction starts to increase from that in the quiescent solution, so that $\mathcal{J}_x(\dot{\gamma}) > \mathcal{J}_z(\dot{\gamma} = 0)$. In this regime, $\dot{\gamma}$ is smaller than the relaxation rate of the molecular deformation and orientation Γ_{MD}: the flow in the weak shear regime with respect to Γ_{MD} would not affect the molecular deformation and orientation as discussed in Sect. 1.4.

The Regime II to Regime IV exist at $\dot{\gamma} > \dot{\gamma}_{c,x}$, where the butterfly SALS pattern appears and $\mathcal{J}_x(\dot{\gamma})$ increases with $\dot{\gamma}$. The increase of the scattered intensity and hence the concentration fluctuations in these Regimes occur almost concurrently with the shear thinning as evidenced by the decrease of η and ψ_1 with $\dot{\gamma}$. The scattered

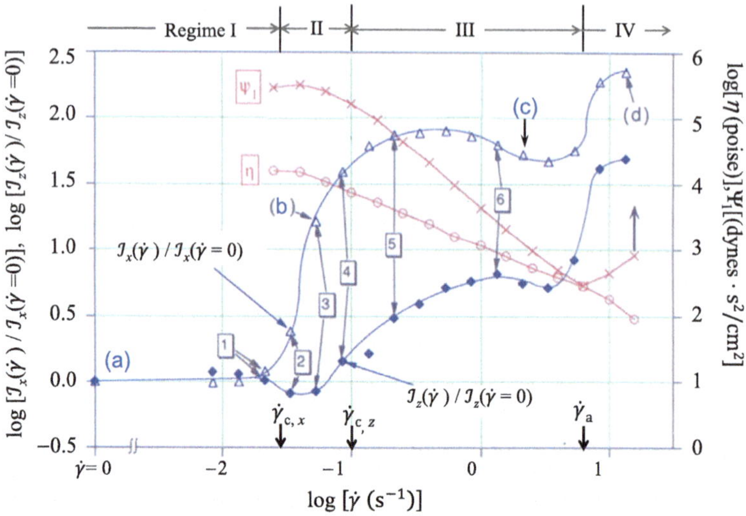

Fig. 5.4 Integrated intensity $\mathcal{J}_j(\dot{\gamma})$ of the scattered SALS intensity presented in Fig. 5.3, $\mathcal{J}_j(\dot{\gamma})$, reduced by that at $\dot{\gamma}=0$, $\mathcal{J}_j(\dot{\gamma}=0)$, along O$x$- or O$z$-axis ($j=x$ or z), respectively, (the ordinate at the left-hand side), and the steady-state shear viscosity η and the coefficient of the first normal stress difference ψ_1 (the ordinate at the right-handside of the figure) as a function of the $\dot{\gamma}$ for the PS548/OP 6.0 wt.% at $T=27\,°\text{C}$, $T_{\text{cl}}(0)=13.8\,°\text{C}$. Annotations (**a**) to (**d**) with the arrows attached on the curve $\mathcal{J}_x(\dot{\gamma})/\mathcal{J}_x(\dot{\gamma}=0)$ versus $\dot{\gamma}$ indicate the shear rates at which the SALS patterns in Fig. 5.5 are taken. Based on [9]

intensity strongly depends on the direction of the observation, $\mathcal{J}_x(\dot{\gamma})$ being much stronger than $\mathcal{J}_z(\dot{\gamma})$.

In Regime II, the dramatic shear-enhanced scattering, reflecting the emergence of the butterfly SALS pattern, occurs along the flow direction, as seen by the increased of $\mathcal{J}_x(\dot{\gamma})$ by the approximately 100-fold relative to $\mathcal{J}_x(\dot{\gamma}=0)$ as shown in Fig. 5.4. However, $\mathcal{J}_z(\dot{\gamma})$ effectively remains equal to $\mathcal{J}_z(\dot{\gamma}=0)$, giving rise to the butterfly SALS pattern shown in Fig. 5.5(b). More specifically, $\mathcal{J}_z(\dot{\gamma})$ tends to be suppressed relative to $\mathcal{J}_z(\dot{\gamma}=0)$, so that $\mathcal{J}_z(\dot{\gamma})/\mathcal{J}_z(\dot{\gamma}=0)<1$. This suppression is theoretically predicted to be due to the effect of the first normal stress difference N_1 on the suppression of the $I(q_x=0, q_z; \dot{\gamma})$ at the small q_z by the numerical analyses [31] of the dynamic equation on the concentration fluctuations based on the HFMO theory [32-34]. It is clarified numerically that at the given N_1 the suppression increases with the increased $\dot{\gamma}$ in Regime II and that at a given $\dot{\gamma}$ in the Regime II the suppression vanishes when $N_1 \to 0$ [31].

In Regime III, $\mathcal{J}_x(\dot{\gamma})$ does not vary much with $\dot{\gamma}$, reaching approximately the plateau value as shown by the high intensity of $\mathcal{J}_x(\dot{\gamma})/\mathcal{J}_x(\dot{\gamma}=0)\sim 100$ as presented in Fig. 5.4. On the other hands, $\mathcal{J}_z(\dot{\gamma})$ starts to gradually increase with increased $\dot{\gamma}$ above the critical shear rate defined as $\dot{\gamma}_{c,z}$ to reach the maximum value which is ~ 1/10 of the plateau value of $\mathcal{J}_x(\dot{\gamma})$. This increase of $\mathcal{J}_z(\dot{\gamma})$ and hence the increase of the concentration fluctuations along the neutral direction is the nonlinear effect brought about by the enhanced concentration fluctuations along the flow direction as manifested by the large increase of $\mathcal{J}_x(\dot{\gamma})$, which cannot be predicted by the linearized HFMO theory (to be described by Eq. 5.7.6 in Sect. 5.7.1 later). The nonlinear effects prevailing at $\dot{\gamma} > \dot{\gamma}_{c,x}$ have not been theoretically solved yet up to today. It may be plausible to predict that this nonlinear effect brings about the **shear-induced phase-separated domains** [35, 36].

The domain formation involves the transformation from the plane wave concentration fluctuations with the wave vectors preferentially oriented along q_x axis in Regime II ($\dot{\gamma}_{c,x} < \dot{\gamma} < \gamma_{c,z}$) into the phase-separated domains with the domain boundary interface developed along q_z axis also, which raises $\mathcal{J}_z(\dot{\gamma})$ in Regime III. The suppression of the integrated intensity $\mathcal{J}_j(\dot{\gamma})(j = x$ or $y)$ in Regime II, i.e., the suppression of $\mathcal{J}_x$ or $\mathcal{J}_y(= \mathcal{J}_x)$ of the scattered butterfly-SALS intensity having uniaxially symmetric intensity distribution around q_x-axis (FD) is predicted and validated as well to be arisen for the "non-equilibrium solutions (with $\dot{\gamma} \neq 0$)" subjected to the applied shear-flow fields at the various $\dot{\gamma}$ values and at the various polymer concentrations c's. That is, "the structures evolved in the nonequilibrium solutions" (or in "the solutions in the dynamic equilibrium states" or at the steady state") under the flow of the energy at the various shear-rates $\dot{\gamma}$'s and concentrations c's are predicted to be arisen by the energy flows applied to the solutions. That is, the nonequilibrium solutions under the flows of the energy applied by the given shear rates are accompanied by the suppression of the scattered butterfly-SALS intensity as clarified by the drop of $\mathcal{J}_x(\dot{\gamma})$ (Eq. 5.3.1) and $\mathcal{J}_y(\dot{\gamma})$ with respect to any given shear rates $\dot{\gamma}$'s in Regime II (Fig. 5.4).

This suppression is presented and validated as well in Fig. 5.3b,c, Fig. 5.4 at $\dot{\gamma}$ specified at point (b), and Fig. 5.5b also. This "increased suppression of $\mathcal{J}_z(\dot{\gamma})$" with respect to the increased shear rates at any given $\dot{\gamma}$'s in Regime II may cause also the net increase of $\mathcal{J}_z(\dot{\gamma})$ (Eq. 5.3.2), as predicted and as validated in Fig. 5.6b,c later, at $\dot{\gamma} = 10.0\ \mathrm{s}^{-1}$ and $89.4\ \mathrm{s}^{-1}$ in Regime III and IV, respectively. This increased suppression of $\mathcal{J}_z(\dot{\gamma})$ is well anticipated also from the butterfly SALS pattern as illustrated in the bottom left corner of the Fig. 5.3.

This suppression described above in Regime II is theoretically predicted and validated as well to be due to the effect of the first normal stress difference N_1 on the suppression of $I(q_z; \dot{\gamma})[\equiv I(q_x = 0, q_y = 0, q_z; \dot{\gamma})]$ by the numerical analyses [31] of the dynamic equation on the concentration fluctuations of the nonequilibrium systems at the $\dot{\gamma}$ under consideration based on the HFMO theory [32–34]. It is clarified and validated as well that, at the given $\dot{\gamma}$ in Regime II, the suppression vanishes when the $N_1 \to 0$ [31]. It should be important to note that the shear-thinning in Regime III is not able to be predicted by "the linearized HFMO theory, $\dot{\gamma}_s$" as will be discussed later in conjunction with Eqs. 5.7.1–5.7.9 in Sect. 5.7.1. It is also important to note that, up to date, there have been no theories to predict the time evolution of the butterfly-SALS pattrns and the relevant time evolution of the dissipation structures in the phase-separating, nonequilibrium systems under the shear-flow-fields quenched into the thermodynamically unstable state.

In Regime IV, an anomaly is observed in both the scattering and rheological behaviors in the increased shear-rate protocol as presented in Fig. 5.4. The scattering

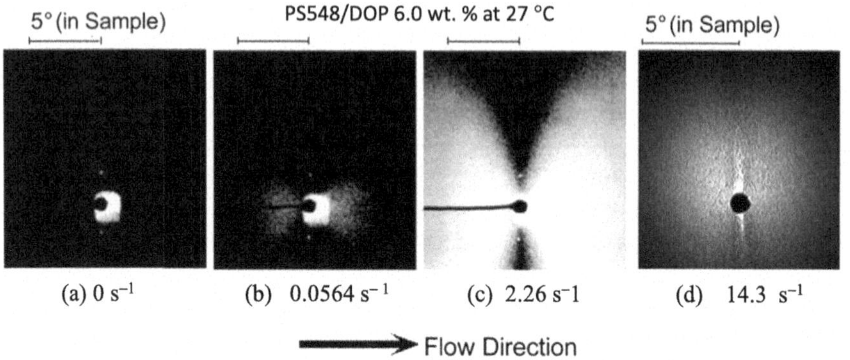

Fig. 5.5 SALS patterns at characteristic shear rates, $\dot{\gamma} = 0\,\mathrm{s}^{-1}$ (**a**), $0.0564\,\mathrm{s}^{-1}$ (**b**), $2.26\,\mathrm{s}^{-1}$ (**c**), and $14.3\,\mathrm{s}^{-1}$ (**d**), as indicated by the arrows marked (**a**) to (**d**) on the curve of $\mathcal{J}_x(\dot{\gamma})/\mathcal{J}_x(\dot{\gamma} = 0)$ in Fig. 5.4. PS548/DOP 6.0 wt.% at $T = 27\,°\mathrm{C}$

anomaly is recognized in the strong uprise of the intensity of both $\mathcal{J}_x(\dot{\gamma})$ and $\mathcal{J}_z(\dot{\gamma})$ at $\dot{\gamma} \geq \dot{\gamma}_a$, the critical shear rate for the onset of the anomaly. This anomaly is accompanied by the rheological anomalies. Upon increasing $\dot{\gamma}$ into the any given shear rates in Regime IV from a lower shear rate in Regime III, the shear viscosity η momentarily increases dramatically with time, but this momentary build-up of the shear stress eventually decays to reach the $\dot{\gamma}$-dependent steady-state values at the any given shear rates in Regime IV, if the solution is kept under the shear flow. The steady-state values of the η at $\dot{\gamma} = 10$ and $14.3\,\mathrm{s}^{-1}$ are still on the same shear-thinning line as observed at the shear rates in Regime II to Regime III given by the Carreau function [37],

$$\eta(\dot{\gamma}) = \eta_0\left[1 + (\dot{\gamma}\tau_\mathrm{m})^2\right]^{-a}, \tag{5.3.3}$$

where the exponent a is the shear-thinning exponent, η_0 is the zero-shear viscosity, and τ_m is the longest relaxation time of the polymer chains in the solution.

The anomaly is also observed in the upturn of ψ_1 (in Regime IV) with increased $\dot{\gamma}$, as typically presented by the broken red-arrow with its head pointing upward in Fig. 5.4; this upturn of ψ_1 at $\dot{\gamma} = 10$ and $14.3\,\mathrm{s}^{-1}$ are the steady-state values, in contrast to the momentary upturns of η found at $\dot{\gamma} = 10$ and $14.3\,\mathrm{s}^{-1}$, as presented also in Fig. 5.4; the momentary upturns of η is found to disappear with time to reach the steady-state values on the shear-thinning line. The steady-state value of ψ_1 is found to increase with increased $\dot{\gamma}$ above $\dot{\gamma}_a$; specifically at the shear rate $\dot{\gamma} = 14.3$ s^{-1} ψ_1 increases as pointed by the broken red-arrow in Fig. 5.4. At $\dot{\gamma} = 10\,\mathrm{s}^{-1}$ the large momentary overshoot was also observed in the ψ_1 before reaching the steady-state value η as in the case of the momentary overshoot η before reaching the steady-state volume. The steady-state value of $\psi_1 = 10^3$ dynes $\mathrm{s}^2/\mathrm{cm}^2$ was attained at the highest shear rate $\dot{\gamma} = 14.3\,\mathrm{s}^{-1}$ after the disappearance of the momentary overshoot as indicated by the broken red-arrow in Fig. 5.4. This steady-state value is the higher limit of the possible value of ψ_1 which could be measured by the rheo-optical apparatus employed in this work, i.e., the steady state ψ_1 may become larger than this value with the increased shear rate, but we stopped the experiment at this shear rate for the safety of the apparatus employed.

Figure 5.5 presents the SALS patterns at the characteristic shear rates as indicated by the arrows marked (a) to (d) on the curve of $\mathcal{J}_x(\dot{\gamma})/\mathcal{J}_x(\dot{\gamma} = 0)$ in Fig. 5.4, each point of which belongs to Regimes I to Regime IV, respectively [20]. Each pattern was taken with an exposure time of 1/30 s, except for the pattern (d), for which the exposure time is reduced to 1/500 s due to the much stronger scattering intensity from the SALS-butterfly pattern than the other SALS-butterfly patterns.

The pattern in part (a) at $\dot{\gamma} = 0$ does not exhibit the appreciable intensity of the SALS pattern, which is typical to that in Regime I. The pattern in Fig. 5.5b shows the typical SALS pattern in Regime II having the butterfly lobes oriented along q_x with the narrow lateral spread along q_z. The pattern in Fig. 5.5c is typical to that in Regime III, having the butterfly lobes with the strong scattered intensity and with increased lateral spread. The pattern in Fig. 5.5d is typical to that in Regime IV. It exhibits the sharp streak-like scattering pattern centered around $q_x = 0$ and oriented along q_z axis, along the neutral direction, with the strong scattered intensity. The sharp streak-like scattering pattern extended along q_x axis, as shown in the 2D SALS pattern in q_x-q_z plane, is superposed to the butterfly pattern compressed along q_x axis but extended along q_z axis (in the 2D SALS-butterfly pattern in q_x-q_z plane) which has the broad scattering maximum along q_x at $q_x = q_{x,\mathrm{m}} \cong 6.5 \times 10^{-4}\ \mathrm{nm}^{-1}$, indicating the periodic domain structure with the spacing of $2\pi/q_{x,\mathrm{m}} \cong 9.7\ \mu\mathrm{m}$ aligned along q_x axis, the flow direction. Here, $q_{x,\mathrm{m}}$ is defined as the magnitude of the scattering vector where the SALS intensity profile along q_x (in the 2D SALS-butterfly pattern in q_x-q_z plane) reaches the maximum intensity, defined as $I_{x,\mathrm{m}}$.

Figure 5.6 presents the simultaneously measured birefringence $\Delta n'$, form dichroism $\Delta n''$, and shear viscosity η (part a) and the normalized integrated scattered intensity along q_x-axis (filled triangles) and q_z-axis (unfilled squares) in the 2D SALS-butterfly pattern observed in q_x-q_z plane where q_x-axis and q_z-axis are the flow direction and the neutral direction, respectively. Those quantities in part a and b are measured as a function of shear rate $\dot{\gamma}$ (with the increased shear-rate protocol, together with the contour plots of the iso-intensity lines of the 2D SALS-butterfly pattern at the characteristic shear rates at $\dot{\gamma} = 0$, 0.0063, 10.0, and 89.4 s^{-1} (part c) for the PS384/DOP 6.0 wt.% at 21 °C [12]. The results on the shear-rate dependence of the shear viscosity and the integrated scattered intensity along the q_x-axis and q_z-axis direction show the same tendency as those shown in Fig. 5.4 for the PS548/DOP 6.0 wt.% at 27 °C, except for the shifts of the critical shear rates. That is, the shear-rate dependence of the relevant results can be classified into the four regimes, Regime I to Regime IV, with the Regime boundaries specified by the critical shear rates $\dot{\gamma}_{\mathrm{c},x}$, $\dot{\gamma}_{\mathrm{c},z}$, and $\dot{\gamma}_{\mathrm{a}}$ as defined at the top of Fig. 5.6a.

The shifts of the critical shear rates depend on the molecular weight, the concentration, and the temperature of the solutions. Each of the critical shear rates $\dot{\gamma}_{\mathrm{c},x}$, $\dot{\gamma}_{\mathrm{c},z}$, and $\dot{\gamma}_{\mathrm{a}}$ for PS/384 DOP 6.0 wt. % presented in Fig. 5.6 is higher than the corresponding critical shear rate for PS548/DOP 6.0 wt. % presented in Fig. 5.4, which

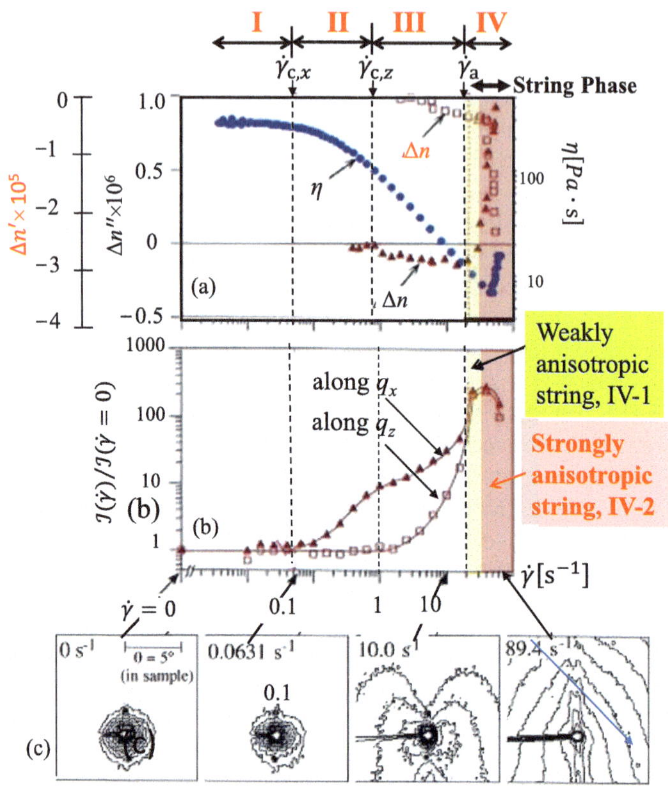

Fig. 5.6 Simultaneously measured birefringent $\Delta n'$, form dichroism $\Delta n''$, and shear viscosity η (part **a**) and normalized integrated scattered intensity $\mathcal{J}(\dot{\gamma}) = \mathcal{J}(\dot{\gamma} = 0)$ of the 2D SALS-butterfly pattern along q_x-axis and q_z-axis in q_x-q_z plane as a function of shear rate where q_x-axis and q_z-axis are taken along the flow direction (FD) and the neutral direction (ND), respectively $\dot{\gamma}$ (part **b**); both data in part (a) and (b) are taken with the increased shear-rate protocol. Contour patterns of the iso-intensity lines of the 2D SALS patterns at the characteristic shear rates at $\dot{\gamma} = 0$, 0.0631, 10.0, and 89.4 s^{-1} (part **c**) for PS384/DOP 6.0 wt.% at 21 °C, $T_{cl}(0) = 13.8$ °C. Based on [12], reprinted with permission, American Chemical Society

is expected to be dominantly due to the difference in the molecular weights. For the solutions at the given temperatures and concentrations under the shear flow it is experimentally predicted that the critical shear rate $\dot{\gamma}_{c,x}$, $\dot{\gamma}_{c,z}$ and $\dot{\gamma}_a$ are increased by the increased molecular weight (12). On the basis of this experimental prediction, we can further proposed, later in conjunction with Fig. 5.25 and Eqs. 5.7.16–5.7.19 in Sect. 5.7.2, that the critical shear rates $\dot{\gamma}_{c,x}$, $\dot{\gamma}_{c,z}$ and $\dot{\gamma}_a$ are controlled by the mechanical energy dissipated (or the work done) per the unit volume of the solution by the applied flow at the critical shear rates $\dot{\gamma}_{c,x}$, $\dot{\gamma}_{c,z}$ and $\dot{\gamma}_a$, respectively.

Figure 5.6a presents newly added information concerning the shear-rate dependence of the birefringence $\Delta n'$ and the form dichroism $\Delta n''$ which are absent in Fig. 5.4. The form dichroism $\Delta n''$ is the dichroism regarding the extinction of the transmitted light due to the anisotropic 2D SALS-butterfly pattern which has the uniaxial symmetry with respect to q_x-axis in q_x-q_z plane. Both quantities ($\Delta n'$ and $\Delta n''$) are negligibly small in Regime I and the II.

The small negative birefringence emerges with the increased $\dot{\gamma}$ in the higher shear-rate part of the Regime III. Figure 5.6c represents a typical butterfly SALS pattern (at $\dot{\gamma} = 10.0\ \mathrm{s}^{-1}$) in this shear rate range. The negative birefringence gradually increases with increasing shear rate up to the lower shear-rate part of Regime IV, specified by the yellow color region designated as the region of the "weakly anisotropic string" and defined as Regime IV-1. It drastically and abruptly increases with the further increase of $\dot{\gamma}$ at the higher shear-rate part of Regime IV as specified by the red color region designated as the region of the "strongly anisotropic string" and defined as Regime IV-2.

Since the intrinsic birefringence of the PS segment, $\Delta_{PS}{}^{o}$, is negative, as will be discussed later in Sect. 5.9 in conjunction with Fig. 5.34, and the $\dot{\gamma}$ dependence of $\Delta n'$ is given by

$$\Delta n' = \Delta_{PS}{}^{o} f_{\text{segment}}, f_{\text{segment}} \equiv \left[3\langle\cos^2\theta_{\text{segment}}\rangle - 1\right]/2, \tag{5.3.4}$$

where θ_{segment} is the polar angle of the segmental axis with respect to q_x-axis, the flow direction (FD), and f_{segment} is the orientation factor which describes the average orientation of the segments with respect to FD under the flow: $f_{\text{segment}} = 1$ for the perfect orientation of the segments along FD and $f_{\text{segment}} = 0$ for the random orientation. Thus, the shear-rate dependence of the birefringence is controlled by the shear-rate dependence of the orientation factor f_{segment}. The gradual increase of the negative $\Delta n'$ suggests that the gradual increase of the segmental orientation f_{segment} along FD, whereas the abrupt and drastic increase of the negative $\Delta n'$ suggests the corresponding increase of the orientation factor of the segments over the narrow range of $\dot{\gamma}$, suggesting the **chain stretching along the FD**.

The negative dichroism $\Delta n''$ emerges from the beginning of Regime III (part a). The small negative value of $\Delta n''$ increases gradually with the increased $\dot{\gamma}$ in Regime III. This dichroism change is due to the increased scattered intensity of the butterfly SALS pattern with the increased $\dot{\gamma}$ as revealed by the results shown in Figs. 5.4, 5.5, and 5.6c. It is also predicted that the anisotropy of the extinction of the intensity of the incident light beam increases due to the increased anisotropy of the SALS-butterfly pattern having the uniaxial symmetry with respect to q_x-axis in q_x-q_z plane. It should be also noted that the increased anisotropy of the SALS-butterfly pattern around q_x-axis (FD) in q_x-q_z plane as described above is accomanied by the increase of the negative value of $\Delta n''$ (12). The result reflects the growth of the phase-separated domains with the increased $\dot{\gamma}$ in Regime III with respect to the increased concentration difference between the domains and the matrix solution and the increased domain size.

On the other hand, the negative $\Delta n'$ emerges only from the later part (higher shear-rate part) of Regime III with increased $\dot{\gamma}$; it increases with the increased $\dot{\gamma}$ as in the case of $\Delta n''$ (part a). The small increase of the negative $\Delta n'$ is revealed to be accompanied by the decreased negative form dichroism, $\Delta n''$, toward zero to positive value with the increased $\dot{\gamma}$'s in Regime IV-1. In Regime IV-1, the strings begin to develop, which contributes to the weak optical anisotropy in this regime, as shown by the yellow color region in part (a) and (b) in Fig. 5.6. The strings in Regime IV-1 are oriented along FD and composed of a series of bundles of oriented domains self organized under the flow with the shear-rate range in this regime, as shown later in part (b-2) in Fig. 5.33b in Sect. 5.10.5. The small increase of the negative value of $\Delta n'$ in Regime III and IV, is predicted to be due to the small increase of the orientational order parameter of the polymer chains, defined herein as f_{d}, within the phase-separated domains. This phase-separated domains as described above are further deformed upon further increase of shear rate $\dot{\gamma}$ at any given shear rate in Regime IV-1. Moreover, the increase of f_{d} is due to the small increase of the chain orientation inside the domains within the string-like dissipative structures as shown in the part (d, d′) in Fig. 5.7 under the flow along FD (q_x-axis, Ox-axis), thereby due to the small increase of the degree of the chain stretching along FD within the deformed domains under the flow.

The orientational birefringence of the domains oreinted along FD, $\Delta n'_{\mathrm{d}}$, is given by $\Delta n'_{\mathrm{d}} = \Delta n^{\circ} f_{\mathrm{d}}$, where Δn° is the intrinsic birefringence of the segments of the chains within the deformed domains oriented uniaxially along FD (q_x-axis, Ox-axis). The a-PS chains in the solutions have $\Delta n^{\circ} < 0$. Here, the orientation factor f_{d} for the chains in the domains is defined as

$$f_d \equiv [3 < \cos^2 \theta_s > -1]/2,$$

where $<\cos^2\theta_{\mathrm{s}}>$ is the average of the $\cos^2\theta_{\mathrm{s}}$ for all the segments of the polymer chains in the phase-separated domains, and θ_{s} is the polar angle of the principal optical axes of the segments of the polymer chains within the domain with respect to FD. Herein, the principal axis of the segments are assumed to have the uniaxial symmetry with respect to FD, which is validated for the polymer solutions at the any given shear rates $\dot{\gamma}$'s in Regime IV-2.

Both the $\Delta n'$ and the $\Delta n''$ strikingly change with increased $\dot{\gamma}$ in Regime IV, although there is a disparity with respect to their shear-rate dependence, as revealed by the pieces of experimental evidence as presented immediately below.

Evidence (1) in Regime IV-1: The increase of the small negative value of $\Delta n''$ with increased $\dot{\gamma}$ in Regime III starts to decrease toward zero and further increase to the positive value with a further increase of $\dot{\gamma}$ in Regime IV-1 as presented in Fig. 5.6a, whereas the negative value of $\Delta n'$ keeps increasing with $\dot{\gamma}$ in the same shear-rate range as presented also in Fig. 5.6a.

Evidence (2) in Regime IV-2: Both the $\Delta n''$ and the $\Delta n'$ abruptly increase toward the positive values and negative values, respectively, with the further increase of $\dot{\gamma}$ in Regime IV-2 over the extremely narrow shear-rate range.

The evolution of $\Delta n''$ in Regime IV is associated with the evolution of the strong streak-like SALS pattern spread over the small q_x range around the $q_x = 0$ but extended along q_z-axis, which predicts itself the evolution of the closely packed demixed domains aligned along FD with the large correlation length. The evolution of the aligned demixed domains with the larger correlation length predicts the increased positive value of $\Delta n''$ with the increased $\dot{\gamma}$, which outweighs the increase of the negative $\Delta n''$ due to the individual domains causing the butterfly patterns with the increased $\dot{\gamma}$ from Regime III to Regime IV-1. This predicts the change of $\Delta n''$ from the negative to positive value with the increased $\dot{\gamma}$ in Regime IV-1 as described above in Evidence (1).

However, this alignment of the domains along FD itself does not yet give the special effect on $\Delta n'$ in Regime IV-1: The negative value of $\Delta n'$ keeps increasing continuously throughout Regime IV-1. The continuous increase of the negative $\Delta n'$ is predicted to arise from the continuous increase of the segmental orientation within the domains subjected to the increased deformation with the increased $\dot{\gamma}$ and/or the increased form birefringence with increased $\dot{\gamma}$ in Regime IV-1. This accounts for the shear-rate dependence on the $\Delta n'$ as described above in the Evidence (1).

The abrupt increase of the positive $\Delta n''$ in Regime IV-2 with the increased $\dot{\gamma}$ reveals the further dramatic growth of the assembly of the aligned domains, over the extremely narrow shear-rate range as presented in the Fig. 5.6. The assembly of the aligned domains is oriented along FD with much larger correlation length than the correlation length for the aligned domains in Regime IV-1.

The abrupt increase of the negative value of $\Delta n'$ in Regime IV-2 with the increased $\dot{\gamma}$ reveals the dramatic increase of f_{segment} over the extremely narrow shear-rate range, so that the segments of the a-PS chains are highly oriented beyond the level of the segmental orientation of the a-PS chains within the domains attained in Regime IV-1. Thus, the abrupt change in $\Delta n'$ may predict the formation of the bundle of the stretched chains interconnecting throughout the aligned domains along FD as presented by the red bundle in model (d′) in row (C) in Fig. 5.7 and by the bundle of stretched chains in model (b-3) in Fig. 5.33(b) to be presented later in Sect. 5.10.5. The striking change as observed in $\Delta n'$ and $\Delta n''$ with the increased $\dot{\gamma}$ described above in the Evidence (2) predicts the formation of the string structures composed of the bundles of stretched chains interconnecting the domains aligned along FD.

Figure 5.7 summarizes the result obtained by the rheo-optical studies presented in Fig. 5.6, regarding the classification of Regime I to IV (row A), the schematic SALS patterns (row B), the contrast-enhanced OM images (row C), the form optical dichroism and the birefringence (row D), and the dissipative structures developed under the steady shear flow (row C) as a function of $\dot{\gamma}$ in the increased shear rate (row E) [12].

In Regime I at $\dot{\gamma} < \dot{\gamma}_{c,x}$ the sheared solution is essentially identical to the homogeneous quiescent semidilute solutions, as evidenced by no SALS pattern and no characteristic OM image as shown in part (a) in row (B) and (C), respectively, as well as no form dichroism and no birefringence as shown in part (a) and row (D). In Regime II to IV at $\dot{\gamma} > \dot{\gamma}_{c,x}$, the shear-induced dissipative structures are formed due to the elastic effects and stress-diffusion coupling effects in the sheared solutions with the dynamical asymmetry, where the entangled polymer coils kept being deformed by when the relevant shear-rate-dependent dissipative structure is formed, as a consequence of essentially no time available for the disentanglements of the polymer chains under the shear flow at any given shear rates satisfying $\dot{\gamma} > \dot{\gamma}_{c,x}$.

In Regime II at $\dot{\gamma}_{c,x} < \dot{\gamma} < \dot{\gamma}_{c,z}$, the butterfly SALS pattern emerges but it's scattered intensity along the neutral direction (Oz-direction) remains as weak as that from the quiescent solution; the butterfly pattern has the two lobes which are oriented

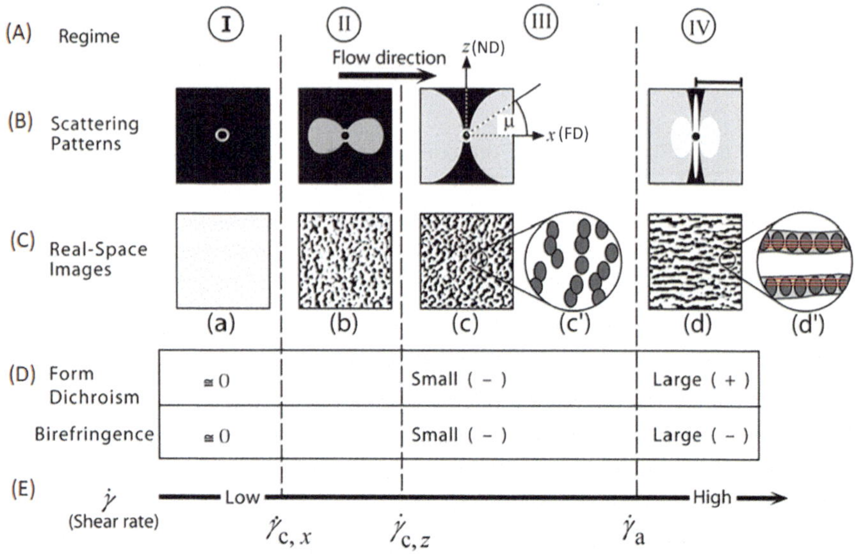

Fig. 5.7 Summary of the results obtained by the rheo-optical studies. (row A): the representation of the classification of the Regime with respect to the shear rate shown in (row E), (row B): the SALS-butterfly patterns (schematics), (row C): the contrast-enhanced real-space images obtained by using the optical microscopy, and (row D): the table showing the behaviors of the form dichroism and birefringence under the increased shear-rate protocol at the various shear rates. The scale bar on the top of part (d) in row (B) indicates the scattering angle of ca 5° in the sample for the SALS-butterfly pattern and the corresponding length scale of ca. 50 μm for the real-space images. The parts (c′) and (d′) are enlarged models for the images in the part (c) and (d) row C), respectively. The model (d) and (d′) in row C in Regime IV specifies the dissipative stuctures developed in Regime IV-2 defined in Fig. 5.6. The model in Regime IV-1 is presented later in part (b-2) in Fig. 5.33b in Sect. 5.10.5. From [12], reprinted with permission, American Chemical Society

along the flow direction (FD, Ox-direction) and have the very narrow spread along the azimuthal angle μ (defined in part c in row B). Moreover, the two lobes of the SALS-butterfly pattern have the uniaxial symmetry around Ox-axis (FD).

The OM image, presented in part (b) in row (C), shows that the dark and bright contrast extends along the neutral direction (ND, Oz-direction). On the other hands, it's contrast varies along the flow direction as shown in part (b) row (C). Here, the Cartesian coordinate Oxyz shown in part (c) and in row (B) is common to all the patterns presented in row (B) and (C) under the flow at the given shear rate. In Sect. 5.6, we shall more precisely clarify in depth and validate as well the experimental evidence regarding this dark-and-bright contrast variation in the OM images along the flow direction (FD, Ox-direction) and the physical significance regarding the structures (the dissipative structures) evolved in the nonequilibrium systems that are open for the flow of the mechanical energy and that are subjected to the dissipation of the applied mechanical energy (see Figs. 5.17, 5.19, and 5.20).

Thus, we can predict that the concentration fluctuations are enhanced along the flow direction (Ox-direction), while they are kept the same as those in the quiescent solution along the neutral direction at the given shear rate. That is, the Fourier modes of the shear-enhanced concentration fluctuations with the wave vectors along the neutral direction (q_z-direction) do not exist. In another words, the enhanced concentration fluctuations evolved can be represented by the superposition of the plane wave concentration fluctuations with their wave vectors preferentially oriented over the narrow azimuthal angle μ centered around $\mu = 0$ [as defined in the part (c) in row (B)].

In Regime III, the butterfly pattern is further developed with the increased $\dot{\gamma}$ in such a way that the butterfly wings spread over the wider range of the azimuthal angle μ and that the scattered intensity along the neutral direction (Oz-axis) increases also. The OM images show the clearer contrast variations along the flow direction (Ox-axis) in Oxz plane (see part c of row C) than those in Regime II (part b of row C), indicating the evolution of the domain-like structures elongated along the neutral axis as schematically shown in part (c′ of row C).

These facts may imply that the phase separation occurs in Regime III, giving rise to the phase-separated domains with the average spacing along q_x of the order of $2\pi/q_{x,\mathrm{m}} = 10\,\mu\mathrm{m}$. The centers of the individual domains are randomly placed along the neutral axis. The domains in the solutions at any given shear rates $\dot{\gamma}$'s tend to grow due to the thermodynamic instability of the solutions driven by the elastic effects at the given shear rates. However, the overgrown domains tend to burst due to the elastic force and viscous drag at the given shear rates. The dynamic balance between the two factors, i.e., (i) the growth driven by the shear-rate dependent thermodynamic driving force and (ii) the burst driven by the shear-rate dependent elastic force and viscous drag, control the shear-rate dependent steady-state domain structures as the dissipative structures evolved under the shear flow at the given shear rates in Regime III. The dichroism is negative and has small absolute values, consistent with the oblate ellipsoidal phase-separated domains extended along the neutral direction (Oz direction) with their major axes oriented along the flow direction (Ox direction). The shape of the oblate ellipsoidal phase-separated domains may reflect their evolution pathway that involves the flow-induced burst of the plane wave concentration fluctuation, brought about by the increased shear rates from Regime II to Regime III. The biregringence has also the small negative values in the higher shear-rate part of Regime III, revealing that the PS chains in the domains are only weakly oriented along the flow direction, or the birefringence may dominantly arise from the form birefringence of the domains in the matrix solution; the negative form birefringence is anticipated for the oblate ellipsoidal phase-separated domains extended along the neutral direction (Oz-direction) with their major axes oriented along the flow direction (Ox-direction). These experimental evidences as described above and the evidences predicted and further validated as well will be more precisely clarified in depth in Sect. 5.6 in conjunction with the Figs. 5.17, 5.19, and 5.20 and in conjunction with Eqs. 5.7.13–5.7.20 in the Sect. 5.7.1.

In Regime IV, the phase-separated domains developed in Regime III as shown in part (c) and (c′) in row (C) may tend to align with their centers of the mass along the flow direction due to the hydrodynamic interactions between the domains and the solvent flow, resulting in the **string-like assembly of the domain structures**, as presented by the dissipative structure shown in the part (d) and part (d′). Consequently,

SALS shows a strong streak-like scattering pattern, extended along the q_z axis and with the small breadth along q_x axis around $q_x = 0$, that arises from the whole string aligned along Ox-axis. The butterfly pattern at the large q_x superposed on the streak pattern at the small q_x may reflect the scattering from the domains within the string with the average interdomain distance of the order of 10 μm. The individual domains themselves cannot be clearly resolved in the contrast-enhanced OM images as shown in the part (c) and part (d) in row (C) in Fig. 5.7, but they are inferred from the butterfly scattering patterns observed in the part (c) and the part (d) in the row (B) in Fig. 5.7.

In the string-like assembly of the domains as presented in the part (d) and (d′) in row (C), the shape and orientation of the individual domains under the flow at the given shear rates are consistent with the oriented butterfly pattern presented in the part (d) in row (B). These experimental evidences described above will be presented and further validated as well in later in conjunction with Figs. 5.30(c), 5.31(c), 5.32, 5.35 (d, d′) and 5.35 (e′) in row (C) in the Sect. 5.8, 5.9, 5.10.4 and 5.11, respectively.

The dissipative structure and the SALS pattern developed in Regime IV are quite reminiscent of the **pearl-necklace structure** and its corresponding SALS-butterfly pattern shown in Fig. 4.21c, d or e and f, respectively. The dissipative structure, having string-like alignment of the phase-separated domains parallel to the flow axis, is anticipated to be the rational optimum structure which are formed at the dynamic equilibrium between the dissipation of the mechanical energy imposed to the systems at the given shear rates under such strong shear flows and thermodynamic force to drive the growth of the dissipative structure under the flows. Although the two systems compared here, i.e., the dynamically symmetric systems (the PS/PB solutions) in Chap. 4 and the dynamically asymmetric systems (the PS/DOP solutions) in Chap. 5. are physically very different, the dissipative patterns formed under the shear flow are universal.

Because of the evolution of the string-like structure oriented along the flow direction in Regime IV (particularly in Regime IV-2), the form dichroism exhibits the large positive value. Correspondingly, the birefringence shows the large negative value, indicating that end-to-end vectors of the a-PS chains in the strings are oriented parallel to the string axis, i.e., along the flow direction (Ox direction), because the single monomeric unit of a-PS chain has the negative intrinsic birefringence. Thus, **the strings have the large optical anisotropy. The bundles of the highly stretched chains** may interconnect the domains aligned along the string axis as schematically illustrated by the red lines in part (d′) in row (C) in Fig. 5.7, specifically in Regime IV-2 defined in conjunction with Fig. 5.6.

If this is the case, the structure developed is strikingly analogous to the "**shish-kebab**" structure developed by shear-induced crystallization of the crystallizable polymers as presented and further validated in Sect. 6.4 in conjunction with Fig. 6.14 [part (f,l)] for the systems composed of UHMW PE/paraffin 5.0 wt. % at 124°C and at $\dot{\gamma} = 2.9\ \text{s}^{-1} (> \dot{\gamma}_a = 1.2\ \text{s}^{-1})$. However, the a-PS used is not the crystallizable polymers, so that crystallization would not occur in Regime IV, i.e., the "shish" and the "kebab" developed are the amorphous bundles of the stretched chains and the amorphous demixed domains in this case, respectively. If the polymers used were crystallizable, they would be crystallized into the so-called **shish-kebab structure (SK)** to be described in the Chap. 6.

The SK is well known as the structure formed due to shear-induced crystallization of the polymer solutions. However, it would be extremely important to recognize the possibility that this shear-induced crystallization may be triggered by the shear-enhanced concentration fluctuations and/or phase separation within the strings, at least in the semidilute solutions of our interest here, simply because the crystallization is anticipated to occur locally in the region or in the phase rich in the polymer concentrations (i.e., in the string phase but not in homogeneous bulk solutions), as inferred from the structure shown by the model (d′) of row (C) in Fig. 5.7. This point will be discussed further in Sect. 6.5 later.

Table 5.2 presents the summary of the steady-state dissipative structures evolved in Regime I to Regime IV in the increased shear-rate protocol.

Table 5.2 Summary of the steady-state dissipative structures evolved in Regime I to Regime IV under increased shear-rate protocol

Regime	Shear rates	Dissipative structures	Remarks
I	$0 < \dot{\gamma} < \dot{\gamma}_{c,x}$ $\dot{\gamma}_{c,x} \sim \tau_m^{-1}$	The homogeneous solutions	o **Dissipation**[a] via disentanglement of the deformed entangle chains o Newtonian flow
II	$\dot{\gamma}_{c,x} < \dot{\gamma} < \dot{\gamma}_{c,z}$	The flow-enhanced **CF**[b] giving rise to the butterfly SALS-pattern and PLWCF	o **Dissipation** via the built-up **CF** o Shear thinning o The superposition of the Fourier modes of the plane wave **CF** with their **k**[c] parallel to FD o The zero butterfly-SALS intensity normal to FD
III	$\dot{\gamma}_{c,z} < \dot{\gamma} < \dot{\gamma}_a$	The flow-induced phase-separated domains and their alignment along FD	o **Dissipation** via the domain formation and domain alignment along FD o Shear thinning o The emerged butterfly-SALS - intensity normal to FD
IV	$\dot{\gamma} > \dot{\gamma}_a$, $\dot{\gamma}_a \sim \tau_e{}^{-1}$	IV-1: the strings with the aligned domains	o **Dissipation** via the string formation o The strong streak-like scattering extended along q_z and localized at around $q_x = 0$
		IV-2: The strings with the bundles of the stretched chains interconnecting the aligned domains	o **Dissipation** via the formation of the bundles of the stretched chains within the strings o Shear thickening o The sharp uprise of the negative $\Delta n'$ and positive $\Delta n''$ with increased $\dot{\gamma}$

τ_m: the longest rheological relaxation time, τ_e: chain retraction time

[a] **Dissipation**: dissipation of the mechanical energy applied to the solutions

[b] **CF**: the concentration fluctuations

[c] **k**: the wave vector for the plane-wave concentration variation in the Fourier modes of the **CF**

5.4 Transient Evolution of Dissipative Structures after Step-Up Shear from $\dot{\gamma} = 0$

In the preceding section, we discussed the various dissipative structures developed at the steady-state shear flow in the increased shear-rate protocol. Let us now focus on how the transient dissipative structures are developed and grown as the function of the time after a step-up shear flow from $\dot{\gamma} = 0$ to the any given shear rates larger than the critical shear rate, i.e., $\dot{\gamma} > \dot{\gamma}_{c,z}$ or $\dot{\gamma} > \dot{\gamma}_a$.

The Fig. 5.8 presents the evolution of the transient rheological properties, the first normal stress difference N_1 and the shear stress σ, after the step-up shear rate from $\dot{\gamma} = 0$ to the constant shear rate $\dot{\gamma} = 14.3\,s^{-1}$ in Regime IV defined in Sect. 5.3 (see Figs. 5.4 and 5.6) for the PS548/DOP 6.0 wt.% solution at 27.5 °C [20, 38]. Both N_1 and σ clearly show the two stress overshoots at least, and even the third overshoot may be discerned at around 70 s on its approach toward a transient steady state at $t \sim 100$ s. The each overshoot in σ appears to occur prior to the corresponding one in N_1. The structural origin of this intriguing rheological behavior was followed by the simultaneously taken butterfly-SALS patterns. For this purpose, the butterfly-SALS patterns are taken at the times marked (a) to (l) on N_1 in Fig. 5.8 in which N_1 undergoes characteristic changes with time. The results are shown in Fig. 5.9 [20, 38], where the patterns (a) and (b) (both recorded at $t = 0$ s) as well as the patterns (n) and (m) (both taken at $t = 150$ s) represent the original pattern and the contour pattern built up from the iso-intensity lines, respectively, for comparison.

The scattered intensity along the flow direction (FD, Ox direction) starts to increase remarkably with time at about 15 s (the pattern shown in Fig. 5.9e), and then after that time, the butterfly pattern is clearly developed. Its intensity parallel to the flow direction (Ox direction) increases with time, whereas the intensity perpendicular to the flow direction (along the neutral direction, Oz direction) stays weak up to about the 25 s (the pattern g), almost at the same intensity level as that for the quiescent solution, giving rise to the dark sector around the neutral axis (Oz-axis) as discussed in the preceding section.

The apex angle (or width) of this dark sector is narrowed with time. The result indicates that the wave vectors of the Fourier modes of the concentration fluctuations, the any given modes of which are represented by the plane wave concentration variation along the any given wave vectors (see Fig. 5.19 to be shown later), spread over the wider azimuthal angle μ centered around $\mu = 0$ (see Fig. 5.7 for the definition of μ) with respect to the flow direction (Ox direction) and the (μ-dependent) amplitude of each Fourie mode increases with time. The scattered intensity along the neutral direction (Oz direction) tends to increase with time at the time longer than 20 s after the onset of the flow, implying the onset of the shear-induced phase separation. Thus, the transient changes in the scattering patterns with time seem to be quite similar to the changes in the steady-state scattering patterns with the increased shear-rate protocol up to $\dot{\gamma} = 14.3$ in Regime IV.

The butterfly pattern has the scattering maximum along FD (Ox direction) at the magnitude of the scattering vector q_{max} with its value slightly decreasing with time

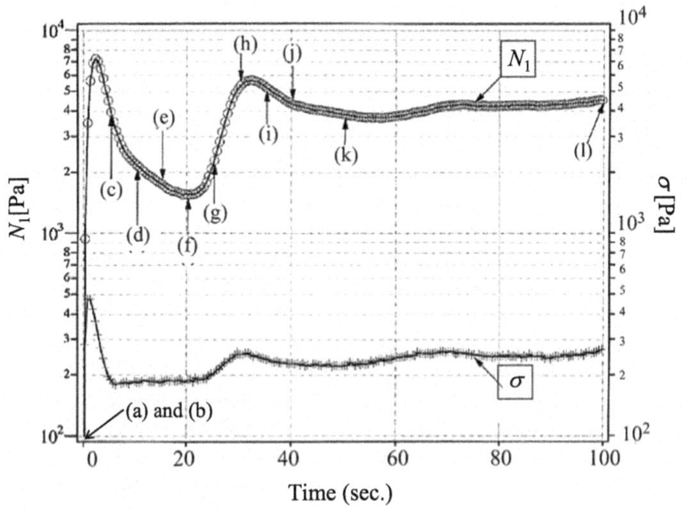

Fig. 5.8 Time evolution of the first normal stress difference (N_1) and shear stress (σ) in the PS548/DOP 6.0 wt.% solution at 27.5 °C after the onset of the step-up shear flow from $\dot{\gamma} = 0\,\text{s}^{-1}$ to the constant shear rate of $\dot{\gamma} = 14.3\,\text{s}^{-1}$ in Regime IV. The annotations **a** to **l** marked on N_1 in the figure indicate the times at which the butterfly-SALS patterns **a** to **l** in Fig. 5.9 are taken. From [38], reprinted with permission, American Chemical Society

from $\sim 0.6 \times 10^{-3}$ to $\sim 0.35 \times 10^{-3}\,\text{nm}^{-1}$ in the q range covered in this experiment. Thus, the characteristic length $2\pi/q_{max}$ of the shear-induced structure along FD varies from ~ 10 to ~ 18 μm with time. In the time span from the onset of the flow (0 s) to 50 s, the scattered intensity along FD is increased by more than 1 order of magnitude [38]. The butterfly-SALS patterns, especially the intensity parallel to FD, showed no appreciable changes with time at $t \gtrsim 100$ s over the time span covered in the experiments presented in Figs. 5.8 and 5.9, indicating the time evolution of the global dissipative structure reached the transient steady state in the time span between 100 s and 150 s.

To facilitate comparison of the results shown in Figs. 5.8 and 5.9, Fig. 5.10 presents the time variation of N_1 after the step-up shear from $\dot{\gamma} = 0$ to 14.3 s^{-1} together with the time evolution of the butterfly-SALS patterns at the particular times of $N_1(t)$ (indicated by the arrows), where $N_1(t)$ exhibits the characteristic changes: (b) 0 s, (c) 5 s, (e) 15 s, (f) 20 s, (h) 30 s, (k) 50 s, and (l) 100 s for the PS548/DOP 6.0 wt.% at 27.5 °C. The data of N_1, butterfly-SALS patterns, and the points (b) to (l) are the same as those in Figs. 5.8 and 5.9.

The comparisons of the two results on the time dependence of N_1 and the SALS-butterfly patterns after the step-up shear flow shown in Figs. 5.8, 5.9 and 5.10 as well reveal the following pieces of evidence.

(1) The butterfly patterns are clearly developed, and hence, the shear-induced dissipative structures (or the concentration fluctuations) are found to be formed after the first stress overshoot of N_1 at ~ 20 s (f) where the stress level reaches the

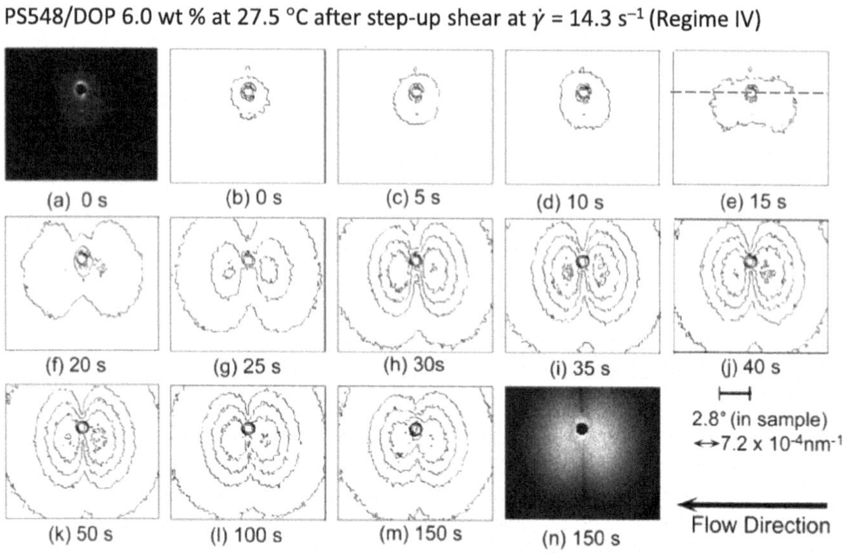

Fig. 5.9 The time evolution of the 2D butterfly-SALS patterns from the same experimental system as shown in Fig. 5.8. The patterns **a** and **n** are the original scattering patterns, while others are contour patterns exhbiting the iso-intensity levels of the 2D scattering patterns. The patterns **a** to **l** correspond to the annotation **a** to **l** with arrows in Fig. 5.8. The asymmetry of each of the 2D scattering patterns with respect to the horizontal axis passing through its center (representatively indicated by the red broken line in part **e**) is due to the artifact brought by the shape of the window employed for the temperature enclosure used in this experiment; the bottom half of each of the patterns presents the correct intensity distribution free from the artifact (see Fig. 3.1 in Ref. [38]). Based on [38], reprinted with permission, American Chemical Society

minimum in between the first and second stress-overshoot maxima at $\sim$ 1.9 s and at $\sim$ 30 s, respectively. In the time span of $0 < t < 5$ s^{-1} where the first stress overshoot is discerned, the SALS intensity remains almost the same as that at $\dot{\gamma} = 0$ as shown in pattern (b) and (c) in Figs. 5.9 and 5.10. This piece of evidence implies that the entangled chains are uniformly deformed and oriented without the disentanglements and thereby without involving the solvent squeeze in this time span under the shear flow at the shear rate satisfying $\dot{\gamma} = 14.3\,\mathrm{s}^{-1} > \dot{\gamma}_a = {\tau_e}^{-1} = 4.5\ \mathrm{s}^{-1}$ thereby this shear rate belonging to Regime IV. After this time span, i.e., $t > 5\,\mathrm{s}^{-1}$, the **solvent squeeze** from the regions rich in polymer concentrations to the regions poor in polymer concentrations, that builds up the enhanced concentration fluctuations against osmotic pressure, may become significant, and the shear-enhanced butterfly-SALS pattern may become appreciable. Thus, this time span ($0 < t < 5\,\mathrm{s}^{-1}$) is predicted to be the one where the stress (mechanical energy per unit volume) is stored in the uniformly deformed entangled networks under the applied flow. That is, the one before the solvent squeeze occurs to relax the stress or to dissipate the energy stored in the deformed swollen entangled networks.

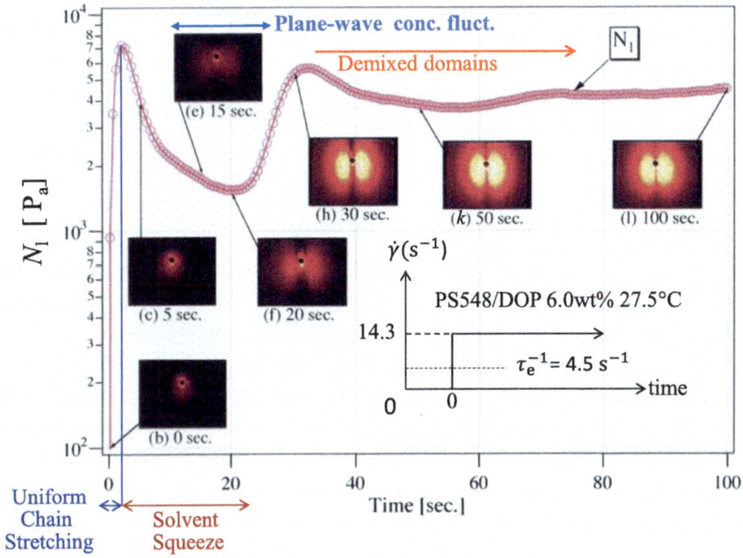

Fig. 5.10 Time variation of N_1 after the step-up shear from $\dot{\gamma} = 0\ \text{s}^{-1}$ to $14.3\ \text{s}^{-1}$ together with the butterfly-SALS-patterns at the particular times of the $N_1(t)$ indicated by the arrows, where $N_1(t)$ exhibits the characteristic changes: **b** 0 s, **c** 5 s, **e** 15 s, **f** 20 s, **h** 30 s, **k** 50 s, and **l** 100 s for the PS548/DOP 6.0 wt.% at 27.5 °C

Thereby, the time span may be assigned to be the **incubation time for the evolution of the butterfly pattern** that reflect the evolution of the first dissipative structure in the sheared solution. We should note that the shear rate at the any given shear rates in Regime IV, defined as $\dot{\gamma}_{\text{Regime,IV}}$, satisfies,

$$\dot{\gamma}_{\text{Regime,IV}} > 1/\tau_{\text{m}} \quad \text{or} \quad 1/\dot{\gamma}_{\text{Regime,IV}} < \tau_{\text{m}}, \tag{5.4.1}$$

where τ_{m} is the longest rheological relaxation time of the entangled chains in the solution. Thus, the shear rate $\dot{\gamma}_{\text{Regime,IV}}$ is faster than the disentanglement rate $1/\tau_{\text{m}}$, so that the entanglements exist as the permanent ones over the shear distortion time $1/\dot{\gamma}_{\text{Regime,IV}}$. Consequently, we cannot ascribe the stress relaxation after the first stress overshoot to the disentanglement of the deformed entangled chains. The solvent-squeezing process is responsible for both the stress relaxation and evolution of the butterfly pattern.

(2) As the time further elapses, both N_1 and σ increase toward the maximum values at the second stress overshoot; the butterfly pattern concurrently increases its scattered intensity, and the butterfly wings spread over the wider azimuthal angle μ. Thus, the uprise of the stress level of N_1 and σ with time after the first stress overshoot is related to the growth of the characteristic structures (the shear-enhanced plane wave concentration fluctuations with the larger amplitude of the fluctuations and with the larger spread of their wave vectors along the

azimuthal angle μ around $\mu = 0$) as discussed in the preceding paragraph in conjunction with Figs. 5.6, 5.7, and Table 5.2 in the case of increased shear-rate protocol.

(3) The growth of the shear-induced structure and deformation of the grown structure with time under the flow would give rise to the second stress overshoot. At around the time 35 s when the N_1 and the σ reach the peak or slightly pass through the peak of the second-overshoot (corresponding to point (i) in Figs. 5.8 and 5.9), the demixing (the formation of the phase-separated domains) may start to occur, which is related to the onset of the increase of the integrated scattered intensity along q_z-axis (Oz-axis). The demixing involves the transformation of the shear-enhanced plane wave concentration fluctuations to the phase-separated domains through the burst of the plane wave concentration fluctuations along the neutral direction (q_z direction, Oz direction). The demixing appears to decrease the stress levels of N_1 and σ toward their steady values through the dissipation of the mechanical energy imposed to the solutions under the flow.

As the stress levels N_1 and σ approach the transient steady values at the time around $100\,\text{s} < t < 150\,\text{s}$ as specified by the values at the point (l) in Fig. 5.8, the phase-separated domain structures also appear to attain the transient steady structures, as shown part (c) in row C in Fig. 5.7 and as indicated by the butterfly-SALS patterns (l) and (m) in Fig. 5.9. The transient steady structures themselves were discussed in the preceding section in conjunction with Fig. 5.7. However, it may be conceivable that the local stress levels and SALS intensity **fluctuate** or even **oscillate** with time around their steady values, because demixed structures will grow with time under the flow but the overgrown structures will be burst by the flow, and a dynamic balance between the growth and burst will give the local **fluctuations** or the **oscillations** with time, though they tend to be averaged out with the increased length scale of observation.

Figure 5.11 presents the time evolution of the scattering profile along the flow direction, $I(q_x; t) \equiv I(q_x, q_z = 0; t)$, after the step-up shear from $\dot{\gamma} = 0$ to $14.3\,\text{s}^{-1}$ for the PS548/DOP 6.0 wt.% at 27.5 °C [38]. The representative times in the figure are selected as the times where the 2D SALS patterns and N_1 undergo the characteristic changes as shown in Figs. 5.8, 5.9, and 5.10. For the sake of convenience, the 2D SALS patterns are also presented in the bottom half of Fig. 5.11. In the incubation period, defined as t_{incub} (between 0 s and ~ 5 s) as evidenced by the SALS pattern in part (c) of Fig. 5.10), $I(q_x; t)$ under the flow remains identical to the one from the quiescent solution ($t = 0\,\text{s}$). The shear-enhanced scattering starts to be clearly discernible at the time longer than the characteristic time, defined as $t_{\text{c},x}$, ($\sim 5\text{s} < t_{\text{c},x} < 15\,\text{s}$). $I(q_x; t)$ at $t > t_{\text{c},x}$ exhibits the broad scattering maximum. The peak intensity of the butterfly-SALS pattern increases with t, and the q value at the peak decreases with t at $t \geq 15$ s as shown by the black arrows with their heads pointing upward. Thus, t_{incub} is specified by $0 < t_{\text{incub}} < t_{\text{c},x} < 15$ s.

Flow-induced growth of dissipative structures. Those trends described above reveal the shear-induced spinodal decomposition (SD) and the **coarsening (growth) of the SD structures** under the flow at any given times after the evolution of the

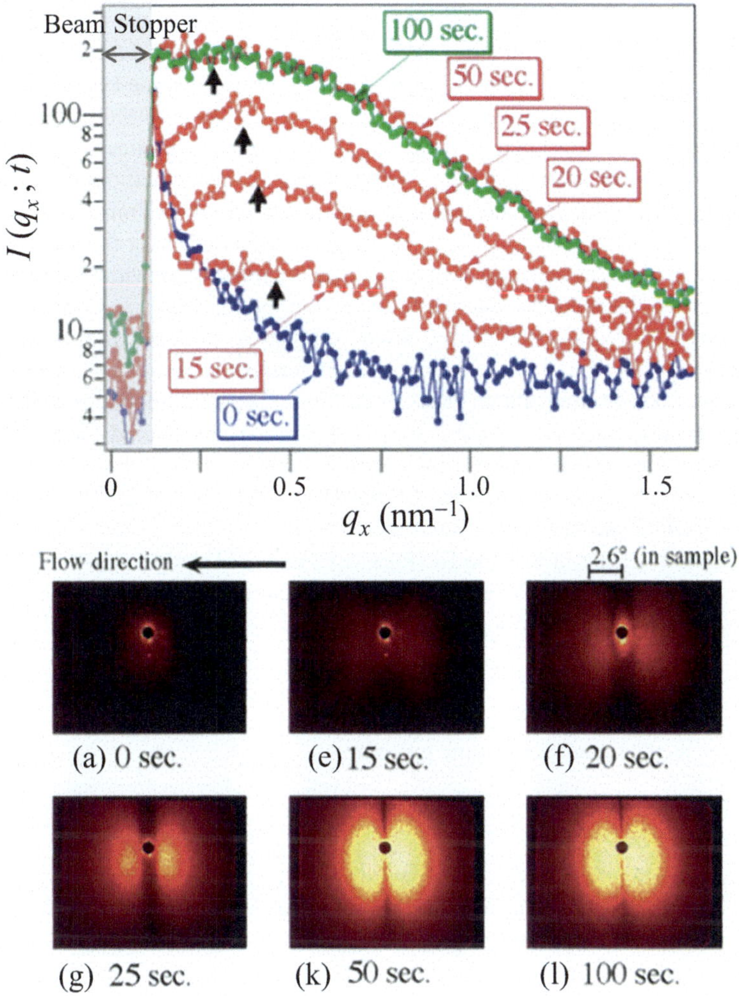

Fig. 5.11 Time evolution of the scattering profile along flow direction, $I(q_x; t)$, after the step-up shear from $\dot{\gamma} = 0$ to $14.3\,\mathrm{s}^{-1}$ for the PS548/DOP 6.0 wt.% at 27.5 °C

early stage SD. The flow-induced SD is driven by the flow-induced thermodynamic instability of the solution under the flow as will be detailed in Sect. 5.7. The thermodynamic driving force for the SD depends on the shear rate, because the spinodal temperature depends on the shear rates. The coarsening occurs until the times where the shear-rate dependent steady state is achieved. The steady state is attained due to a dynamic equilibrium between the growth of the structures driven by the thermodynamic driving force and the burst of the grown structures, i.e., the dissipation of the relevant energy driven by the imposed flow.

Up to 25 s at least after the step-up flow, the dark sector of the butterfly pattern clearly exists along the Oz direction (the neutral direction), so that the shear-induced dissipative structures can be described by the superposition of the Fourier modes of the shear-enhanced plane wave concentration fluctuations with their wave vectors oriented over the narrow polar angle μ with respect to the flow direction: The wavelength and amplitude of the dominant Fourier modes of the fluctuations increase with time, as manifested by the decreased $q_{x,\mathrm{m}}$ and the increased maximum intensity $I_{x,\mathrm{m}}$ with time, and the correlation length of the Fourier modes along the wave vectors increases with time, as manifested by the narrowing of the peak width with the time.

At $t > 30$ s, the scattered intensity $I(q_z; t)$ and the integrated scattered intensity $\mathcal{J}_z(t)$ along q_z direction (the neutral direction) start to increase from those in the quiescent solution (Fig. 5.9), revealing the transformation from the shear-enhanced plane wave concentration fluctuations to the phase-separated domains at the characteristic time defined as $t_{\mathrm{c},z}$ ($25\,\mathrm{s} < t_{\mathrm{c},z} < 30\,\mathrm{s}$). The scattered intensity along FD (q_x direction), $I(q_x; t)$, further increases with t up to 50 s and reaches the transient steady-state over the time span from 100 to 150 s covered in this experiment. The steady-state attained here at the time span of $t_{\mathrm{c},z} < t < t_\mathrm{a}$ is the transient (or apparent) steady state in Regime III (Figs. 5.12 and 5.13) before reaching the final stage at $t > t_\mathrm{a}$, in Regime IV where t_a is the characteristic time, defined as

$$t_\mathrm{a} \equiv 1/\dot{\gamma}_\mathrm{a} \sim \tau_\mathrm{e}^{-1}. \tag{5.4.2}$$

Here, τ_e is the chain retraction time given by

$$\tau_\mathrm{e} = 2\tau_R, \tag{5.4.3}$$

where τ_R is the longest Rouse relaxation time [39]. That is, the evolution of the SALS and the stress is predicted to be continued to occur even after the apparent steady state being reached at the time span between $\sim$ 100 and 150 s: The final stage is predicted to be attained much later time, although the experiments presented here in Figs. 5.8, 5.9, 5.10, and 5.11 do not cover the corresponding time span.

Note: the solvent-squeeze time or incubation time as observed by **LAOS**. The relaxation time for the solvent-squeeze process or the incubation time for the shear-enhanced concentration fluctuations was clearly elucidated by large amplitude oscillatory shear flow experiments (**LAOS**), as described in Sect. 3.2.2, on the same semidilute solutions [40]. In this experiment, the butterfly pattern was observed at the strain amplitude γ_0 greater than the critical value defined as $\gamma_{0,\mathrm{c}}$. At $\gamma_0 > \gamma_{0,\mathrm{c}}$, the butterfly pattern was observed at the angular frequency ω satisfying $\omega_{\mathrm{c},1} < \omega < \omega_{\mathrm{c,u}}$, where $\gamma_0\omega_{\mathrm{c},1}$ corresponds to $\dot{\gamma}_{\mathrm{c},x}$ and $\gamma_0\omega_{\mathrm{c,u}}$ corresponds to the rate for the solvent squeezing [40]. At $\omega \geq \omega_{\mathrm{c,u}}$ the entangled polymer solutions are uniformly stretched and relaxed under the oscillatory shear flow, giving rise to the homogeneous solution without enhanced concentration fluctuations, simply because there would be no time available for solvent molecules to be squeezed out from polymer-rich regions subjected to larger stress than the polymer-poor regions to relax the stress for the deformed swollen entangled networks. It is clarified that $\omega_{c,l}$ and $\omega_{c,u}$ depend on γ_0: upon increasing γ_0, $\omega_{c,l}$ and $\omega_{c,u}$ decreases and increases, respectively for the given PS/DOP solution at a given concentration and temperature.

Figure 5.12 presents a comparison of the evolution of dissipative structures under the following two shear protocols: (a) the shear flow with step-up shear rate from $\dot{\gamma} = 0$ to $\dot{\gamma} > \dot{\gamma}_a$ and (b) the shear flow with increased shear rate from $\dot{\gamma} = 0$ to $\dot{\gamma} > \dot{\gamma}_a$. The structural evolution process in the case (a) where the step-up shear flow with increased time is imposed to the system until reaching the steady-state string-like structure at $t > t_a$ is the same as that in the case (b) where the shear flow with increased shear rate is imposed to the system until reaching the steady-state string-like structure at $\dot{\gamma} > \dot{\gamma}_a$, if the characteristic times, $t_{c,x}$, $t_{c,z}$ and t_a in the case (a) are replaced by the characteristic shear rates $\dot{\gamma}_{c,x}$,, $\dot{\gamma}_{c,z}$, and $\dot{\gamma}_a$ in the case (b), respectively. Both protocols (a) and (b) commonly involve Regime [I] to [III] before reaching Regime [IV]: The characteristic times, $t_{c,x}$, $t_{c,z}$, and t_a and the characteristic shear rates, $\dot{\gamma}_{c,x}$, $\dot{\gamma}_{c,z}$, and $\dot{\gamma}_a$ which specify the respective regimes are related to the work dissipated per unit volume W by the flow imposes to the systems to reach the critical times and critical shear rates,

$$W = \sigma\gamma = \eta\dot{\gamma}^2 t. \tag{5.4.4}$$

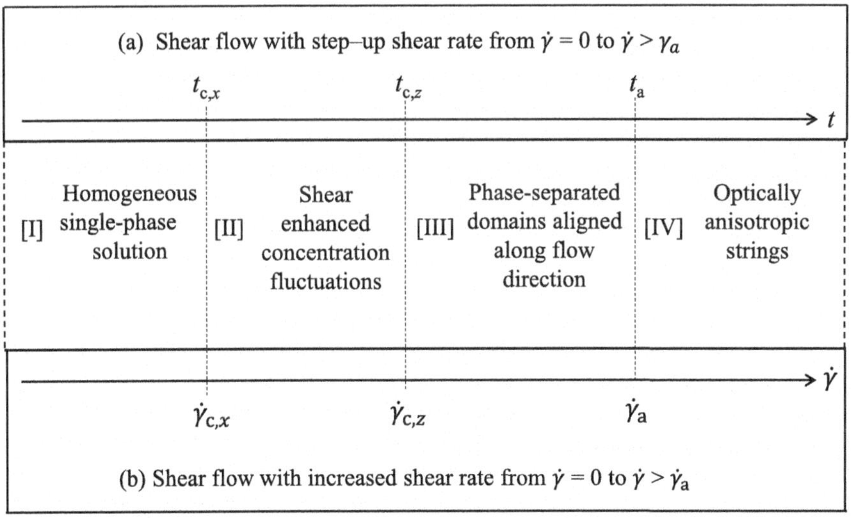

Fig. 5.12 Evolution of dissipative structures under shear flow with step-up shear rate from $\dot{\gamma} = 0$ to $\dot{\gamma} > \dot{\gamma}_a$ (part **a**) and under shear flow with increased shear rate protocol from $\dot{\gamma} = 0$ to $\dot{\gamma} > \dot{\gamma}_a$ (part **b**). The two processes **a** and **b** involve the common evolution pathway classified by Regime I to Regime IV with the characteristic times, $t_{c,x}$, $t_{c,z}$ and t_a, and the corresponding characteristic shear rates $\dot{\gamma}_{c,x}$, $\dot{\gamma}_{c,z}$, and $\dot{\gamma}_a$, respectively

Figure 5.13 summarizes time evolutions of the SALS patterns (schematic patterns), the OM images (the contrast-enhanced real images), and N_1 after the step-up $\dot{\gamma}$ from zero to a constant value larger than the $\dot{\gamma}_a$. The series of the events induced by the step-up shear flow, which lead to the dissipative structures, is summarized as follows [20, 38]:

(1) the incubation period $t < t_{c,x}$, (part a), corresponding to the period of the first stress overshoot, where the entangled polymer chains in the solutions are uniformly deformed without involving the solvent-squeeze process;
(2) the early stage of the structure formation in the time span of $t_{c,x} < t < t_{c,z}$ (part b), corresponding to the period between the first and second stress overshoot

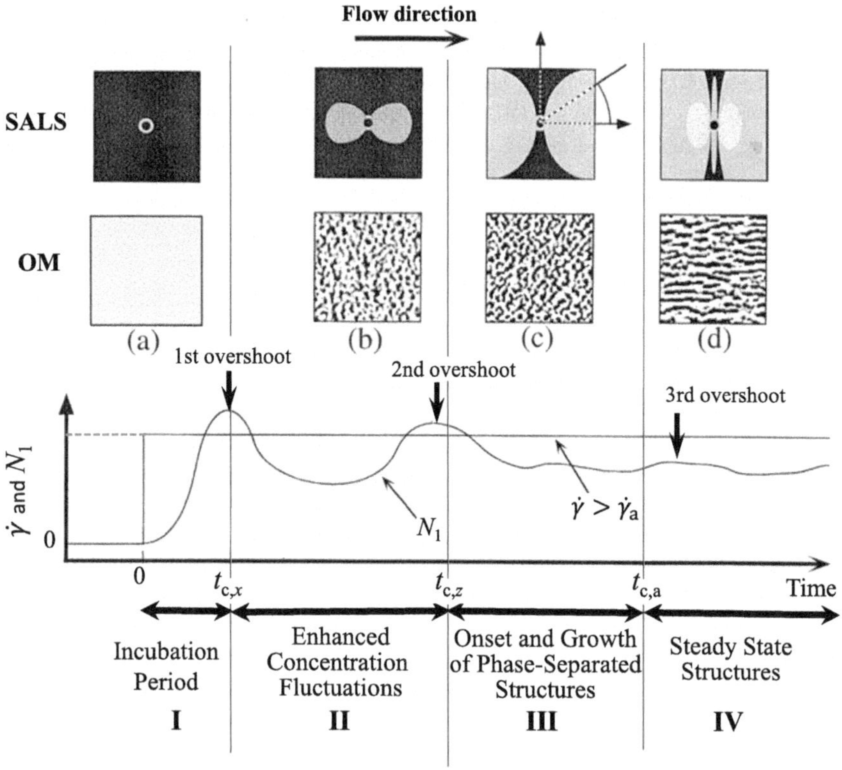

Fig. 5.13 Representation of the time evolutions of the SALS (schematics); butterfly patterns *b* and *c*, and the superposed butterfly pattern and streak pattern *d*; the corresponding measured contrast-enhanced OM (real-space) images, and normal stress N_1 after the onset of the flow from $\dot{\gamma} = 0\ s^{-1}$ to a constant shear rate $\dot{\gamma}$ larger than $\dot{\gamma}_a$. The time evolutions of the scattering patterns and the dissipative structures involve the series of the processes which are composed of an incubation period (Regime I), the enhanced concentration fluctuations (Regime II), onset of phase separation and growth of phase-separated structures (Regime III), and the formation of the string-like structure as the steady-state structure (Regime IV). The critical time $t_{c,a}$ in this figure is identical to the critical time t_a in Figs. 5.6 and 5.12

where the shear-induced concentration fluctuations with the characteristic length of ~ 10 μm are observed to develop along q_x in the q_x-q_z plane due to the solvent-squeeze process;

(3) the late stage of the structure formation at $t > t_{c,z}$ composed of the following two stages:

(3-1) part (c), corresponding to the period between the time $t_{c,z}$, in which the second stress overshoot starts relaxing toward the average steady-state stress level, and the time $t_{c,a}$, in which the third stress overshoot starts to uprise, i.e., the time span where the phase-separated domains start being developed, the phase-separated domains grow and overgrown domains burst, and the domains evolved in dynamically equilibrium are aligned align along the flow direction;

(3-2) part (d), corresponding to the time span at $t > t_a$ where the optically anisotropic string-like structures are eventually formed as the transient steady-state dissipative structures. This stage (3-2) is not covered by the experimental results shown in Figs. 5.8, 5.9, 5.10, and 5.11 which are attained under the step-up shear protocol with the limited time scale up to 100 s. However, this stage is covered by the results shown in Figs. 5.4, 5.5, and 5.6, which are attained under the increased shear-rate protocol.

5.5 Dissipative Structures Observed over the Wide Range of Length Scale

A combined SALS and SANS experiment was conducted under the steady-state shear flow with the increased shear protocol in order to explore the flow-induced dissipative structures developed over the wide range of the length scale of about 3 orders of magnitude (~ 10 μm to 3 nm) [41]. For this purpose, the experiments were conducted with the semidilute solution of the 8.0 wt.% ($c/c^* = 6.4$) deuterated PS ($M_W = 2.0 \times 10^6$) solution in DOP, designated as dPS200/DOP 8.0 wt.%.

The SANS experiments were performed with the instrument D11 of the Institute of Laue-Langevin, Grenoble, France. Figure 3.14 shows the schematic diagram of the sample cell and the geometry of the experimental setup [14, 15, 41]. The sample cell uses a Couette cell made of quartz. The temperature of the solution is controlled by the circulation of the temperature-controlled water in the compartments placed inside the inner cylinder on either side of the beam path. The incident neutron beam passes through the sample along Oy-axis parallel to the velocity gradient direction, as in the case of the SALS using the cone-plate cell. The sample-to-detector distance of $L = 2.5$, 10, and 35.7 m with the wavelength of 1 nm was employed to cover the q range $6.76 \times 10^{-3} \leq q\left(\text{nm}^{-1}\right) \leq 7.75 \times 10^{-1}$: The q range covered by the SANS is larger than that covered by the SALS ($6 \times 10^{-4} \leq q\left(\text{nm}^{-1}\right) \leq 4 \times 10^{-3}$). The 2D SANS intensity data $I(q_x, q_z)$, the scattering cross section per unit volume irradiated by the incident beam, were obtained in the absolute unit (cm^{-1}), and sector-averaged over an apex angle of 30° and 15° along the direction parallel and perpendicular to the flow direction to obtain $I(q_x)$ and $I(q_z)$, respectively (see ref. [48] in the Reference and Notes as for the scattering cross section).

Figure 5.14 presents the shear-rate dependence of the SALS patterns (top row) and the SANS patterns at $L = 35.7$ m (2nd row), 10 m (3rd row), and 2.5 m (bottom row) at the steady-state shear flow for the dPS200/DOP 8.0 wt.% at 22 °C [41]. The patterns in each column are taken at the common shear rate: The shear rate is increased from 0 s^{-1} (part a) to 2.6 s^{-1} (part f) from the left column to the right column of the figure.

In the quiescent state ($\dot{\gamma} = 0\,\text{s}^{-1}$, part a), both the SALS pattern and the SANS patterns at different L values are isotropic and have the weak scattered intensity. At $\dot{\gamma} = 0.1\,\text{s}^{-1}$ (part b), there is no change in the shape and intensity of the SALS pattern and the SANS patterns at the different L values: Therefore, the shear flow does not affect the concentration fluctuations over the wide q-range covered in this experiment (from 6×10^{-4} to 7.75×10^{-1} nm^{-1}). Thus, $\dot{\gamma} = 0.1\,\text{s}^{-1}$ belongs to the flow of Regime I (the shear-rate Regime $[\dot{\gamma}_1]$) at all the length scales covered by the SALS and the SANS.

At $\dot{\gamma} = 0.2\,\text{s}^{-1}$ (part c), only the SANS pattern at $L = 35.7$ m (the second row and part c, covering the q-range of $6.76 \times 10^{-3} \leq q\,(\text{nm}^{-1}) \leq 6.00 \times 10^{-2}$),

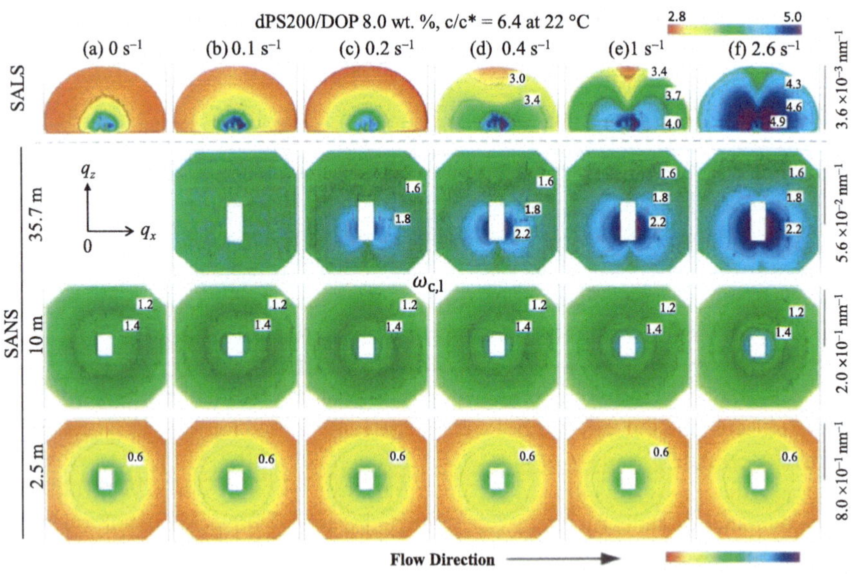

Fig. 5.14 Shear-rate dependence of SALS patterns (top row) and SANS patterns at sample-to-detector distance $D = 35.7$ m (2nd row), 10 m (3rd row), and 2.5 m (bottom row) at steady-state shear flow for dPS200/DOP 8.0 wt.% with $T_{cl}(0) \leq 2$ °C at 22 °C. The patterns in each column are taken at a common shear rate: The shear rate is increased from 0 (part a) to 2.6 s^{-1} (part f). SANS experiments are performed at the D11 instrument of ILL, Grenoble, France, with $\lambda = 1$ nm. Based on [41], reprinted with permission, American Chemical Society

exhibits the butterfly pattern with the appreciable intensity excess to that at $\dot{\gamma} = 0$ or $0.1\,\mathrm{s}^{-1}$. The pattern is anisotropic with the strong intensity along q_x but the weak intensity along q_z, as weak as the intensity at $\dot{\gamma} = 0$ or $0.1\,\mathrm{s}^{-1}$. Thus, $\dot{\gamma} = 0.2\,\mathrm{s}^{-1}$ belongs to the flow in Regime II (the shear-rate Regime $[\dot{\gamma}_{\mathrm{II}}]$ (see Fig. 1.8d)). This indicates the large amplitude of the concentration fluctuations is developed only along the q_x-direction and only in the q-range covered by SANS at $L = 35.7$ m in the solutions under the flow at the any given flow rates $\dot{\gamma}'$s in Regime II and at the given concentrations. In the larger q-range covered by the SANS at $L = 10$ and 2.5 m ($2.5 \times 10^{-2} \leq q\,(\mathrm{nm}^{-1}) \leq 8.0 \times 10^{-1}$) in the solutions under the flow at the any given flow rates $\dot{\gamma}'$s in Regime II and at the given concentrations, the scattering pattern is almost same as that in the quiescent solution, indicating that the structures at the small length scale corresponding to the q-range covered by the SANS at $L = 10$ and 2.5 m are not affected much by the flow (the 3rd and 4th row in Fig. 5.14). It is important to note that the SALS pattern at $\dot{\gamma} = 0.2\,\mathrm{s}^{-1}$ (shown in part c in the 1st row in Fig. 5.14) does not exhibit the butterfly pattern. This is due to the fact that the q_x-value at the maximum intensity of the butterfly pattern along q_x, defined as $q_{x,\mathrm{m}}$, exists in between $q_{\mathrm{u},SALS}$ and $q_{\mathrm{l,SANS}}$, where $q_{\mathrm{u},SALS}$ and $q_{\mathrm{l,SANS}}$ are the upper and lower limit of q accessible by the SALS apparatus and the SANS apparatus at $L = 35.7$ m, respectively, as evidenced by the result shown below in Fig. 5.15.

At $\dot{\gamma} = 0.4\,\mathrm{s}^{-1}$ (part d), the SALS pattern also begins to exhibit the butterfly pattern from the solution under the flow (as presented in part d in the 1st row in Fig. 5.14). This experimental evidence clarifies and further validates also the shift of the magnitude of the scattering vector $q_{x,\mathrm{m}}$ at the maximum scattering intensity along q_x direction (FD, Ox direction) from $q_{x,\mathrm{m}}$ for the butterfly-SANS pattern defined as, $q_{x,\mathrm{m,SANS}}$, to that for the SALS-butterfly patterns defined as, $q_{x,\mathrm{m,SALS}}$. This shift described above satisfies $q_{x,\mathrm{m,SANS}} > q_{x,\mathrm{m,SALS}}$. This relation reveals also that $\Lambda_{\mathrm{SANS}} < \Lambda_{\mathrm{SALS}}$, where Λ_{SALS} and Λ_{SANS} are the characteristic wavelength of the Fourier modes of the concentration fluctuations corresponding to the maximum intensity of the SALS and SANS butterfly pattern, respectively. Nevertheless, the scattered intensity along q_z-direction (ND) from the SALS pattern (part d in the first row) and the SANS pattern taken at $L = 35.7$ m (part d) in the second row) are as weak as the intensity from the thermal concentration fluctuations along q_z-direction in the single-phase solution, although the intensity from the single-phase solution at $\dot{\gamma} = 0.4\,\mathrm{s}^{-1}$ is higher than that at $\dot{\gamma} = 0.2\,\mathrm{s}^{-1}$, due to the shear-rate dependence of the critical temperature. Thus, the solution at $\dot{\gamma} = 0.4\,\mathrm{s}^{-1}$ still belongs to that in Regime II.

The $q_{x,\mathrm{m}}$ of the SALS butterfly pattern shifts toward the smaller q with increased $\dot{\gamma}'$s relative to $q_{x,\mathrm{m}}$ of the SANS butterfly pattern at $\dot{\gamma} = 0.2\,\mathrm{s}^{-1}$ (as presented in the part c in the second row in Fig. 5.14), thereby revealing the concentration fluctuations covered by the SALS butterfly patterns extend over the larger length scale than those covered by the SANS. That is, the combined SALS and SANS under the flow at the any given shear rates $\dot{\gamma}'$s in Regime II covered the concentration fluctuations in Regime II over the wide range of the length scale from ~ 10 μm to 8 nm, as indicated later by the scale of q_x covered in Fig. 5.15.

At the higher shear rates of $\dot{\gamma} = 1$ and $2.6\,\mathrm{s}^{-1}$ (the part e and f in Fig. 5.14, respectively), both the SALS butterfly pattern and the SANS butterfly pattern at $L = 35.7$ m evolve in such a way that the scattered intensity along the flow direction further increases and the butterfly wings expand toward the larger azimuthal angle μ, resulting in the narrowed dark sector perpendicular to the flow direction, as presented in the 1st and 2nd row in Fig. 5.14. It is also noted that the scattered intensity along q_z becomes higher than that in the quiescent solution. Thus, the $\dot{\gamma} = 1$ and $2.6\,\mathrm{s}^{-1}$ belong to the flow of Regime III (the shear-rate Regime $[\dot{\gamma}_{\mathrm{II}}]$). On the contrary, the SANS patterns at $L = 10$ and 2.5 m (at any given $\dot{\gamma}$'s of $\dot{\gamma} = 1$ and $2.6\ \mathrm{s}^{-1}$) remain equal to those at $\dot{\gamma} = 0$: The concentration fluctuations in this q range $(2.5 \times 10^{-2} \leq q\,(\mathrm{nm}^{-1}) \leq 8.0 \times 10^{-1})$ are not affected by those $\dot{\gamma}$ values.

Figure 5.15 presents the combined SALS and SANS scattering profiles parallel to the flow $I(q_x)$ (part a) and perpendicular to the flow $I(q_z)$ (part b) obtained at the various $\dot{\gamma}$ values at the given temperature $T = 22\,°\mathrm{C}$ where the dPS200/DOP 8.0 wt.% solution in quiescent state is in the single-phase state above the cloud point which is lower than 2 °C [41]. In part (a) and (b) in Fig. 5.15, the SANS intensity profiles are shown over the q range marked by the arrow-labeled SANS in the top of the figure: The SANS intensity assessed by the absolute intensity unit should be referred to the ordinate scale in the right-hand side of each figure. The SALS intensity profiles are shown over the q range marked by the arrow-labeled SALS in the top of the figure: The intensity is displayed in the ordinate scale of left-hand side of each figure. Although the measured intensity of the SALS profiles is in arbitrary unit, it is converted to the absolute intensity unit of the SANS profiles (cm^{-1}) by the vertical shift of the given set of the SALS profiles at the different shear rates in such a way that they are smoothly interconnected to the corresponding set of the SANS profiles,

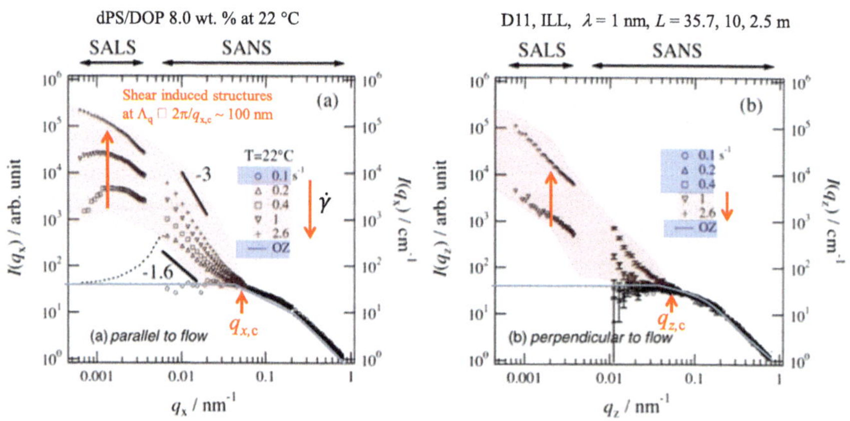

Fig. 5.15 SALS intensity (referred to the ordinate scale on the left-hand side) and SANS intensity (referred to the ordinate scale on the right-hand side) profiles parallel, part (**a**), and perpendicular, part (**b**), to the flow direction, measured at the various shear rates and at the given temperature 22 °C. The SALS intensity profiles were equally multiplied by the shift factor in order to bring their intensity in arbitrary unit into the absolute intensity unit of the SANS profiles (cm^{-1})

as shown by the ordinate scale in the right-hand side of each figure. The combined SALS and SANS experiment covers over a wide range of wave number (q_x, q_z) scale of $\sim$ 3 orders of magnitude, corresponding to the length scale from 6 nm to 6 μm, and a large dynamic range of the intensity of $\sim$ 6 orders of magnitude.

It is interesting to note that the combined SALS and SANS scattering profile for $I(q_x)$ and $I(q_z)$, presented respectively in Figs. 5.15(a) and (b), are generally comprised of the following three scattering components:

(i) the scattering from the flow-induced dissipative structures in the large length scale which appears in the SALS q-range,

(ii) the scattering from the dissipative structures in the local length scale, which appears in the SANS q range at $q_{x,1} < q < q_{x,c}$ and at $q_{z,1} < q < q_{z,c}$ where $q_{x,1}$ and $q_{z,1}$ are the lower limit of q_x and q_z covered by SANS, respectively. The scattering functions $I(q_x)$ and $I(q_z)$ from the local structures are approximately given by the power-law given, respectively, as follows,

$$I(q_x) \sim q_x^{-p_x}, \; I(q_z) \sim q_z^{-p_z}, \tag{5.5.1}$$

(iii) the OZ scattering function from thermal concentration fluctuations of the entangled polymer solutions under the flow which appears in the SANS q range at $q > q_{x,c}$ and $q > q_{z,c}$ where $q_{x,c}$ and $q_{z,c}$ are the critical q's above which the OZ functions along the q_x-axis and q_z-axis, respectively, become independent of $\dot{\gamma}'s$. The critical value of $q_{x,c}$ and $q_{z,x}$ are highlighted by the red arrow in part (a) and (b), respectively.

The OZ functions at $q < q_{x,c}$ and $q < q_{z,c}$ are independent of the flow-rates at any $\dot{\gamma}'s$ as shown by the gray horizontal straight line and the symbols. That is, all the SANS scattering function taken at the different shear rates are merged into the single OZ scattering function in this q range. The merged scattering function is identical to the OZ function in the quiescent solutions as presented by the gray OZ scattering curve at zero shear rate, implying that the imposed shear flow does not utterly affect the thermal concentration fluctuations of the solution at $q > q_{x,c}$ and $q > q_{z,c}$ under the flow at any shear-rates $\dot{\gamma}'s$ in Regime I to III. This is because the relaxation rates of the Fourier modes of the thermal concentration fluctuations Γ_{conc} at $q_x > q_{x,c}$ and at $q_z > q_{z,c}$ satisfy,

$$\Gamma_{\text{conc}}\left(q_x > q_{x,c}\right) > \dot{\gamma}, \; \Gamma_{\text{conc}}\left(q_z > q_{z,c}\right) > \dot{\gamma}, \tag{5.5.2}$$

in any shear-rates range of the $\dot{\gamma}'s$ covered in this experiment, so that the flow does not affect the fluctuations at $q_x > q_{x,c}$ and at $q_z > q_{z,c}$, giving rise to $I(q_x) = I(q_z)$ and $q_{x,c} = q_{z,c} = 6.0 \times 10^{-2}\,\text{nm}^{-1}$, both being universal with $\dot{\gamma}'s$.

The exponent p_x for $I(q_x)$ in the power-law region (Eq. 5.5.1 and Fig. 5.15a), which appears in the SANS q region at $q_{x,1} < q < q_{x,c}$, tends to increase from $\sim$ 1.6 to $\sim$ 3 with the increased $\dot{\gamma}$ from 0.2 to 2.6 s^{-1}, whereas the corresponding exponent p_z for $I(q_z)$ is constant independent of $\dot{\gamma}'s$ equal to $\sim$ 4 as presented in Fig. 5.15(b). The fact that $p_z \sim 4$ reveals that the dissipative structures evolved at these shear rates ($\dot{\gamma} = 1.0$ and $2.6\,\mathrm{s}^{-1}$) have the sharp interface against the matrix solution in the direction perpendicular to the flow direction. This point, regarding the exponent p_z is equal to $\sim$4, shall be discussed later in Sect. 5.11 in conjunction with Fig. 5.36. Thus, these shear rates belong to the Regime III where the phase-separated domains are formed and tend to be aligned along the flow direction.

The scattering intensity profile $I(q_x)$ excess to the OZ scattering is observed at $\dot{\gamma}$'s $> 0.1\,\mathrm{s}^{-1}$, whereas $I(q_z)$ excess to the OZ is observed at $\dot{\gamma}$'s $> 1\,\mathrm{s}^{-1}$, as highlighted with the pink-color background in the figure and figure legend. Both intensity of $I(q_x)$ and that of $I(q_z)$ increase with the increased $\dot{\gamma}$'s, as highlighted by the direction of the red arrows. More specifically, $I(q_x)$ excess to the OZ scattering is observed at $\dot{\gamma}$'s $\geq 0.4\,\mathrm{s}^{-1}$ for the SALS intensity profiles and $\dot{\gamma}$'s $\geq 0.2\,\mathrm{s}^{-1}$ for the SANS intensity profiles (part a). This means that the butterfly pattern is observed only by the SANS intensity profiles in the shear-rate range at $0.2\,\mathrm{s}^{-1} < \dot{\gamma} < 0.4\,\mathrm{s}^{-1}$; in this shear-rate range, the SALS intensity profiles still tends to look at the system as the homogeneous solutions. This reveals that $I(q_x)$ at $\dot{\gamma} = 0.2\,\mathrm{s}^{-1}$ has the maximum intensity in the q-range between $(q_{x,c})_{\mathrm{SALS}}$ and $(q_{x,l})_{\mathrm{SANS}}$, as indicated by the expected profile drawn with the dotted line for the guide of eye. Here, $(q_{x,c})_{\mathrm{SALS}}$ and $(q_{x,l})_{\mathrm{SANS}}$ are the highest limit of q in the q-range covered by the SALS intensity profiles and the lowest limit of q in the q-range covered by the SANS intensity profiles, respectively. This scattering maximum increases with respect to the scattering intensity and shifts toward small q's with the further increase of the shear rates.

Growth and coarsening of the dissipative structures with the increased shear rates. The trend described above regarding the change of the scattering maximum with the shear rates as revealed by Fig. 5.15a elucidates the shear-induced spinodal decomposition (SD) and the coarsening of the dissipative structures via the SD under the flow at any shear rates $\dot{\gamma}$'s in Regime III. The flow-induced SD is driven by the flow-induced enhancement of the thermodynamic instability of the solution under the flow (Sect. 4.7). On the one hand, in Regime III increased shear rates from $\dot{\gamma}_{\mathrm{l}}$ to $\dot{\gamma}_{\mathrm{h}}$ involve the corresponding increase of the thermodynamic driving force for the growth of the phase-separated structure via the SD. This is because the increased shear rate in causes the increased quench depth at the given working temperature T, $\Delta T(\dot{\gamma}) = T_{\mathrm{c}}(\dot{\gamma}) - T$, due to the shift of the critical temperature $T_{\mathrm{c}}(\dot{\gamma})$ toward the higher temperature with the increased $\dot{\gamma}$. On the other hand, the increased shear rate causes the increased shear force for deformation and burst of the structures. Thus, the shear rate controls the steady-state dissipative structure attained by the dynamic balance between the growth of the structures and deformation and the burst of them, due to the two opposing forces.

One must also note that the assessment of $\dot{\gamma}_{\mathrm{c},x}$ may depend on the q range covered experimentally, as elucidated from the results shown in Figs. 5.14 and 5.15. However, at any shear rates in Regime III satisfying $\dot{\gamma}$'s $\geq 0.4\,\mathrm{s}^{-1}$, the excess scattering $I(q_x)$ is discerned over the SALS and SANS q-ranges as presented in Fig. 5.15, which reflects the shear-induced structure extending over the wide length scale in this regime. These are clearly indicated also in Fig. 5.14 which directly compared the 2D SALS and 2D SANS patterns over the range of shear rates at $\dot{\gamma}$'s $\geq 0.4\,\mathrm{s}^{-1}$. The OZ scattering shows the power-law behavior as given by Eq. 5.5.1 with the exponent p_x and p_z equal to 2 at the high q-limits of q_x and q_z larger than 0.3 nm^{-1} as obviously seen in Fig. 5.15 [41].

Figure 5.16 presents comparisons of $I(q_x)$ and $I(q_z)$ at the various shear rates: (a) $\dot{\gamma} = 0.2\,\mathrm{s}^{-1}$ and (b) 0.4 s^{-1} (both in Regime II), (c) 1.0 s^{-1} and (d) 2.6 s^{-1}(both in Regime III) for the dPS200/DOP 8.0 wt.% at 22 °C [41]. To compare $I(q_x)$ and $I(q_z)$ at the same $\dot{\gamma}$'s, the scattering profiles shown in Fig. 5.15 are replotted in Fig. 5.16. It is clearly shown that both $I(q_x)$ and $I(q_z)$ are identical at $q_x = \ \ q_z > 0.06\,\mathrm{nm}^{-1}$ (marked by the arrows) where both are described by the OZ scattering function which is identical at all shear rates in Regime II and III (as shown in the Fig. 5.16).

At all $\dot{\gamma}$'s in Regime II and III and at the any q's satisfying $q < 0.06\,\mathrm{nm}^{-1}$, $I(q_x)$ exceeds $I(q_z)$, indicating the evolution of the anisotropic structures that have the amplitude of the characteristic Fourier modes of the concentration fluctuations along the q_x direction is larger than those along q_z direction.

At $\dot{\gamma} = 0.2\,\mathrm{s}^{-1}$ (part a) and 0.4 s^{-1} (part b), only $I(q_x)$ exhibits the scattered intensity excess to the OZ scattered intensity level at the small q range: the OZ intensity level at the small q range is shown by the horizontal dotted line. This result reveals the evolution of the plane wave concentration fluctuations with the wave vectors essentially oriented along q_x direction.

When $\dot{\gamma}$ is increased, the increment of $I(q_z)$ with $\dot{\gamma}$ at a given q_z ($< q_{zc}$) from the OZ intensity level is larger than that of $I(q_x)$ with $\dot{\gamma}$, resulting in the gradual approach of $I(q_z)$ to $I(q_x)$, as clearly shown by the decreases of the intensity ratio of $I(q_x)$ to $I(q_z)$ in part (c) and (d) in Fig. 5.16. At $\dot{\gamma} = 1.0\,\mathrm{s}^{-1}$ in Regime III (part c), the scattering excess to the OZ function is observed for $I(q_z)$ at $q_z \leq 0.02\,\mathrm{nm}^{-1}$. This value of $q_z = 0.02\ \mathrm{nm}^{-1}$ is the critical value of q where the excess scattering $I(q_z)$ to the OZ function crosses the OZ intensity level with the increased $\dot{\gamma}$. This

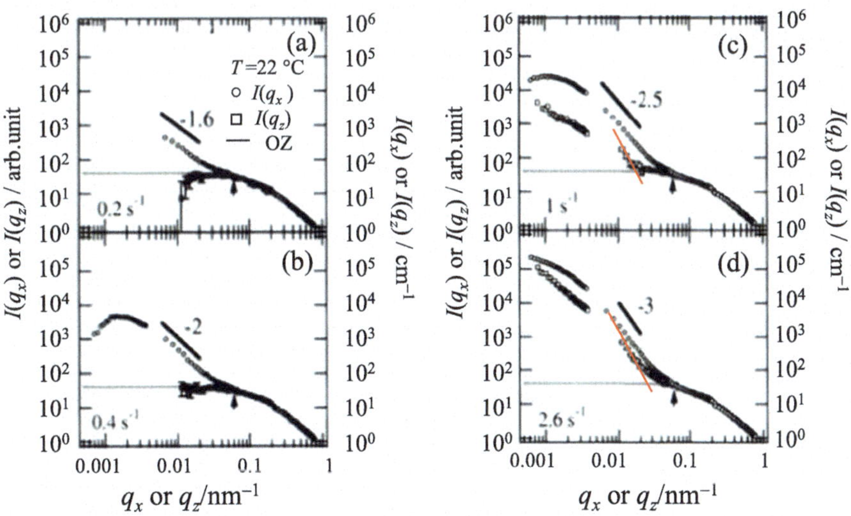

Fig. 5.16 Comparisons of $I(q_x)$ and $I(q_z)$ at various shear rates: **a** $\dot{\gamma} = 0.2\,\mathrm{s}^{-1}$ and **b** 0.4 s^{-1} in Regime II: **c** 1.0 s^{-1} and **d** 2.6 s^{-1}in Regime III for the dPS200/DOP 8.0 wt.% at 22 °C. Based on [41], reprinted with permission, American Chemical Society

critical value is much smaller than $q_{z,c} = q_{x,c} \leq 0.06\,\text{nm}^{-1}$ specified by the arrow. The excess scattering of $I(q_z)$ at the small q_z-range of $q \leq 0.06\,\text{nm}^{-1}$ is due to the phase separation which develops a sharp interface along q_z-axis, giving rise to

$$I(q_z) \sim q_z^{-4} \text{ (Porod law)}, \tag{5.5.3}$$

at the large q-range, as shown by the red straight line, where the excess scattering appears in SANS q-range. At $\dot{\gamma} = 2.6\,\text{s}^{-1}$in Regime III (part d), the scattering $I(q_z)$ which obeys the Porod law extends up to much higher q_z close to $q_{x,c} = 0.06\,\text{nm}^{-1}$ in contrast to the case at $\dot{\gamma} = 1.0\,\text{s}^{-1}$. The excess scattering of $I(q_z)$ at the small q_z range of $q_z \leq 0.02\,\text{nm}^{-1}$ exhibits the Porod law as shown by the red straight line.

5.6 Origin of Flow-Induced Concentration Fluctuations in Regime II

Figure 5.17 schematically represents the origin of the butterfly scattering pattern [9, 20, 42]. The thermal concentration fluctuations in the single-phase solution in quiescent state are isotropic as schematically illustrated by the real-space image shown in part (a) where the light blue regions and the white regions present polymer-rich and polymer-poor region, respectively. The fluctuations are small with respect to the amplitude and the thermal correlation length as illustrated schematically by the wavy pattern representing the spatial variation of the concentration fluctuations along Ox-axis (flow direction) and Oz-axis (neutral direction) as presented underneath and right-hand side of the image in part (a), respectively. Consequently, the corresponding OM image is featureless and isotropic, and the scattering intensity is weak and circularly symmetric around the incident beam axis (Oy-axis) as shown in part (b). The scattering can be predicted by the OZ scattering or more rigorously by the random phase approximation (RPA) [43].

The shear flow with the shear rate large than the critical shear rate $\dot{\gamma}_{c,x}$ but smaller than the critical shear rate $\dot{\gamma}_{c,z}$ enhances the concentration fluctuations along the flow direction (Ox direction) but not much along the neutral direction (Oz direction), as schematically shown in the real-space image of the fluctuations in part (c). The dark-blue and white regions in the image are polymer-rich and polymer-poor regions with enhanced and depressed polymer concentration in relative to the space-time wise average concentration of the polymers in the solutions, respectively. The spatial variation of the fluctuations along the Ox-axis has a large amplitude and correlation length, as shown by the schematic diagram underneath the image (c), whereas that along Oz-axis has a small amplitude and correlation length, as small as the concentration fluctuations of the quiescent solution, as shown by the schematic diagram in the right-hand side of the image (c). The enhanced fluctuations are anisotropic and are represented by a superposition of the Fourier modes of the concentration fluctuations, each mode of which is given by the plane wave concentration variation

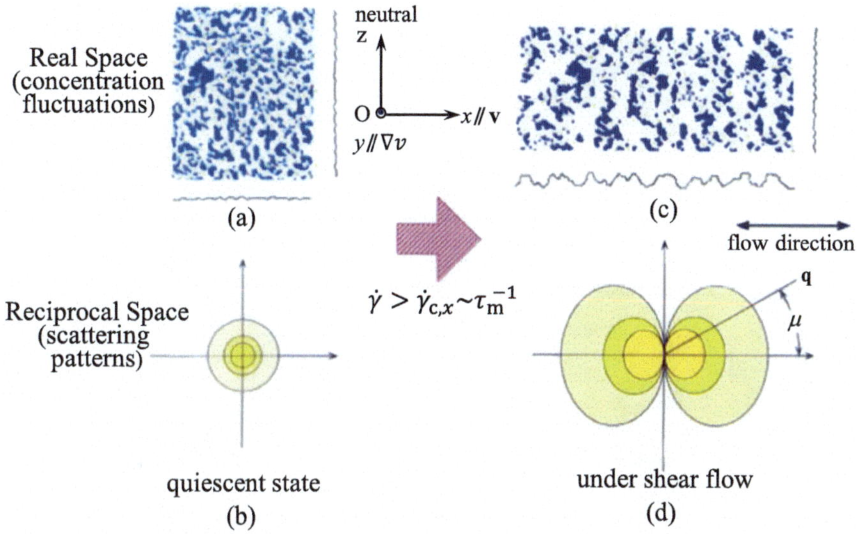

Fig. 5.17 Sketches of shear-enhanced concentration fluctuations (represented by the change from part **a** to **c**) and the evolution of the butterfly-type SALS pattern (represented by a change from part **b** to **d**). The patterns **a** to **d** are observed in the Oxz plane set normal to both the propagation direction of the incident beam and the velocity-gradient direction of the shear flow (Oy-axis). Spatial concentration fluctuations along Ox-axis and Oz-axis are schematically illustrated, respectively, underneath and right-hand side of real-space images **a** and **c**. From [42], permission with Oxford University Press

in space with the given wave vector and amplitude of the concentration variation (see Fig. 5.19 below), with their wave vectors preferentially oriented along the flow direction having a narrow orientation distribution around $\mu = 0$.

These shear-enhanced concentration fluctuations give rise to the strong scattered intensity along Ox-axis but not much along Oz-axis, giving rise to the butterfly-type anisotropic scattering pattern as shown in part (d) and the contrast variation in the transmission OM image as shown in part (c). Upon increasing $\dot{\gamma}$, the wave vectors of the plane wave fluctuations have a broader orientation distribution with respect to the Ox-axis, so that the butterfly wings are spread over a wider polar angle μ: The OM image undergoes corresponding changes [17]. However, the intensity along the Oz-axis is kept unchanged from that for the quiescent solution up to a certain critical shear rate $\dot{\gamma}_{c,z}$, resulting in the dark streak or sector along the Oz-axis in the butterfly pattern. The shear rate $\dot{\gamma}_{c,z}$ is the critical shear rate above which the scattering intensity emerges and starts to increase even along Oz-axis [35], as will be detailed later.

Figure 5.18 schematically represents a "**solvent-squeeze**" model for the shear-enhanced concentration fluctuations [20, 40, 42]. The single-phase solution in the quiescent state is subjected to thermal concentration fluctuations and thus has regions rich in polymers having more entanglements as shown in the gray regions in part

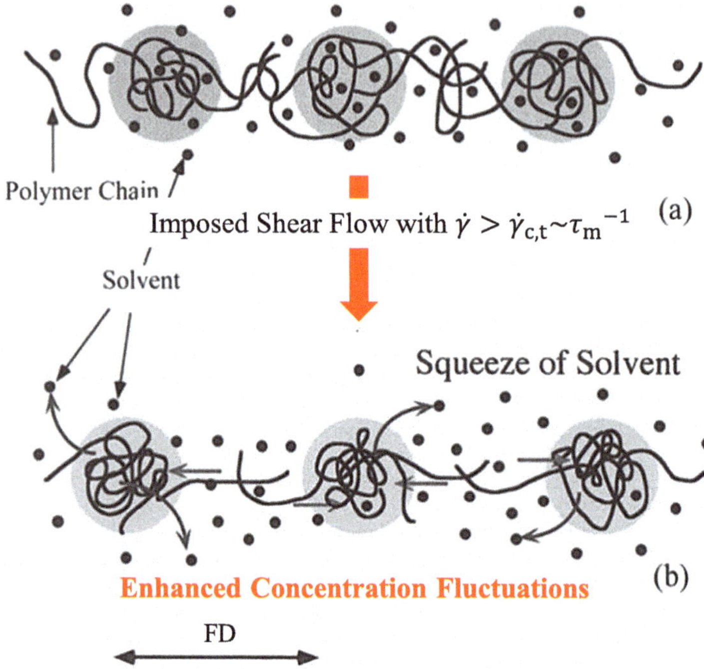

Fig. 5.18 Solvent-squeeze model for the shear-enhanced concentration fluctuations driven by the stress-diffusion coupling. FD: flow direction. Based on [40], reprinted with permission, American Chemical Society

(a) and regions poor in polymers having fewer entanglements (bright matrix). Under the imposed shear flow, the regions rich in polymers are accompanied by the larger stress than the other regions through the concentration dependence of viscosity η and the coefficient of first normal stress difference ψ_1, giving rise to the local stress variations. The excess stress born by the polymer-rich region, which is brought about by the thermal fluctuations, can be relaxed via the disentanglements, if $\dot{\gamma}$ is smaller than the disentanglement rate, $\Gamma_{\text{disent}} \sim 1/\tau_m$. In this case (defined as the weak shear flow in the shear regime $[\dot{\gamma}_{\text{I}}]$ (Table 5.2, Fig. 5.12 in Sect. 5.4), the concentration fluctuations under the flow can be also thermally relaxed, and the solution remains homogeneous even under the flow.

However, in the case where $\Gamma_{\text{disent}} < \dot{\gamma}$, the stress developed in the deformed swollen entangled networks cannot have the sufficient time to relax through the disentanglements. In this case, defined as the strong shear flow regime $[\dot{\gamma}_{\text{II}}]$, the entanglements exist as if they were permanent ones under the strong flow. Consequently, the stress or the deformation and orientation of the entangled polymers can be relaxed only by the squeezing solvent from the regions rich in polymers, having a higher elastic energy stored, into those poor in polymers against osmotic pressure as shown in part (b) in Fig. 5.18, resulting in regions rich in polymer becoming even richer and those poor in polymer becoming even poorer. Upon squeezing, the swollen deformed chains can have relaxed conformations, hence resulting in the dissipation of the elastic free energy and stress relaxation. Since deformations of the swollen entangled chains are anisotropic under the flow, the squeeze of the solvents is also anisotropic and occurs dominantly along the flow direction (FD), and thereby, the shear-enhanced concentration fluctuations are also anisotropic. The wave vectors **k** of Fourier modes of the shear-enhanced concentration fluctuations spread over a limited range of the polar angle μ around $\mu = 0°$ in the Oxz plane, as shown in Fig. 5.17d.

Figure 5.19 presents one of the Fourier modes of the thermal concentration fluctuations viewed in Oxz plane (part a) and in Oxy plane (part b) with the wave vector **k** oriented parallel to Ox-axis, the flow direction. The Fourier mode at a given time t is described by a plane wave concentration variation in space, $\delta\phi_{\mathbf{k}}(\mathbf{r}, t)$, at a given t with the wave vector $\mathbf{k} = k\widehat{\mathbf{X}}$ (k: the wave number, $\widehat{\mathbf{X}}$: the unit vector parallel to Ox-axis), the amplitude of the concentration variation $\eta_{\mathbf{k}}$ and wavelength $\Lambda_{\mathbf{k}} = 2\pi/k$ as shown in part (b). $\delta\phi_{\mathbf{k}}(\mathbf{r}, t)$ is given by

$$\delta\phi_{\mathbf{k}}(\mathbf{r}, t) = \eta_{\mathbf{k}}(t)\exp[\mathrm{i}(\mathbf{k}\cdot\mathbf{r})]. \tag{5.6.1}$$

The red and blue region in Fig. 5.19 are the region rich in polymer chains and the region poor in polymer chains, respectively. In this plane wave concentration variation, the concentration varies only along x-axis; the concentration along y-axis and z-axis are constant.

The elastic energy density (or the stress) stored by the deformed entangled chains in the solution under the externally applied flow field is large in the polymer-rich region. Under the strong-flow condition, the deformation of the entangled chains and therefore the stress can be relaxed by the squeeze of solvent molecules from the regions rich in polymer chains (light-red regions) to the regions poor in polymer

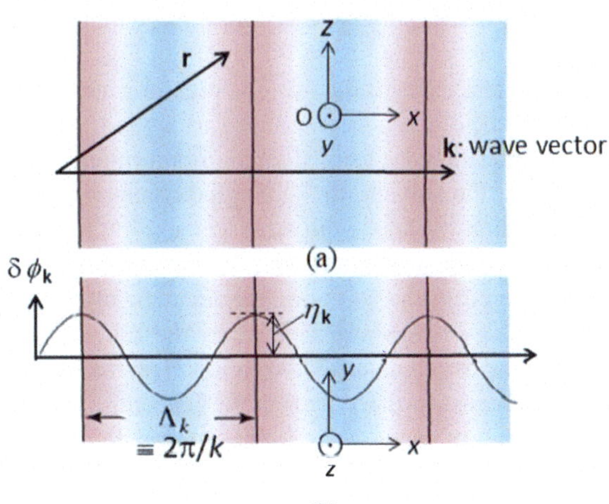

Fig. 5.19 One of the Fourier modes of the thermal concentration fluctuations with wave vector **k** oriented parallel to Ox-axis, flow direction, in Oxz plane (part **a**), and in Oxy plane (part **b**). The Oy-axis and Oz-axis are parallel to the velocity gradient direction and neutral direction, respectively

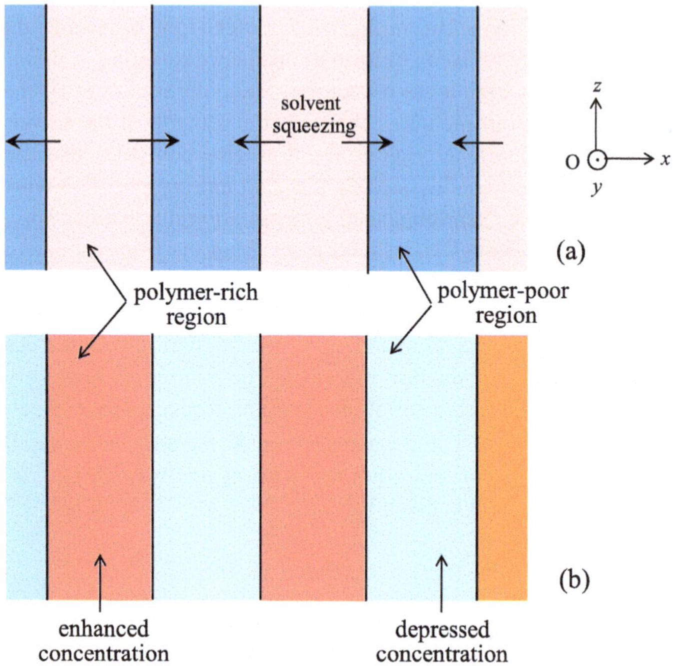

Fig. 5.20 Solvent squeeze from the polymer-rich region to the polymer-poor region along Ox-axis, flow direction, as illustrated by the arrows marked by "solvent squeezing" (part **a**) brings about enhanced (light-red to thick-red regions) and depressed concentration from blue to light-blue regions in the plane wave concentration fluctuation with wave vector **k** parallel to Ox-axis (part **b**)

chains (blue regions) as shown by the arrows marked by the solvent squeezing in Fig. 5.20a, so that the polymer-rich regions have the enhanced polymer concentration (with the higher concentration illustrated by thick red) and the polymer-poor regions has the depressed polymer concentration (with the lower concentration illustrated by light blue) as schematically illustrated in Fig. 5.20b. The squeeze occurs when the built-up stress exceeds the osmotic pressure. The squeeze brings about the increased amplitude $\eta_{\mathbf{k}}$ of the concentration variation of the Fourier mode.

After the first cycle of the chain-deformation/chain-relaxation process via the squeeze of the solvent, the entangled chains are brought in the field of the higher concentration variation with the larger amplitude $\eta_{\mathbf{k}}$. The chains are further subjected to the flow, causing the deformation of the entangled chains and the increased elastic energy stored in the solution, as schematically illustrated in Fig. 5.21. The stored elastic energy is again dissipated by the squeeze of the solvent molecules, which in turn relaxes the chain deformation. The dissipation of the stored energy via the squeeze process further enhances the concentration fluctuations of the solutions under the flow at any given $\dot{\gamma}$'s in Regime II, as illuminated by the increased amplitudes of $\eta_{\mathbf{k}}$ for the concentration variations of the Fourier modes of any wave vectors, $\mathbf{k}'$s, existing in the concentration fluctuations of the solutions. The relaxed entangled

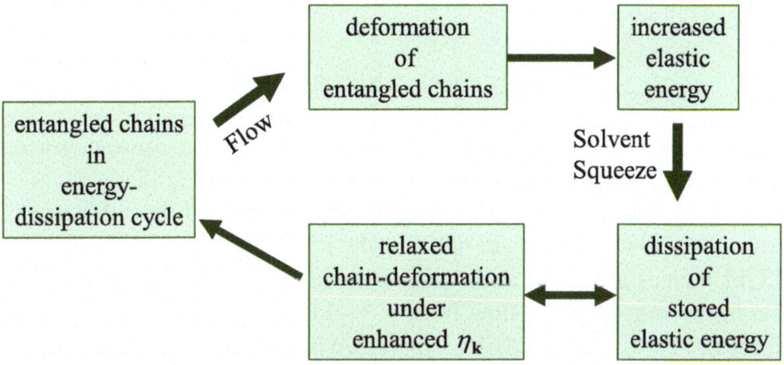

Fig. 5.21 Mechanical energy dissipation cycle in the repeated chain-deformation and chain-relaxation process via the solvent squeeze. In each cycle, the amplitude of $\eta_{\mathbf{k}}$ of the **k**th Fourier mode increases

chains are placed under the given concentration variation field with the amplitude $\eta_{\mathbf{k}}$ larger than that in the preceding cycle. The amplitude of the concentration variation $\eta_{\mathbf{k}}$ is kept being increased with the increased number of cycles via the repeated energy dissipation process with increased time in the step-up shear protocol or increased shear rate in the increased shear-rate protocol in Regime II (Fig. 5.12).

5.6.1 Analysis with Phase-Contrast Optical Microscopy

The flow-enhanced concentration fluctuations in Regime II, which gives rise to the butterfly SALS patterns with the characteristic size along the flow direction of a few μm can be analyzed in real space with phase-contrast optical microscopy (PCOM). In the PCOM, the local image contrast of the object under consideration at the given position vectorr is described by the difference between the brightness of the object at $\mathbf{r}$ and that of its surrounding medium, defined as $I(\mathbf{r})$ and I_0, respectively. The local image contrast linearly depends on the difference between the local refractive index of the object and refractive index of the medium, defined as $n(\mathbf{r})$ and n_0, respectively,

$$\{I(\mathbf{r}) - I_0\}/I_0 = 4\pi\{n_0 - n(\mathbf{r})\}d\, D^{1/2}/\lambda, \tag{5.6.2}$$

when a phase-contrast objective lens with dark (or positive) contrast is used [44]. d, D, and λ are the thickness of the object along the optical axis of the PCOM (Oy-axis, parallel to both the velocity gradient axis and thepropagation direction of the incident light beam), the absorption coefficient of the phase filter, and the wavelength of incident beam in the medium, respectively. According to Eq. 5.6.2, when the observed object has a higher refractive index than that of the medium, the object looks darker than the medium. The difference of the brightness reduced by I_0, $\Delta I(\mathbf{r}) \equiv |I(\mathbf{r}) - I_0|/I_0$, is proportional to the difference of the refractive index

$\delta n \equiv |n_0 - n(\mathbf{r})|$, i.e.,

$$\Delta I(\mathbf{r}) \sim \delta n(\mathbf{r}) \simeq (\partial n/\partial c)\ \Delta c(r), \tag{5.6.3}$$

where $(\partial n/\partial c)$ is the refractive-index increment which is a constant for a given solution at a given temperature.

Figure 5.22 presents the PCOM images (upper half) and their 2D fast Fourier transform (2DFFT) images (lower half) for the PS288/DOP 5.0 wt.% at 25 °C [18]. The PCOM images are taken at the time defined as t_{obs}, i.e., the time t_{ac} after the cessation of the step-up shear flow from $\dot{\gamma} = 0$ to the constant shear rates of $\dot{\gamma} = 27\ s^{-1}$(part a), $54\ s^{-1}$(part b) and $61\ s^{-1}$(part c), where t_{ac} is defined as $t_{ac} \equiv t_{obs} - t_{cess}$, and t_{cess} is the time span over which the given shear rate is imposed to the solution (see top-left corner of Fig. 5.22 for definitions of t_{cess}, t_{ac}, and t_{obs}). These shear rates imposed to the solution belong to those in Regime II. The set of (t_{cess}, t_{obs}) employed in part (a) to (c) is (2.3, 3.8), (1.1, 3.0), and (1.0, 3.3) in unit of second, respectively. First, t_{cess} is set in such a way that the PCOM images are taken for the solution

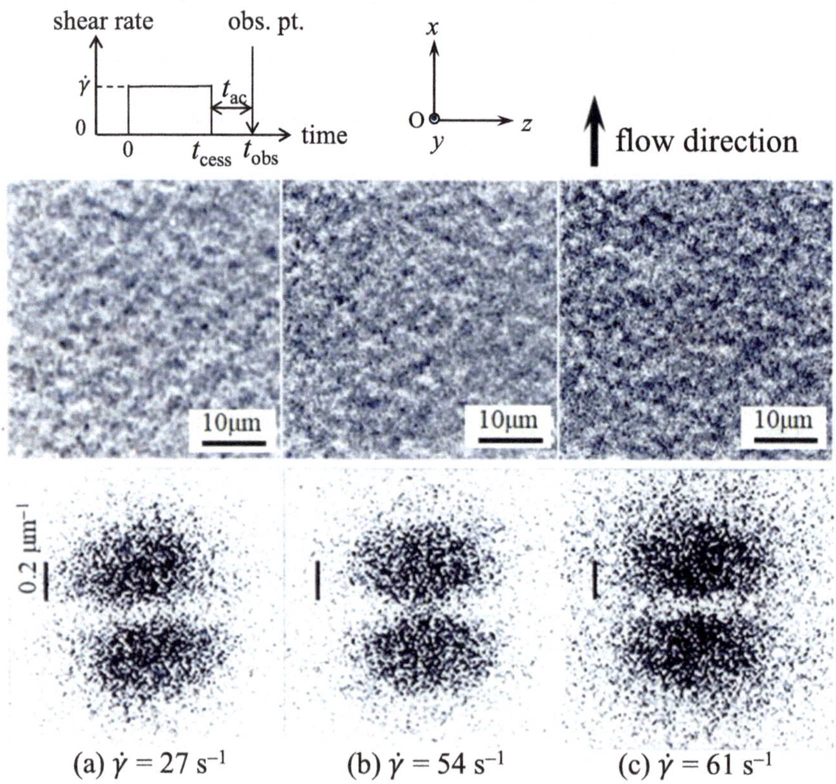

Fig. 5.22 PCOM images (upper half) and their 2D fast Fourier transform (2DFFT) images (lower half) for PS288/DOP 5.0 wt.% at 25 °C. Based on [18]

subjected to nearly equal total strain $\dot{\gamma} t_{cess} \cong 60.8$ under the protocol of the step-up shear/step-down shear between $\dot{\gamma} = 0$ and $27\ s^{-1}$ (part a), $\dot{\gamma} = 0$ and $54\ s^{-1}$ (part b), or $\dot{\gamma} = 0$ and $61\ s^{-1}$ (part c). Subsequently, t_{obs} is set at the time when the integrated scattered intensity along the flow direction (Ox-axis) after the cessation of shear, $\mathcal{J}_x(t_{ac}, \dot{\gamma})$, reaches the maximum intensity with respect to the time t_{ac} elapsed after the cessation, as shown in Fig. 5.23 [18]. The 2DFFT images are the frequency domain images corresponding to $F_a(q)$ in the second line of Eq. 5.6.5 to be described below, but they are not the power spectrum $P(\mathbf{q})$ given by Eq. 5.6.4.

Even after the cessation of the shear, the SALS intensity of the butterfly pattern is found to increase with t_{ac} to the maximum intensity at t_{obs} and then decreased gradually to the SALS intensity for the quiescent solution, as highlighted in Fig. 5.23 where the integrated intensity $\mathcal{J}_x(t_{ac}; \dot{\gamma})$ along q_x reduced with $\mathcal{J}_x(t_{ac} = 0; \dot{\gamma})$ is shown as a function of the time elapsed t_{ac} after the cessation for the three different shear rates [18]. The reduced integrated intensity linearly increases first with t_{ac}, followed by the gradual increase to the maximum intensity, and then decay gradually to the integrated intensity $\mathcal{J}_x(t = 0)$ inherent to the quiescent solution.

The initial rate of the intensity increase ($1.1\ s^{-1}$) is independent of $\dot{\gamma}$: but the maximum intensity attained with the increased t_{ac} depends on $\dot{\gamma}$, the larger the $\dot{\gamma}$ is, the larger is the maximum intensity. These results imply that the amount of the squeeze of solvent molecules after the cessation of the shear depends on $\dot{\gamma}$, although the same total strain $\dot{\gamma} t_{cess} \cong 60.8$ is set for the experiments with the three different shear rates. This may be reasonable from the view point of the work dissipated per unit volume of the systems imposed to shear flow during the period t_{cess} is $\eta\dot{\gamma}(\dot{\gamma} t_{cess}) \cong 60.8\eta\dot{\gamma} \propto \dot{\gamma}$. It may be predicted that the increased intensity after the cessation is due to the retarded solvent squeeze which causes the rapid stress relaxation and the rapid intensity uprise after the cessation due to increased concentration fluctuations: This process is against osmotic pressure of the solutions. The increased concentration fluctuations are then relaxed slowly to the concentration fluctuations inherent to the quiescent solution via the collective diffusion under the relaxed stress state after the rapid stress relaxation: This process is driven by osmotic pressure inherent to the solutions at rest.

To obtain the highest image contrast, the PCOM image are taken at the time t_{obs} or t_{ac} at which the intensity of the butterfly SALS pattern or more specifically the integrated intensity $\mathcal{J}_x(t_{ac}; \dot{\gamma})$ reaches maximum, and thereby, the highest image contrast is captured.

Based on the spatial variation of the brightness of the image $I(\mathbf{r})$ measured by the PCOM, we can obtain its power spectrum $P(\mathbf{q})$,

$$P(\mathbf{q}) = \int \langle I(\mathbf{r}_1)\, I(\mathbf{r}_2)\rangle_{\mathbf{r}} \exp(-\mathrm{i}\,\mathbf{q}\cdot\mathbf{r})d\mathbf{r}, \tag{5.6.4}$$

where $\langle XY\rangle_{\mathbf{r}}$ designates the average of XY overall space keeping the separation vector $\mathbf{r} = \mathbf{r}_2 - \mathbf{r}_1$ constant. $P(\mathbf{q})$ is related to the structure amplitude $|F_a(\mathbf{q})|$ given by the equation below,

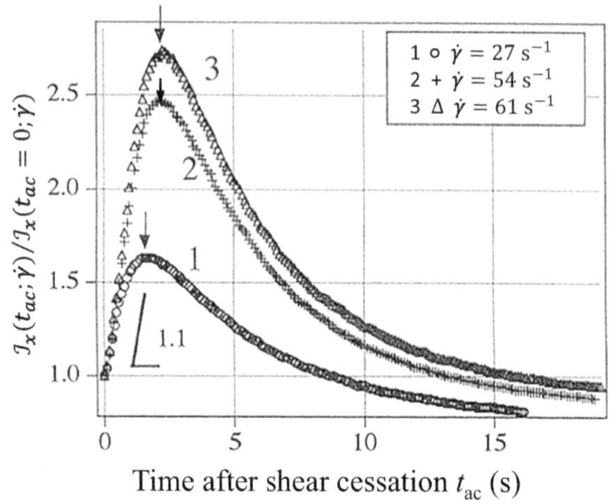

Fig. 5.23 Integrated SALS intensity $\mathcal{J}_x(t_{ac};\dot{\gamma})$ along q_x reduced with $\mathcal{J}_x(t_{ac}=0;\dot{\gamma})$ is shown as a function of time t_{ac} elapsed after cessation for the three different shear rates. Based on [18]

$$P(\mathbf{q}) = |F_a(\mathbf{q})|^2,\ F_a(\mathbf{q}) = \int I(\mathbf{r})\ \exp(-i\ \mathbf{q}\cdot\mathbf{r})d\mathbf{r}, \tag{5.6.5}$$

where $F_a(\mathbf{q})$ is Fourier transform of $I(\mathbf{r})$ associated with the FFT pattern presented in the lower half of Fig. 5.22. Using Eqs. 5.6.2 and 5.6.4, $P(\mathbf{q})$ is given by

$$P(\mathbf{q}) \sim \int \langle n(\mathbf{r}_1)n(\mathbf{r}_2)\rangle_{\mathbf{r}}\exp(-i\ \mathbf{q}\cdot\mathbf{r})d\mathbf{r} \tag{5.6.6}$$

$$\sim I_s(\mathbf{q}), \tag{5.6.7}$$

where $I_s(\mathbf{q})$ is the light scattering intensity distribution from the object. Thus, $P(\mathbf{q})$ assessed from the PCOM image reflects the light scattering ntensity distribution from the object.

5.6.2 Filtered PCOM Image Emphasizing Butterfly SALS Pattern

To capture the PCOM image which emphasizes the butterfly SALS pattern, we can reconstruct the filtered PCOM image using the raw PCOM image and its 2DFFT pattern (frequency domain pattern). The frequency domain pattern obtained by 2DFFT of the raw PCOM image is designated hereafter as $F_{OM,FFT}(\mathbf{q})$. The reconstruction of the filtered PCOM image can be conducted based on the free software of [NIH Image] [45] as follows.

The bottom half of Fig. 5.22 presents the frequency domain pattern, $F_{\mathrm{OM,FFT}}(\mathbf{q})$, calculated by the 2DFFT of the raw PCOM images shown in the upper half of the figure [18]. The results indicate that the frequency domain pattern well reproduce the butterfly scattering pattern measured by the SALS experiments under the shear flow: The brightness variation of the PCOM image well reproduces the shear-enhanced concentration fluctuations in the solution in Regime II.

5.6.3 Image Processing of the PCOM Image

The PCOM image reflects the superposition of all the Fourier modes of the concentration fluctuations which give rise to the butterfly pattern. We try reconstructing the PCOM image which reflects the low-q-Fourier modes, reflecting the butterfly SALS pattern, but which suppresses high q-Fourier modes; the high q modes are superposed on the low-q modes as the high frequency noises. For this purpose, we first smooth the measured 2DFFT (frequency domain) patterns, $F_{\mathrm{OM,FFT}}(\mathbf{q})$, shown in the bottom half of Fig. 5.22 with respect to the speckled noise and then obtain the maximum intensity of the patterns, $I_{\max}$. Then, we define the domain D in the low-q region which is enclosed by the contour line of iso-intensity of $0.99 \times I_{\max}$. We define the frequency-filtered function $F_{\mathrm{OM,filter}}(\mathbf{q})$ such that

$$F_{\mathrm{OM,filter}}(\mathbf{q}) = \begin{cases} 1.0, & \text{if } \mathbf{q} \in D \\ 0.3, & \text{if } \mathbf{q} \notin D. \end{cases} \tag{5.6.8}$$

The filtered frequency domain pattern, designated as $F^{\mathrm{filter}}_{\mathrm{OM,FFT}}(\mathbf{q})$, is given by

$$F^{\mathrm{filter}}_{\mathrm{OM,FFT}}(\mathbf{q}) = F_{\mathrm{OM,filter}}(\mathbf{q})\, F_{\mathrm{OM,FFT}}(\mathbf{q}). \tag{5.6.9}$$

The filtered real-space image is reconstructed from $F^{\mathrm{filter}}_{\mathrm{OM,FFT}}(\mathbf{q})$ through the inverse 2DFFT of $F^{\mathrm{filter}}_{\mathrm{OM,FFT}}(\mathbf{q})$.

Figure 5.24 presents a close-up image of the original PCOM image, shown in the upper part of Fig. 5.22b, obtained at $\dot{\gamma} = 54\,\mathrm{s}^{-1}$ (part a) (the frequency domain pattern), $F^{\mathrm{filter}}_{\mathrm{OM,FFT}}(\mathbf{q})$ obtained from $F_{\mathrm{OM,FFT}}(\mathbf{q})$ and $F_{\mathrm{OM,filter}}(\mathbf{q})$ (part b), the reconstructed PCOM image obtained from $F^{\mathrm{filter}}_{\mathrm{OM,FFT}}(\mathbf{q})$ through its inverse 2DFFT (part c), and the contour map of the reconstructed PCOM image showing various iso-intensity levels of the reconstructed image (c) (part d) [18]. The number attached to the contour lines indicates the brightness level of the image such that **the darkness increases with the increased number**: Number 1 represents the average brightness of the image.

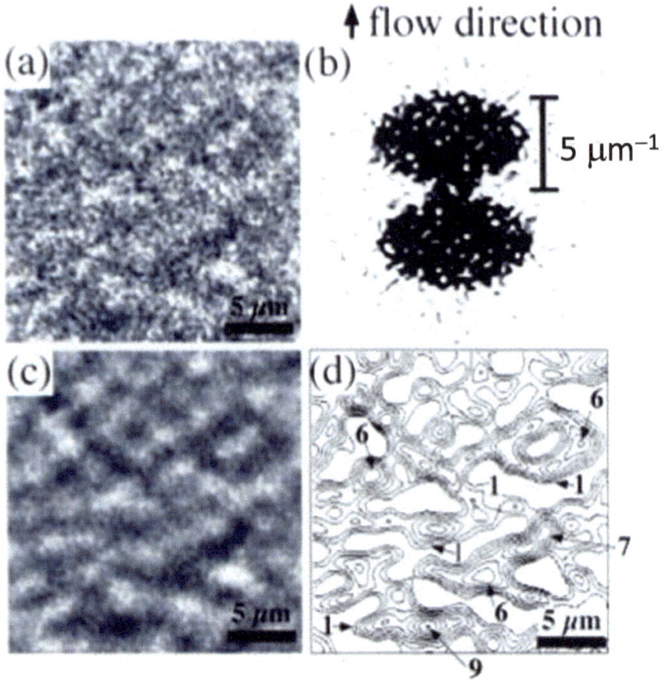

Fig. 5.24 A close-up of the original PCOM image, shown in upper part of Fig. 5.22b, obtained at $\dot{\gamma} = 54\,\mathrm{s}^{-1}$ (part **a**), $F^{\mathrm{filter}}_{\mathrm{OM,FFT}}(\mathbf{q})$ obtained from $F_{\mathrm{OM,FT}}(\mathbf{q})$ and $F_{\mathrm{filter}}(\mathbf{q})$ (part **b**), the reconstructed PCOM image obtained from the inverse 2DFFT of $F^{\mathrm{filter}}_{\mathrm{OM,FFT}}(\mathbf{q})$ (part **c**), and the contour map of reconstructed PCOM image showing various iso-intensity levels of reconstructed image (**c**) (part **d**). The number attached to the contour lines in part **d** indicates the brightness level of the image. Based on [18]

This experiments use a dark contrast lens for the objective lens. Thus, in the given image, the dark part has a higher refractive index than the bright part. The PS/DOP solution has $(\partial n/\partial c) = 0.119$ ml/g at 25 °C [46], and the darker regions correspond to the PS-rich regions. The bright region in the image shown in part (d) has the brightness higher than the average brightness (specified by the contour line with number 1), reflecting the PS-poor phase, whereas the region with contour lines with number 1-9 corresponds to the PS-rich phase. The dark and bright regions tend to have a large extension along the neutral direction perpendicular to the flow direction and align periodically along the flow direction with the average spacing $\Lambda_{\mathrm{m}} \sim 3\,\mu\mathrm{m}$. The results reveal a superposition of the plane wave concentration variation with the wavelength ~ 3 μm and with their wave vectors preferentially oriented along the flow direction.

5.7 Flow-Induced Shift of Spinodal Temperature in Regime II

5.7.1 Theoretical Background for the Flow-Induced Spinodal Decomposition

We discuss here the flow-induced spinodal decomposition. The theory for the shear-induced concentration fluctuations in the semidilute solutions under the flow is developed by Helfand, Fredrickson, Milner, and Onuki (HFMO) based on the two fluid model [32–34]. According to the theory, the dynamics of the concentration fluctuations $\delta\phi(\mathbf{r};t)$ is described by the following time-dependent Ginzburg-Landau equation which incorporates the coupling between the stress and the diffusion for the dynamic asymmetric system:

$$\left(\frac{\partial}{\partial t}+\mathbf{v}\cdot\nabla\right)\delta\phi(\mathbf{r};t)=\nabla\cdot L_{\mathrm{eff}}\left(\nabla\frac{\delta F}{\delta\phi}-\frac{1}{\phi}\nabla\cdot\boldsymbol{\sigma}^{(n)}\right)+\theta_{\phi}, \tag{5.7.1}$$

where $\mathbf{v}\equiv\phi\mathbf{v}_{\mathrm{p}}+(1-\phi)\mathbf{v}_{\mathrm{s}}$ is the volume average velocity of the polymer ($\mathbf{v}_{\mathrm{p}}$) and the solvent ($\mathbf{v}_{\mathrm{s}}$), ϕ is the volume fraction of the polymer, L_{eff} is the effective Onsager coefficient for the dynamically asymmetric system (Eq. 2.4.21 in Sect. 2.4), F is the free energy functional of the spatial concentration fluctuation $\delta\phi(\mathbf{r};t)$, $\boldsymbol{\sigma}^{(n)}$ is the stress tensor on the entangled network chains given by Eq. 5.7.2 below, and θ_{ϕ} is the thermal noise term.

$$\boldsymbol{\sigma}^{(n)}=\boldsymbol{\sigma}-\frac{1}{3}\left(\mathrm{Tr}\boldsymbol{\sigma}\right)\mathbf{I}, \tag{5.7.2}$$

where

$$\boldsymbol{\sigma}\equiv\left\{\sigma_{ij}\right\}(i.j=x,y,z) \tag{5.7.3}$$

is the shear stress tensor, and $\mathbf{I}$ is the unit matrix. $\nabla\cdot\nabla\cdot\boldsymbol{\sigma}^{(n)}$ in Eq. 5.7.1 is given by

$$\begin{aligned}\nabla\cdot\nabla\cdot\boldsymbol{\sigma}^{(n)}&=2\,\frac{\partial^2}{\partial x\partial y}\sigma_{xy}\\&+\frac{1}{3}\left\{\left(2\frac{\partial^2}{\partial x^2}-\frac{\partial^2}{\partial y^2}-\frac{\partial^2}{\partial z^2}\right)N_1+\left(\frac{\partial^2}{\partial x^2}+\frac{\partial^2}{\partial y^2}-2\frac{\partial^2}{\partial z^2}\right)N_2\right\},\end{aligned} \tag{5.7.4}$$

where N_1 and N_2 are the first and second normal stress difference defined by

$$N_1=\sigma_{xx}-\sigma_{yy},\quad N_2=\sigma_{yy}-\sigma_{zz}, \tag{5.7.5}$$

where σ_{ij} $(i,j=x,y,z)$ is the component of the stress tensor.

The linearized Eq. 5.7.1 with respect to $\delta F/\delta\phi$, $\sigma_{xy}(\phi)$, and $N_1(\phi)$ to the linear order in $\delta\phi$ is given by

$$\left(\frac{\partial}{\partial t}+v\cdot\nabla\right)\delta\phi = L_{\mathrm{eff}}\left\{\nabla^2\left(r_0 - C_g\,\nabla^2\right) - \frac{1}{\phi}\right.$$
$$\left.\times\left[2\frac{\partial^2}{\partial x\partial y}\left(\frac{\partial\sigma_{xy}}{\partial\phi}\right)+\frac{1}{3}\left(2\,\frac{\partial^2}{\partial x^2}-\frac{\partial^2}{\partial y^2}-\frac{\partial^2}{\partial z^2}\right)\left(\frac{\partial N_1}{\partial\phi}\right)\right]\right\}\delta\phi+\theta_\phi, \quad (5.7.6)$$

where the term related to N_2 is dropped, because $N_2 = 0$ (Weissenberg hypothesis). r_0 and C_g are given by Eqs. 2.4.13 and 2.4.14 in Sect. 2.4.

The Fourier transform of Eq. 5.7.6 gives the dynamical equation for the **q**-Fourier mode of the concentration fluctuations,

$$\delta\phi_{\mathbf{q}} = \phi_{\mathbf{q}}\exp(i\mathbf{q}\cdot\mathbf{r}), \quad (5.7.7)$$

of the solutions under the shear flow as follows.

$$\left(\frac{\partial}{\partial t}-\dot\gamma\,q_x\,\frac{\partial}{\partial q_y}\right)\delta\phi_q = -L_{\mathrm{eff}}(q)\times$$
$$\left\{q^2\left(r_0 + C_g\,q^2\right)-\frac{1}{\phi}\left[2\,q_x\,q_y\left(\frac{\partial\sigma_{xy}}{\partial\phi}\right)+\frac{1}{3}\left(2\,q_x^2-q_y^2-q_z^2\right)\left(\frac{\partial N_1}{\partial\phi}\right)\right]\right\}\delta\phi_q+\theta_q. \quad (5.7.8a)$$

The term $\theta_{\mathbf{q}}$ is the **q**-Fourier modes of the thermal noises. The second term in the left-hand side of Eq. 5.7.8a,

$$\dot\gamma\,q_x\,\frac{\partial\delta\phi_{\mathbf{q}}}{\partial q_y}, \quad (5.7.8b)$$

represents the convection of the **q**-Fourier mode of the fluctuations under the shear flow. The wave vector **q** of the **q**-Fourier mode is convected by $\dot\gamma\;t\;q_x\;\mathbf{e}_y$ over the time t to change into $\mathbf{q}(t)$, so that

$$\mathbf{q}(t) = \mathbf{q}+\dot\gamma\;t\;q_x\;\mathbf{e}_y = \left(q_x, (q_y+\dot\gamma\;t\;q_x), q_z\right), \quad (5.7.8c)$$

where $\mathbf{e}_y$ is the unit vector along Oy-axis, i.e., the velocity gradient axis.

Equation 5.7.8a gives the relaxation rate $\Gamma(\mathbf{q})$ of the **q**-Fourier modes of the concentration fluctuations,

$$\Gamma(\boldsymbol{q}) = L_{\mathrm{eff}}(q)\times$$
$$\left\{q^2\left(r_0 + C_g\,q^2\right)-\frac{1}{\phi}\left[2\,q_x\,q_y\left(\frac{\partial\sigma_{xy}}{\partial\phi}\right)+\frac{1}{3}\left(2\,q_x^2-q_y^2-q_z^2\right)\left(\frac{\partial N_1}{\partial\phi}\right)\right]\right\}. \quad (5.7.9)$$

Note that

$$\sigma_{xy} = \eta_0 \dot{\gamma}, \quad N_1 = \psi_{1,0} \dot{\gamma}^2, \tag{5.7.10}$$

where η_0 and $\psi_{1,0}$ are the zero-shear viscosity and zero-shear coefficient for the first normal stress difference N_1, respectively. The critical shear rate $\dot{\gamma}_{c,x}$ above which the shear-enhanced concentration is observed is very small, e.g., $\dot{\gamma}_{c,x} < 0.0564\ \mathrm{s}^{-1}$ for the PS548/DOP 6.0 wt.% at $T = 27\ °\mathrm{C}$ (Fig. 5.3). Thus, the contribution of N_1 to $\Gamma(q)$ at $\dot{\gamma}$ close to $\dot{\gamma}_c$ may be neglected in comparison with that of σ_{xy} based on Eq. 5.7.10. Under this condition, let us consider $\Gamma(\mathbf{q})$ at $\mathbf{q}$ satisfying $q_x = q_y = q/\sqrt{2}$ in the Oq_xq_y plane. In this case, $\Gamma(q)$ is given by

$$\Gamma(q) = L_{\mathrm{eff}}(q)\, q^2 \left\{ \left[r_0 - \frac{1}{\phi} \left(\frac{\partial \eta_0}{\partial \phi} \right) \dot{\gamma} \right] + C_{\mathrm{g}}\, q^2 \right\}. \tag{5.7.11}$$

$\Gamma(q)$ is given by the product of the q-dependent transport term, $L_{\mathrm{eff}}(q)\, q^2$, and the term of q- and $\dot{\gamma}$-dependent thermodynamic driving force for the relaxation, $F_{\mathrm{relax}}(q;\, \dot{\gamma})$, where $F_{\mathrm{relax}}(q;\, \dot{\gamma})$ is given by

$$F_{\mathrm{relax}}(q;\, \dot{\gamma}) = r_0 - \frac{1}{\phi} \left(\frac{\partial \eta_0}{\partial \phi} \right) \dot{\gamma} + C_{\mathrm{g}}\, q^2. \tag{5.7.12}$$

In the thermodynamic limit at $q \to 0$, where the convection does not play a role (Eq. 5.7.8b), the thermodynamic driving force for the relaxation of the concentrations is given by

$$\lim_{q \to 0} F_{\mathrm{relax}}(q;\, \dot{\gamma}) \equiv F_{\mathrm{relax}}(q = 0;\, \dot{\gamma}) = r_0 - \frac{1}{\phi} \left(\frac{\partial \eta_0}{\partial \phi} \right) \dot{\gamma}. \tag{5.7.13}$$

Equation 5.7.13 predicts the followings

(1) When $F_{\mathrm{relax}}(q = 0;\, \dot{\gamma}) > 0$, the thermally activated fluctuation with $q = 0$ is relaxed, so that the solution is thermodynamically stable.
(2) When $F_{\mathrm{relax}}(q = 0;\, \dot{\gamma}) < 0$, the thermally activated fluctuation does not relax but rather grows, so that the solution is thermodynamically unstable and undergoes SD.
(3) When $F_{\mathrm{relax}}(q = 0;\, \dot{\gamma}) = 0$, the solution is in thermodynamic stability limit, i.e., the so-called spinodal point.

Figure 5.25 presents the thermodynamic driving force for the relaxation of the concentration fluctuations $F_{\mathrm{relax}}(q;\, \dot{\gamma})$ as a function of q at representative $\dot{\gamma}$ values for the thermodynamically stable solution with $r_0 > 0$ at $\dot{\gamma} = 0$ and at a given T. Here, r_0 is given by

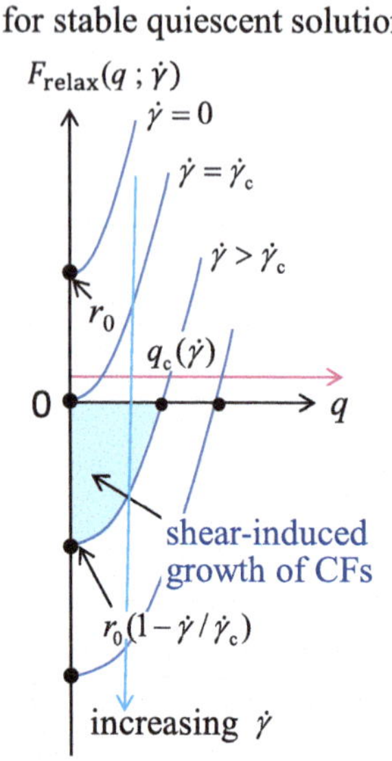

Fig. 5.25 Thermodynamic driving force for relaxation of concentration fluctuations $F_{\text{relax}}(q;\dot{\gamma})$ as a function of q at representative $\dot{\gamma}$ values for the thermodynamically stable solution with the given $r_0 > 0$ at a given T and at $\dot{\gamma} = 0$

$$r_0 = \left(\frac{\partial^2 f}{\partial \phi^2}\right)_0 = \frac{k_B T}{v_0} 2(\chi_s - \chi), \tag{5.7.14}$$

where v_0 is the segmental volume, χ is the thermodynamic interaction parameter between polymer segments and solvent molecules, and χ_s is the χ parameter at spinodal point. For the thermodynamically stable solution, $F_{\text{relax}}(q;\dot{\gamma})$ at $\dot{\gamma} = 0$ versus q is the concave downward parabolic function with a minimum value r_0 (> 0) at $q = 0$ as shown in Fig. 5.25. Since the coefficient C_g of the gradient-free energy term,

$$C_g = \frac{k_B\, T}{v_0} \frac{a^2}{18\,\phi}, \tag{5.7.15}$$

is independent of $\dot{\gamma}$, where a is the length of the segments, the effect of $\dot{\gamma}$ on $F_{\text{relax}}(q;\dot{\gamma})$ at the given r_0 gives a vertical shift of the parabolic curve as predicted by Eq. 5.7.12: The curve shifts downward with increased $\dot{\gamma}$, as shown in Fig. 5.25. At the spinodal point, $\dot{\gamma}$ satisfies the equation given below,

$$F_{\text{relax}}(q = 0;\dot{\gamma}) = r_0 - \frac{5\,\eta_0}{\phi^2}\,\dot{\gamma} = 0. \tag{5.7.16}$$

Equation 5.7.16 is obtained by using the experimental result on concentration dependence of η for the semidilute solutions at $\phi > \phi_e$, the entanglement concentration, as given by

$$\eta_0 \sim \phi^5. \tag{5.7.17}$$

The shear rate $\dot{\gamma}$ which satisfies the flow-induced spinodal point is defined as $\dot{\gamma}_c$. Then, $\dot{\gamma}_c$ is given by

$$\dot{\gamma}_c = \frac{r_0\,\phi^2}{5\,\eta_0}. \tag{5.7.18}$$

Equations 5.7.13 and 5.7.18 lead to

$$F_{relax}(q = 0; \dot{\gamma}) = r_0\left(1 - \frac{\dot{\gamma}}{\dot{\gamma}_c}\right). \tag{5.7.19}$$

As shown in Fig. 5.25, the single-phase solution at a given T, which satisfies $r_0 > 0$ and thereby $\chi < \chi_s$ (Eq. 5.7.14), is

(1) stable at $\dot{\gamma} < \dot{\gamma}_c$ for all the Fourier modes with q,
(2) in the spinodal point at $\dot{\gamma} = \dot{\gamma}_c$, where only the Fourier mode with $q = 0$ is faced to the stability limit, whereas all other Fourier modes with $q \neq 0$ is stable, and
(3) unstable at $\dot{\gamma} > \dot{\gamma}_c$ for the Fourier modes with $q \leq q_c(\dot{\gamma})$, where $q_c(\dot{\gamma})$ is given from Eqs. 5.7.12 and 5.7.18 by

$$q_c(\dot{\gamma})^2 = \left(\frac{r_0}{C_g}\right)\frac{\dot{\gamma} - \dot{\gamma}_c}{\dot{\gamma}_c}, \tag{5.7.20}$$

although the Fourier modes with $q > q_c(\dot{\gamma})$ are stable. The region marked by the blue color in Fig. 5.25 indicates the q-region having the thermodynamic driving force for the growth, defined as $F_{growth}(q)$, where $F_{growth}(q)$ is given by

$$F_{growth}(q) = -F_{relax}(q) = -\left\{r_0 - \frac{1}{\phi}\left(\frac{\partial \eta_0}{\partial \phi}\right)\dot{\gamma}\right\} - C_g\,q^2. \tag{5.7.21}$$

The driving force for the growth $F_{growth}(q)$ increases with $\dot{\gamma}$ such that both the maximum $r_0\,(\dot{\gamma}/\dot{\gamma}_c - 1)$ at $q = 0$ and the cut-off wave number $q_c(\dot{\gamma})$ increases and extends to the larger q with increased $\dot{\gamma}$, respectively.

$F_{growth}(q)$ at $\dot{\gamma} > \dot{\gamma}_c$ has a maximum at $q = 0$ and decreases parabolically with increased q and becomes zero at $q = q_c(\dot{\gamma})$. The Fourier modes with $q > q_c(\dot{\gamma})$ relax rather than grow with time. The growth rate $R(q)$ of the modes with $q \leq q_c(\dot{\gamma})$ is given by

$$R(q) = -\,\Gamma(q) = L_{eff}(q)\,q^2 F_{growth}(q). \tag{5.7.22}$$

The competing effects of the transport term, $L_{eff}(q)\,q^2$, and the thermodynamic term $F_{growth}(q)$ on the growth rate $R(q)$ gives rise to a maximum growth rate at $q = q_m$ which is the selected wave number of the dominant Fourier mode of the concentration the fluctuations for the growth in the early stage of SD and gives rise to the maximum

SALS intensity at $q = q_m$. As time elapses after the onset of SD, the q_m shifts toward smaller q values in the coarsening process of SD which involves the mode-mode coupling in the Fourier mode of the fluctuations to lower the free energy of the shear-induced unstable solution, as shown in Fig. 5.11 in Sect. 5.4.

5.7.2 Flow-Induced Shift of the Spinodal Line

$F_{\text{relax}}(q = 0, \dot{\gamma})$ under the shear flow at a given shear rate $\dot{\gamma}$, given by Eq. 5.7.13, can be rewritten as

$$F_{\text{relax}}(q = 0, \dot{\gamma}) = \frac{k_B T}{v_0} 2(\chi_s - \chi) - \frac{5\,\eta_0}{\phi^2}\dot{\gamma}, \tag{5.7.23}$$

using Eq. 5.7.14 for r_0 and Eq. 5.7.17 for η_0. Here, χ_s is the χ parameter at the spinodal point inherent to the quiescent solution at $\dot{\gamma} = 0$,

$$\chi_s = \frac{1}{2}\left(\frac{1}{N\phi} + \frac{1}{1-\phi}\right), \tag{5.7.24}$$

where N is degree of polymerization of the polymer in the solution. χ varies with T, as given by

$$\chi = A + \frac{B}{T}. \tag{5.7.25}$$

Here, A and B are constants associated with entropic and enthalpic part of the χ parameter, respectively.

Equations 5.7.23 and 5.7.25 predict that the solution under a given $\dot{\gamma}$ reaches the spinodal temperature, defined as $T_{s,\text{shear}}(\dot{\gamma})$, by changing temperature. $T_{s,\text{shear}}(\dot{\gamma})$ or the χ parameter at the spinodal temperature at the given $\dot{\gamma}$, defined as $\chi_{s,\text{shear}}(\dot{\gamma})$, satisfies $F_{\text{relax}}(q = 0, \dot{\gamma}) = 0$ in Eq. 5.7.23 and thereby the equation given below,

$$\frac{k_B T_{s,\text{shear}}(\dot{\gamma})}{v_0} 2\left[\chi_s - \chi_{s,\text{shear}}(\dot{\gamma})\right] - \frac{5\,\eta_0}{\phi^2}\dot{\gamma} = 0. \tag{5.7.26}$$

Thus, the shift of the spinodal line with the shear rate, relative to the spinodal line in the quiescent solution, defined by $\Delta\chi_{s,\text{shear}}(\dot{\gamma})$, is given by

$$\Delta\chi_{s,\text{shear}}(\dot{\gamma}) \equiv \chi_s - \chi_{s,\text{shear}}(\dot{\gamma}) = \frac{5\,v_0}{2\,k_B T_{s,\text{shear}}(\dot{\gamma})}\frac{\eta_0}{\phi^2}\dot{\gamma} \tag{5.7.27}$$

$$\propto \begin{cases} \phi^3\,\dot{\gamma} & \text{at } \phi > \phi_e, \\ \phi^{-1}\,\dot{\gamma} & \text{at } \phi < \phi_e, \end{cases} \tag{5.7.28}$$

where ϕ_e is the entanglement concentration at which the concentration dependence of η_0 undergoes the following change,

$$\eta_0 \propto \begin{cases} \phi & \text{at } \phi < \phi_e, \\ \phi^5 & \text{at } \phi > \phi_e. \end{cases} \tag{5.7.29}$$

Equation 5.7.27 predicts that $\Delta\chi_{s,\text{shear}}(\dot{\gamma})$ is proportional to $\dot{\gamma}$, provided that the shift of $T_{s,\text{shear}}(\dot{\gamma})$ with $\dot{\gamma}$ is not large, as is the case encountered by the experiments. That is, an increase of $\dot{\gamma}$ proportionally increases $\Delta\chi_{s,\text{shear}}$ for any given sets of (ϕ, η_0), resulting in the **shear-enhanced thermodynamic instability** of the solution.

Figure 5.26 schematically presents the predicted shift of the spinodal line with increased shear rate from $\dot{\gamma} = 0$ to the given $\dot{\gamma} \neq 0$ for the solutions with the upper critical solution temperature (UCST) (part a-1) and for the solutions with the lower critical solution temperature (LCST) (part b-1). The shift is highlighted by the red arrows pointing upward and the blue arrows pointing downward, respectively. For the solution with UCST, χ increases with increased $1/T$ as shown by the red straight line in part (a-2), so that the increased $\Delta\chi_{s,\text{shear}}(\dot{\gamma})$, i.e., increased drop of $\chi_{s,\text{shear}}(\dot{\gamma})$, accompanied by the increased $\dot{\gamma}$, as shown in part (a-2) by the thick downward black arrows, results in the decreased $1/T_{s,\text{shear}}(\dot{\gamma})$ from $1/T_s(\dot{\gamma} = 0)$, as shown by the horizontal thin red arrow pointing left in part (a-2), where $T_s(\dot{\gamma} = 0)$ shown by the red solid line in part (a-1) is the spinodal temperature for the quiescent solution in the UCST solution.

Thus, the spinodal line shifts upward, as shown by the upward shift of the red solid line at $\dot{\gamma} = 0$ into the broken red line at the given $\dot{\gamma}$ in part (a-1).

On the other hand, for the LCST solution, χ decreases with increased $1/T$ as shown by the blue solid straight line in part (b-2), so that the increase of $\Delta\chi_{s,\text{shear}}(\dot{\gamma})$, i.e., increased drop of $\chi_{s,\text{shear}}(\dot{\gamma})$, accompanied by the increased $\dot{\gamma}$ as shown in part (b-2) by the thick black arrows, results in the increased $1/T_{s,\text{shear}}(\dot{\gamma})$ from $1/T_s(\dot{\gamma} = 0)$, as shown by the horizontal thin blue arrow in part (b-2). Thus, the spinodal line shifts downward, as shown by the downward shift of the blue solid line at $\dot{\gamma} = 0$ into the broken blue line at $\dot{\gamma} \neq 0$ in part (b-1).

In either the UCST solution or LCST solution, the shear commonly expands the thermodynamically unstable region in each phase diagram and thereby enhances phase separation. The shift of the spinodal temperature with shear rate depends also on the concentration ϕ. At a given shear rate, the shift of $T_{s,\text{shear}}(\dot{\gamma})$ from $T_s(\dot{\gamma} = 0)$ increases with the increased concentration ϕ in the region of $\phi > \phi_e$, whereas the shift increases with decreased concentration in the region of $\phi < \phi_e$, as predicted by Eq. 5.7.28 and as illustrated by the change in the length of the red or blue arrows with ϕ in part (a-1) and (b-1), respectively. The shift of the spinodal temperature depends on η_0 also, so that the larger the molecular weight is, the larger is the shift for a given set of ϕ and $\dot{\gamma}$.

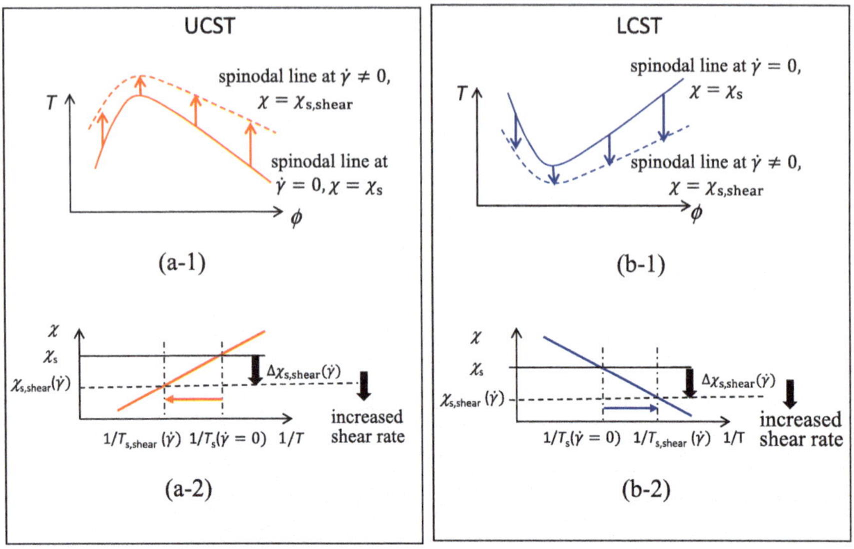

Fig. 5.26 Schematic presentation of the shift of spinodal line with the increased shear rate from $\dot{\gamma} = 0$ to any given $\dot{\gamma}$'s $\neq 0$ for solutions with the upper critical solution temperature (UCST) (part **a-1**) and for the solutions with lower critical solution temperature (LCST) (part **b-1**), as highlighted by the upward-pointing red arrows (part a-1) and downward-pointing blue arrows (part b-1). For the solution with UCST χ increases with increased $1/T$ as shown in part (**a-2**), so that the increased $\Delta\chi_{s,shear}(\dot{\gamma})$, i.e., increased drop of $\chi_{s,shear}(\dot{\gamma})$ relative to χ_s at $\dot{\gamma} = 0$ accomanied by the increased $\dot{\gamma}$, as shown in part (**a-2**) by the thick downward pointing arrows, results in the decreased $1/T_{s,shear}(\dot{\gamma})$ from $1/T_s(\dot{\gamma} = 0)$ as shown by the horizontal thin arrow in part (**a-2**), where $T_s(\dot{\gamma} = 0)$ is the spinodal temperature for the quiescent solution. Thus, the spinodal line shifts upward, as shown by the upward shift of the solid line at $\dot{\gamma} = 0$ into the broken line at the given $\dot{\gamma}$ in part (**a-1**). Similar arguments are applicable to the shift of the LCST phase diagram with any given $\dot{\gamma}$'s

5.7.3 *Comparison of the Experimental and Theoretical Critical Shear Rate for the Spinodal Decomposition*

We compare $\dot{\gamma}_{c,x}$ measured experimentally with $\dot{\gamma}_c$ predicted by the theory [34] based on the assumption,

$$\dot{\gamma}_{c,x} \propto \dot{\gamma}_c = \frac{K_{OS}}{5\,\eta_0} \tag{5.7.30}$$

where K_{OS} is the osmotic modulus of the solutions at $\dot{\gamma} = 0$, defined by

$$K_{OS} \equiv \phi\,\frac{\partial \Pi}{\partial \phi} = \phi^2\left(\frac{\partial^2 f}{\partial \phi^2}\right) = r_0\,\phi^2. \tag{5.7.31}$$

Equation 5.7.30 is obtained by noting that $F_{\mathrm{relax}}(q=0;\dot{\gamma}) = 0$ at $\dot{\gamma} = \dot{\gamma}_{\mathrm{c}}$ in Eq. 5.7.13 and $\partial\eta_0/\partial\phi = 5\eta_0/\phi$ at $\phi > \phi_e$ from Eq. 5.7.29. Based on the Flory-Huggins mean-field theory, K_{OS} in Eq. 5.7.31 is given by

$$K_{\mathrm{OS}} = \frac{k_{\mathrm{B}}\,T}{v_0}\phi^2\left[\phi + (1-2\chi) + \frac{1}{N\,\phi}\right]. \tag{5.7.32}$$

Figure 5.27 presents $\dot{\gamma}_{\mathrm{c},x}$ plotted as a function of K_{OS}/η_0 based on Eq. 5.7.30 for the 6.0 wt.% PS548 in DOP at 27, 35, and 50 °C, in trecresyl phosphate (TCP) at 27 °C, in cyclohexane (CH) at 35 °C, in dibutyl phthalate (DBP) at 27 °C, and in diethylmalonate (DEM) at 50 °C [21]. Here, DOP at 22 °C, CH at 35 °C, and DEM at 50 °C are the Θ solvent for polystyrene (PS), whereas DBP at 27 °C and TCP at 27 °C are the good solvent for PS. The χ parameter, K_{OS} (Pa), and K_{OS}/η_0 (s^{-1}) estimated are summarized in Table 2 of Ref. [21]. The experimental results reveal that the effect of the temperature and the solvent quality on $\dot{\gamma}_{\mathrm{c},x}$ is validated by the theorical prediction on $\dot{\gamma}_{\mathrm{c}}$ shown by the dotted line. The results elucidate the following trends.

(1) $\dot{\gamma}_{\mathrm{c}}$ and $\dot{\gamma}_{\mathrm{c},x}$ depend on the solvents used: They increase in the order of DOP, TCP, CH, DBP, and DEM. The χ-parameters of the solvents employed here decrease in this order. Thereby, the critical shear rates for the SD increase with the increased K_{OS} and solvent quality with respect to the thermodynamic interactions between the polymer and the solvents employed here. The solvent quality increases with decreasing the χ-parameter.

(2) $\dot{\gamma}_{\mathrm{c}}$ and $\dot{\gamma}_{\mathrm{c},x}$ increase with the increased temperature for the given solvent: DOP (27 °C), DOP (35 °C), and DOP (50 °C), revealing that they increase with the decreased χ parameter and η_0.

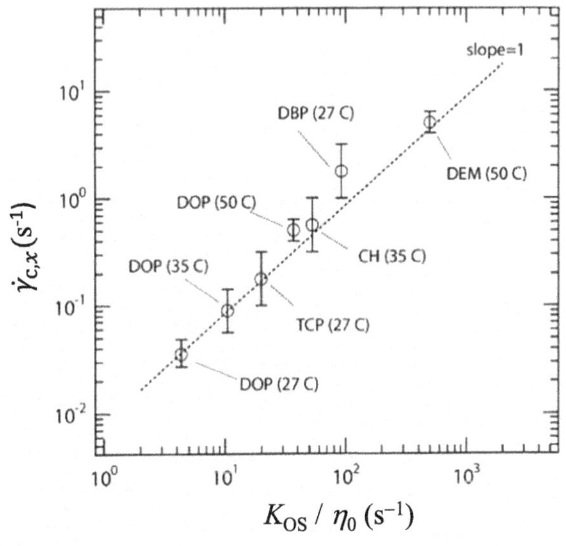

Fig. 5.27 $\dot{\gamma}_{\mathrm{c},x}$ plotted as a function of K_{OS}/η_0 based on Eq. 5.7.30 for 6.0 wt.% the PS548 in DOP at 27, 35, and 50 °C, in trecresyl phosphate (TCP) at 27 °C, in cyclohexane (CH) at 35 °C, in dibutyl phthalate (DBP) at 27 °C, and in diethylmalonate (DEM) at 50 °C. $\dot{\gamma}_{\mathrm{c},x}$ is predicted to be proportional to $\dot{\gamma}_{\mathrm{c}}$ as shown by the dotted line. From [21], reprinted with permission, American Chemical Society

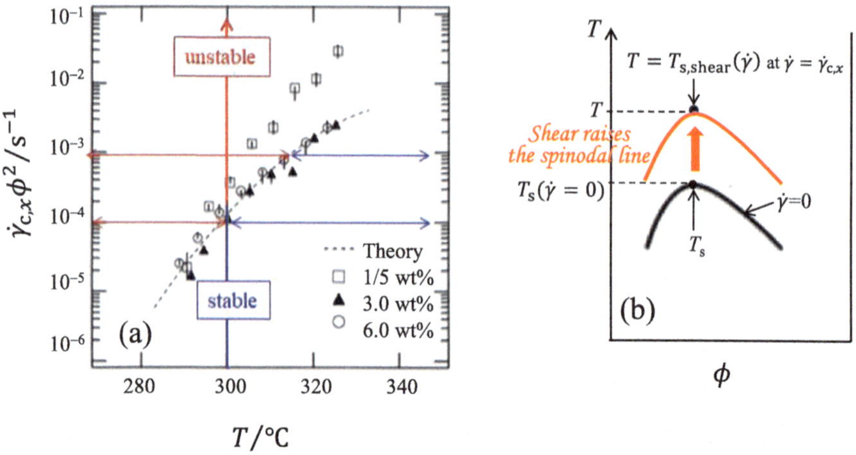

Fig. 5.28 Part **a** presents $\dot{\gamma}_{c,x}$ reduced with ϕ^{-2} for difference in concentrations of the solution at 1.5, 3.0, and 6.0 wt.% plotted as a function of T for PS548/DOP. Part **b** presents the uprise of the spinodal (SD) line with the increased shear rate as predicted from the results shown in part **a**. Part **b** presents the uprise of the SD line from the black line to the red line together with the red arrow pointed upward with the increased $\dot{\gamma}$ from $\dot{\gamma} = 0$ to the given $\dot{\gamma}$

Figure 5.28a presents $\dot{\gamma}_{c,x}$ for the PS548/DOP reduced with ϕ^{-2} for the difference in the concentrations of the solutions at 1.5, 3.0, and 6.0 wt. % plotted as a function of T [35]. In Eq. 5.7.32 for K_{OS}, the concentration dependence of K_{OS} is given by ϕ^3 when $\chi \to 1/2$ (Θ solvent), because $N\phi \gg 1$. Furthermore, $K_{OS} \propto \phi^3$and $\eta_0 \sim \phi^5$ at $\phi > \phi_e$, lead to the concentration dependence of $\dot{\gamma}_{c,x}$ and $\dot{\gamma}_c$ as given by

$$\dot{\gamma}_{c,x} \propto \dot{\gamma}_{c,\text{theory}} \sim \phi^{-2}. \tag{5.7.33}$$

Therefore, $\phi^2\ \dot{\gamma}_{c,x}$ is predicted to become independent of the concentration ϕ and thereby depends only on T, as long as the condition of $\phi > \phi_e$ is satisfied.

The temperature dependence of K_{OS} and η_0 is given by

$$K_{OS} = a_1T - a_2, \quad \eta_0 \sim \exp(a_3/T),$$

where a_1, a_2, and a_3 are the constant independent of $\dot{\gamma}$ and estimated at the three concentrations as presented in Table III in Ref. [35].

In Fig. 5.28a, the data obtained for the solution at 3.0 and 6.0 wt.% are fall onto the universal curve independent of the concentration in the reduced plot. However, the data obtained for the solution at 1.5 wt.% deviate from the universal curve: The deviation increases with the increased temperature above 300°C. It is plausible that the approximation of $\eta_0 \sim \phi^5$ at $\phi > \phi_e$ and $K_{OS} \sim \phi^3$ for the 1.5 wt.% solution is less reliable in comparison with that for the 3.0 wt.% and 6.0 wt.% solutions. The discrepancy in the validity of the approximation between the low (1.5 wt.%) and high (3.0 and 6.0 wt.%) concentration increases with the increased T. The piece of evidence that this discrepancy increases with increasing temperature is due to the fact that K_{OS} is proportional to T.

The results shown in Fig. 5.28a reveal the following pieces of evidence:

(1) at a given ϕ, $\dot{\gamma}_{c,x}$ increases with the increased T;
(2) at a given ϕ and T, the lower and upper side of the curve correspond to the thermodynamically stable and unstable region, respectively;
(3) at a given ϕ and $\dot{\gamma}_{c,x}(\phi)$, the left-hand and right-hand side of the curve correspond to the unstable and stable region, respectively.

Evidence (1) indicates that, for the solution at a given ϕ, the higher the working temperature T above the spinodal temperatures at $\dot{\gamma} = 0$ (defined as $T_s(\dot{\gamma} = 0)$) is, the larger is $\dot{\gamma}_{c,x}$ required for attaining $T = T_{s,\mathrm{shear}}(\dot{\gamma})$ (the spinodal temperature under the shear) as shown in part (b). The shear raises the spinodal line from black line to red line, as shown by the red arrow pointing upward in Fig. 5.28b.

Evidence (2) indicates that the solution at a given T and ϕ changes from the stable solution to the unstable one with increased $\dot{\gamma}$ across $\dot{\gamma}_{c,x}$, as shown by the vertical arrows with blue and red color, respectively, in part (a).

Evidence (3) indicates that the solution at a given $\dot{\gamma}_{c,x}(\phi)$ and ϕ changes from the unstable to stable one with increased temperature across the spinodal temperature under the shear flow, defined as $T_{s,\mathrm{shear}}(\dot{\gamma})$, as shown by the horizontal arrows with red and blue color in Fig. 5.28a: The larger the $\dot{\gamma}$ is, the higher is the $T_{s,\mathrm{shear}}(\dot{\gamma})$, as shown in part (a); the increased shear rate expands the unstable region, as also shown in part (b).

The critical shear rate $\dot{\gamma}_{c,x}$ can be estimated from intensity distribution of the SALS pattern in the Oq_xq_z plane. From Eq. 5.7.9, $\Gamma(\mathbf{q})$ along q_x, i.e., $q_x = q$ and $q_y = q_z = 0$, is given by

$$\Gamma(q, 0, 0; \dot{\gamma}) = L_{\mathrm{eff}}(q)\, q^2 \left\{ \left[r_0 - \frac{2}{3\phi} \left(\frac{\partial N_1}{\partial \phi} \right) \right] + C_g\, q^2 \right\}. \tag{5.7.34}$$

Thus, the relaxation rate at a given $\dot{\gamma}$ along q_x depends only on N_1 but not on σ_{xy} regarding the mechanical properties relevant to the system. The thermodynamic force for the relaxation of the concentration fluctuations along the q_x direction is given from Eq. 5.7.34 by

$$F_{\mathrm{relax}}(q_x = 0, 0, 0; \dot{\gamma}) = r_0 - \frac{2}{3\phi} \left(\frac{\partial N_1}{\partial \phi} \right), \tag{5.7.35}$$

where the second line in Eq. 5.7.10 can be used for the $\dot{\gamma}$ dependence of N_1. Since at $\dot{\gamma} = \dot{\gamma}_{c,x}$ where the solution reaches the spinodal point, $F_{\mathrm{relax}}(q = 0; \dot{\gamma}) = 0$. Thus, $\dot{\gamma}_{c,x}$ is given from the second line of Eqs. 5.7.10 and 5.7.30 by

$$\dot{\gamma}_{c,x} = \left\{ K_{OS} \Big/ \left[\frac{2}{3}\, \phi \left(\frac{\partial \psi_{1,0}}{\partial \phi} \right) \right] \right\}^{1/2}. \tag{5.7.36}$$

It is worth investigating if this theoretical prediction accounts for the experimentally measured $\dot{\gamma}_{c,x}$. The flow suppresses the thermodynamic force for the relaxation of the concentration fluctuations along q_x through the second term in the rhs of Eq. 5.7.35. That is, this suppression depends on $\dot{\gamma}^2$ and $\frac{1}{\phi}\left(\frac{\partial \psi_{1,0}}{\partial \phi}\right)$. It enhances the growth of the concentration fluctuations along q_x at $\dot{\gamma} > \dot{\gamma}_{c,x}$ (see Figs. 5.25, 5.26 (a-1) and (a-2), and 5.28)

$\Gamma(\mathbf{q})$ along q_z, i.e., $q_x = q_y = 0$ and $q_z = q$, is given from Eq. 5.7.9 by

$$\Gamma(0, 0, q; \dot{\gamma}) = L_{\text{eff}}(q)\, q^2 F_{\text{relax}}(0, 0, q; \dot{\gamma}), \tag{5.7.37}$$

$$F_{\text{relax}}(0, 0, q; \dot{\gamma}) = \left[r_0 + \frac{1}{3\phi}\left(\frac{\partial \psi_{1,0}}{\partial \phi}\right)\right] + C_g\, q^2. \tag{5.7.38}$$

Equation (5.7.38) predicts that the thermodynamic force for the relaxation of the concentration fluctuations along q_z is always positive and never goes down below zero at any given shear rates: Thus, the concentration fluctuations along q_z are relaxed to form a stable solution but never grown because the flow cannot attain the condition of $F_{\text{relax}}(0, 0, q; \dot{\gamma}) < 0$. This is the physical origin of the dark sector of the butterfly pattern along the q_z direction in Regime II at $\dot{\gamma} > \dot{\gamma}_{c,x}$ which is accounted for by the linearized version of the HFMO theory. However, the uprise of the SALS intensity along $q_z$in Regime III at $\dot{\gamma} > \dot{\gamma}_{c,z}$ cannot be accounted for in the context of the linearized theory. The verification of the uprise of the SALS intensity demands the analyses which incorporate the nonlinear terms of the theory given by Eqs. 5.7.1 to 5.7.3.

The difference between $F_{\text{relax}}(q \to 0, 0, 0; \dot{\gamma})$ and $F_{\text{relax}}(0, 0, q \to 0; \dot{\gamma})$ in Eqs. (5.7.35) and (5.7.38) predicts the anisotropy in the bulk property (the osmotic modulus K_{OS}) of the solution under the shear flow. The anisotropy vanishes as $\dot{\gamma} \to 0$.

5.8 Flow-Induced Evolution of Phase-Separated Domains and Their Alignment along Flow Direction in Regime III

Figure 5.29 presents the normalized integrated intensity $\mathcal{J}_x(\dot{\gamma})/\mathcal{J}_x(\dot{\gamma} = 0)$ and $\mathcal{J}_z(\dot{\gamma})/\mathcal{J}_z(\dot{\gamma} = 0)$ along the q_x and q_z axis, respectively, as a function of the shear rate (part a) and the SLAS pattern (1) and (2) taken at the shear rate 1.59×10^{-2} and $2.51 \times 10^{-2}\ \text{s}^{-1}$, respectively (part b) for PS548/DOP 6.0 wt. % at 15.7°C [35]. The SALS pattern (1) and (2) are taken at the shear rate below and above the critical shear rate $\dot{\gamma}_{c,z}$, respectively, as shown by the point (1) and (2) with the arrows (part a), respectively. The SALS pattern (1) shows the strong scattered intensity along q_x but the weak intensity along q_z as weak as that from the quiescent solution. On the other hand, the SALS pattern (2) exhibits the strong intensity along q_z excess to that from the quiescent solution. The uprise of the scattered intensity $\mathcal{J}_z(\dot{\gamma})/\mathcal{J}_z(\dot{\gamma} = 0)$ at $\dot{\gamma} >$

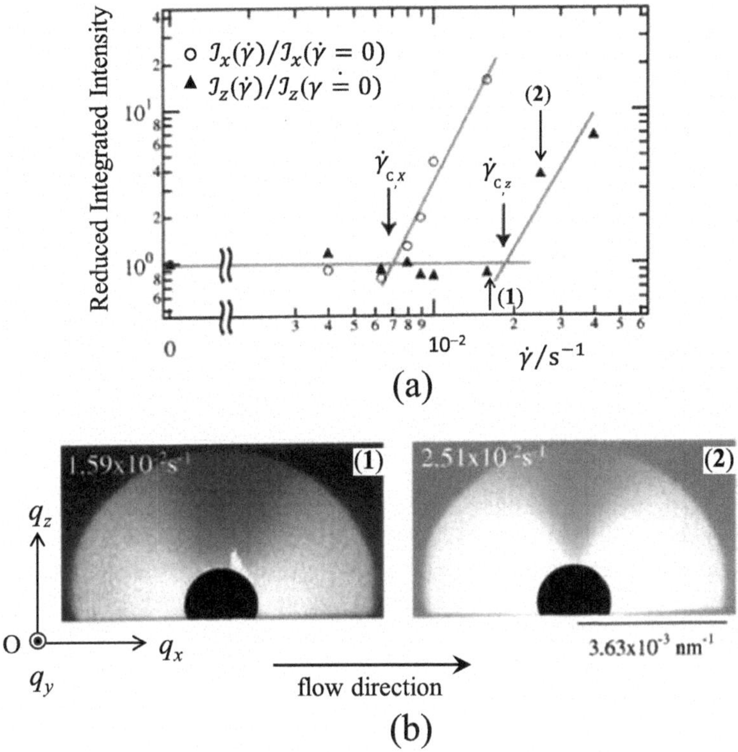

Fig. 5.29 Normalized integrated intensity along q_x and q_z axis as a function of shear rate (part **a**) and the SALS pattern (1) and (2) taken at $\dot{\gamma} = 1.59 \times 10^{-2}$ and 2.51×10^{-2} s^{-1}, respectively, which correspond to the shear rate (1) and (2) marked by the arrows in part (**a**), respectively (part **b**)

$\dot{\gamma}_{c,z}$ cannot be taken into account by the linearized dynamical equation (Eqs. 5.7.6 and 5.7.8a); on the contrary, the linearized theory predicts that the scattered intensity along q_z is suppressed, so that it becomes lower than that from the quiescent solution as discussed above in conjunction with Eqs. 5.7.37 and 5.7.38.

The uprise of the scattered intensity along the q_z axis implies the transformation of the plane wave concentration fluctuations (**PLWCF**) with their wave vectors preferentially oriented along the flow direction (q_x direction) into the phase-separated domains randomly aligned in space, as shown by a plausible model regarding the change of the flow-induced structure from part (a) to part (b) in Fig. 5.30 where the blue arrows schematically illustrate the solvent flow. With the increased shear rate or the time after the step-up shear, in Regime II, the amplitude of the PLWCF increases with the shear rate or the time after the step-up shear, respectively, and thereby, the stored energy increases in the polymer-rich regions shown by the red-color region in part (a). Eventually, the polymer-rich regions become mechanically unstable and are burst into the domains shown by the red color in part (b) driven by increased

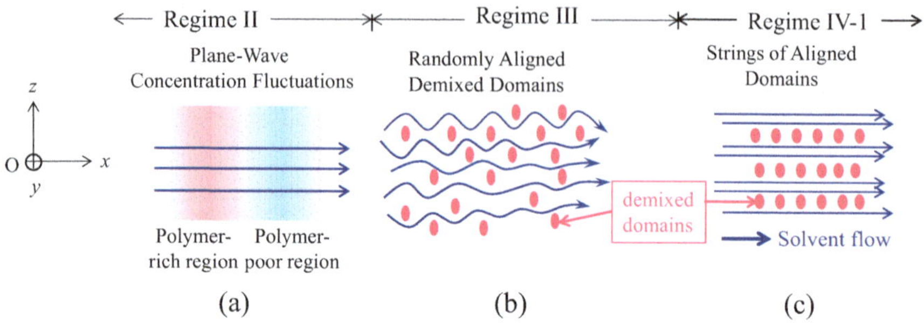

Fig. 5.30 Change of the dissipative structures from Regime II (part **a**) to III (part **b**), and from Regime III to Regime IV-1 (part **c**) with the increased shear rate or the increased time after step-up shear from $\dot{\gamma} = 0$ to $\dot{\gamma} > \dot{\gamma}_a$

hydrodynamic interactions between the PLWCF and the solvent flow. This transformation from the PLWCF to the domains dissipates the energy stored in the solution to stabilize the solution under the shear. The domains evolved by the transformation give rise to the increase of the Fourier modes of the concentration fluctuations along the q_z (neutral) direction, excess to those in the quiescent solution: Thereby, they cause the uprise of $I(q_z; \dot{\gamma})$ and $\mathcal{J}_z(\dot{\gamma})$. The transformation process is evidenced by the second stress-over shoot (Figs. 5.10 and 5.13) and the uprise of the scattered intensity along the q_z axis (Figs. 5.6 and 5.29).

The hydrodynamic interactions between the solvent flow (shown by the blue arrows in part (b)) and the domains increase with the increased shear rate or with the increased time after the step-up shear in Regime III as schematically illustrated in part (b), so that the energy stored in the solution increases with the shear rate or the time elapsed after the step-up shear flow. The stored energy may be dissipated by aligning the domains along the flow direction as schematically presented from part (b) to (c). The gradual alignment of the domains along the flow direction may cause the broad stress overshoot with the time after the step-up shear. Finally, the domains are aligned into the strings oriented along the flow direction in Regime IV-1 as shown in part (c) where eventually, the domains within a given string are partially interconnected by the stretched chains (in Regime IV-2), although the partial interconnection is not shown in part (c).

5.9 String Structure in Regime IV

Figure 5.31 presents typical SALS patterns (part a and b) and OM image (part c) for the PS548/DOP solution in Regime IV [47]. The flow direction (FD) is set in the horizontal direction of the figure as indicated in part (b). The pattern (a) is taken for the 6.0 wt.% solution at $\dot{\gamma} = 50\,\text{s}^{-1}$ and at 27 °C: The upper half of the pattern (a) shows the contour pattern constructed from the iso-intensity lines of the lower half

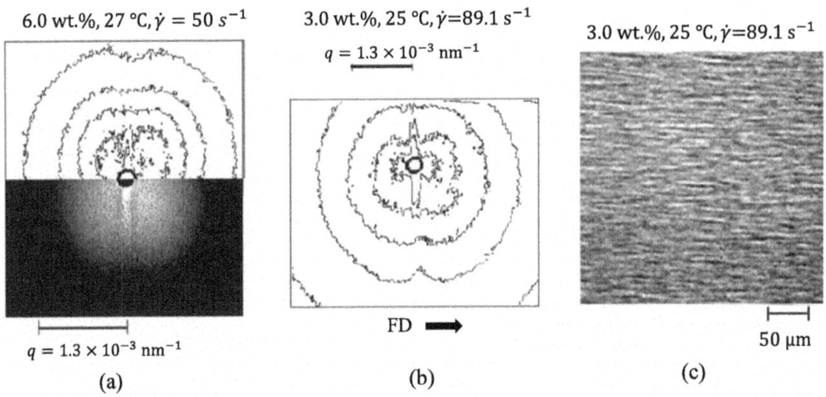

Fig. 5.31 Typical SALS patterns (part **a** and **b**) and OM image (part **c**) for the PS548/DOP solution in the Regime IV. The flow direction (FD) is horizontal as shown by the arrow in part (b). The SALS pattern is composed of the two type of patterns superposed one another: the streak-like pattern centered at $q_x = 0$ and extended along q_z and the butterfly pattern composed of the two elliptical lobes separated by the streak-like pattern at $q_x = 0$.

of the pattern taken with the CCD as described in Chap. 3. The contour pattern (b) and the OM image (c) are taken for the 3.0 wt.% solution at $\dot{\gamma} = 89.1\ s^{-1}$ and at 25 °C. The OM image is the snapshot taken at the shutter speed of 10^{-3} s: The image in part (c) is taken simultaneously with the SALS pattern in part (b).

The OM image reveals the strings composed of the phase-separated domains aligned in a row and oriented along the flow direction as schematically illustrated in Fig. 5.30c. Each of the individual strings with the average persistence length L is expected to give rise to the strong horizontal streak-like pattern centered at $q_x = 0$ and spread laterally along q_z, the direction normal to the flow direction. The width of the streak along the q_x axis depends on $2\pi/L$.

The internal structure of the strings, which is characterized by the domains aligned along the flow direction with an average spacing d, gives rise to the butterfly scattering pattern superposed to the streak pattern. The butterfly pattern has the following features:

(1) at the small scattering angles, the butterfly scattering pattern has a broad maximum arising from the average spacing d between the neighboring d domains in a given string aligned along the flow direction, at $q \equiv q_{x,m} = 2\pi/d$;
(2) whereas at the large scattering angles, the butterfly pattern is composed of the elliptical lobes with their long axes oriented along the q_z axis (normal to FD), as evidenced by the shape of the iso-intensity lines in the contour pattern in part (a) and (b) in Fig. 5.31.

Interpretation of the SALS pattern in Regime IV composed of butterfly pattern and the streak pattern. Figure 5.32 presents the physical interpretation of the SALS pattern observed in Regime IV. We propose the SALS pattern arises from the strings composed of the oriented domains aligned along the flow direction (FD) (part c):

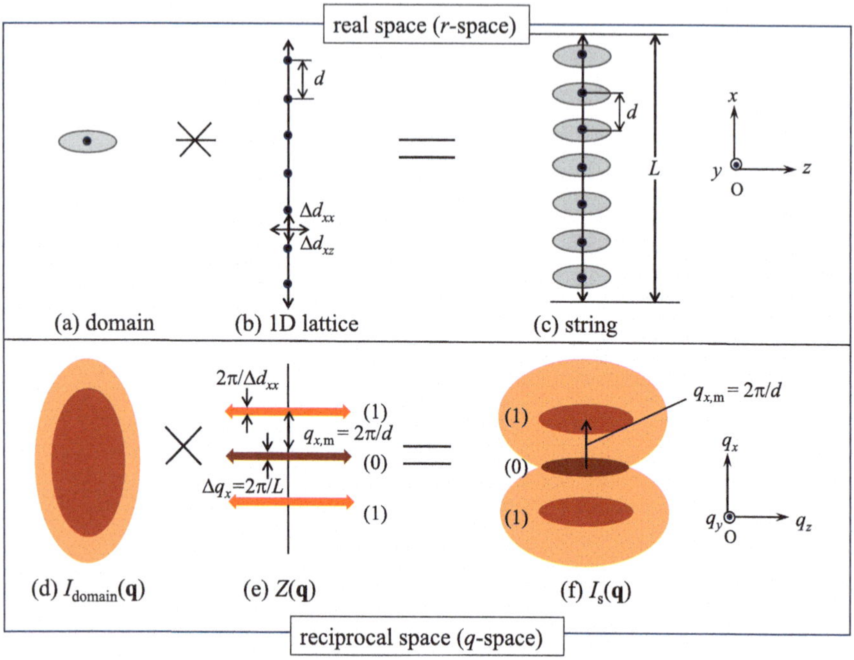

Fig. 5.32 Physical interpretation of the SALS pattern observed in Regime IV

One of the oriented individual domains within the string is shown in part (a). It has the shape of an oblate ellipsoid with its major axis (revolution axis) oriented parallel to Ox direction (FD).The domains are aligned in a row in one of the string with an average spacing d and persistence length L, as shown in part (c): The strings are oriented along FD (Ox direction). In other words, the domains in the string are arranged in one dimensional (1D) lattice with the lattice distortion tensor $\Delta\mathbf{d} = (\Delta d_{xj})$ where $\Delta d_{xj} (j = x, y, z)$ is the component of the lattice distortion tensor along Ox-, Oy-, and Oz-axis, respectively, as shown in part (b). In real space, the string (part c) is described by the convolution of the domain (part a) and the 1D lattice (part b). The operator "*" in between the domain (part a) and the 1D lattice (part b designates the convolution (of the domain and 1D lattice), while the operator "=" in between the 1D lattice (part b) and the string (part c) designates the equality between the convolution of the domain and 1D lattice (part a and b, respectively) and the string (part c).

The SALS pattern in the q space, $I_s(\mathbf{q})$, (part f) is given by the product of the scattering pattern from one of the the individual domains (part d), $I_{\text{domain}}(\mathbf{q})$, and the lattice factor $Z(\mathbf{q})$ (part e), as designated by the operator "×" in between part (d) and (e) as given below by Eq. 5.9.1. The lattice factor $Z(\mathbf{q})$ is given by a convolution of $Z_\infty(\mathbf{q})$ and $S(\mathbf{q})$ as given below by Eq. 5.9.2. Here, $Z_\infty(\mathbf{q})$ is the lattice factor for the 1D lattice with infinitely large extension along the q_x-axis. $S(\mathbf{q})$ is the shape factor which controls the finite size of the string L along Ox direction [48].

$$I_s(\mathbf{q}) \sim I_{\text{domain}}(\mathbf{q})Z(\mathbf{q}), \tag{5.9.1}$$

$$Z(\mathbf{q}) \sim Z_\infty(\mathbf{q}) * S(\mathbf{q}) \equiv \int Z_\infty(\mathbf{u})S(\mathbf{q}-\mathbf{u})d\mathbf{u}, \tag{5.9.2}$$

where the shape factor $S(\mathbf{q})$ along q_x-axis is given by

$$S(\mathbf{q}) = S\left(q_x,\ q_y = 0,\ q_z = 0\right) = (\text{const})\frac{\sin^2(q_x\,L/2)}{(q_x\,L/2)^2}. \tag{5.9.3}$$

$Z(\mathbf{q})$ gives only a broad first order maximum along q_x-axis at $q_x \equiv q_{\text{x,m}} = 2\pi/d$, when $0.18 < \Delta d_{xx}/d < 0.35$ [48], as indicated by the thick red arrows marked (1) in part (e). This first-order maximum of $Z(\mathbf{q})$ extends infinitely along the q_z-axis as shown by the thick red arrows with the double heads labeled by (1) pointing toward $q_z \to \pm\infty$ in part (e). The product of $I_{\text{domain}}(\mathbf{q})$ and $Z(\mathbf{q})$ gives rise to a broad scattering maximum in $I_s(\mathbf{q})$ along q_x-axis at around $q_x = q_{x,\text{m}} = 2\pi/d$ as shown by the purple pattern marked (1) in part (f). The scattering maximum extends along q_z over a wide range of q_z because the maximum of the lattice factor $Z(\mathbf{q})$ at around $q_x = q_{x,\text{m}}$ extends infinitely along q_z axis as presented by the thick double-head red arrows (1) in part (e).

$Z(\mathbf{q}) \to 1$ at $q_x > q_{x,\text{m}}$, so that $I_s(\mathbf{q}) \to I_{\text{domain}}(\mathbf{q})$ at the large range of $q = ({q_x}^2 + {q_z}^2)^{1/2}$(Eq. 5.9.1). In other words, $I_s(\mathbf{q})$ at the large q-range depends only on the scattering from the individual domains, thereby providing the form factor of the domains which gives information on the shape of the individual domains themselves.

$Z(\mathbf{q})$ along q_x and at around $q_x \to 0$ depends on the shape factor $S(\mathbf{q})$. The shape factor $S(\mathbf{q})$ depends on the size of the string (L) along the q_x as indicated in Fig. 5.32c. Thus, $S(\mathbf{q})$ gives rise to the scattering maximum at $q_x = 0$ as predicted by Eq. 5.9.3, i.e., the so-called **zeroth-order scattering maximum**. The breadth of the zeroth-order scattering maximum along q_x, defined as Δq_x, is given by $\Delta q_x = 2\pi/L$, whereas that along q_z axis, Δq_z, is infinitely large as shown in Fig. 5.32e, by the thick purple arrows with the double heads pointing toward $q_z \to \pm\infty$ and labeled by (0). Thus, $I_s(\mathbf{q})$ also exhibits the zeroth-order scattering maximum, as shown by the streak-like pattern colored purple and labeled by (0) in part (f), with the breadth along q_x and q_z axis controlled by those of $S(\mathbf{q})$ and $Z(\mathbf{q})$, respectively; $S(\mathbf{q})$ depends on L. The zeroth-order scattering maximum of $S(\mathbf{q})$ is the **origin of the streak-like scattering pattern** at around $q_x = 0$.

Figure 5.33 presents the experimental results on the shear-rate dependence of the birefringence Δn_{xz}, the reduced integrated SALS intensity parallel and perpendicular to the flow direction, $\mathcal{J}_x(\dot{\gamma})/\mathcal{J}_x(\dot{\gamma} = 0)$ and $\mathcal{J}_z(\dot{\gamma})/\mathcal{J}_z(\dot{\gamma} = 0)$, respectively, and the viscosity η (part a) and the OM image (part b-1 in part b) for the PS384/DOP 6.0 wt.% solution at 21 °C in the increased shear-rate protocol [49]. Here, we focus only on the results in the stages 1 and 2 in Regime IV, defined as Regime IV-1 and IV-2 in Fig. 5.6 in Sect. 5.3, respectively. The stages 1 and 2 are highlighted by the light-gray and dark-gray background color in Fig. 5.33a, respectively.

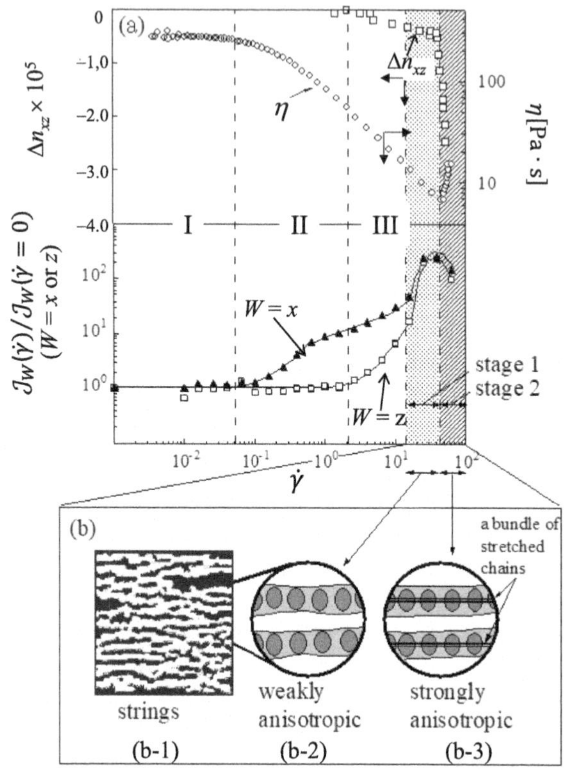

Fig. 5.33 Experimental results on the shear-rate dependence of the birefringence Δn_{xz}, reduced integrated SALS intensity parallel and perpendicular to the flow direction, $\mathcal{J}_x(\dot{\gamma})/\mathcal{J}_x(\dot{\gamma}=0)$ and $\mathcal{J}_z(\dot{\gamma})/\mathcal{J}_z(\dot{\gamma}=0)$, respectively, and the viscosity η (part **a**) and the OM image shown in (**b-1**) (part **b**) for the PS384/DOP 6.0 wt.% solution at 21 °C in increased shear-rate process. Based on [49]

The experimental results revealed in Regime IV-1 and IV-2 validates the followings.

(1) The negative birefringence starts to increase with the increased shear-rate in the higher shear-rate part of Regime III. The negative value continues to increase into a sizably large value in the stage 1 in Regime IV-1;

(2) The $\mathcal{J}_x(\dot{\gamma})$ and $\mathcal{J}_z(\dot{\gamma})$ exhibit the remarkable second-step uprise with the increased $\dot{\gamma}$;

(3) The negative birefringence abruptly increases in Regime IV-2 with the increased shear rate;

(4) The evidence (3) is accompanied by the abrupt increase of the shear viscosity, causing the shear thickening. Those pieces of evidence described above further elucidate the followings.

(i) The result (2) elucidates the emergence of the zeroth-order scattering maximum in the SALS pattern, i.e., the strong streak-like scattering centered at $q_x = 0$ and extended along q_z as shown by the purple pattern marked by (0) in Fig. 5.32f, and thereby the evolution of the strings, as shown in the observed OM image shown in (b-1) in part (b) and by the schematic illustration in (b-2) in part (b) in Fig. 5.33. The result (1) further elucidates the domains in the strings are partially linked by the bundles of the stretched chains. That is, the bundles of the stretched chains interconnect some fractions of the domains in a given string, as illustrated in part b-2 in part b, to exhibit the weak optical anisotropy with the small negative birefringence Δn_{xz} as presented in Regime IV-1 in part (a). Although the partial links interconnecting the domains are not explicitly shown in part (b-2) in part (b), it is explicitly shown later in part (d') in row (C) in Fig. 5.35.

(ii) The results (3) and (4) elucidate the domains in the strings in Regime IV-2 are linked by the bundles of stretched chains over the large persistence length L, giving rise to the strong optical anisotropy with the large negative birefringence Δn_{xz}, as schematically illustrated in part (b-3) in Fig. 5.33b.

The birefringence Δn_{xz} of PS chains in the solution under the shear flow is given by Eq. 5.3.4 in Sect. 5.3, where $\Delta n_{xz} = \Delta n'$. The birefringence Δn_{xz} depends on the orientation factor of the segment $f_{segment}$ and the intrinsic birefringence of the segment $\Delta_{PS}{}^{o}$. The PS/DOP solution is predicted to have the negative intrinsic birefringence $\Delta_{PS}{}^{o}$ due to the reason described below.

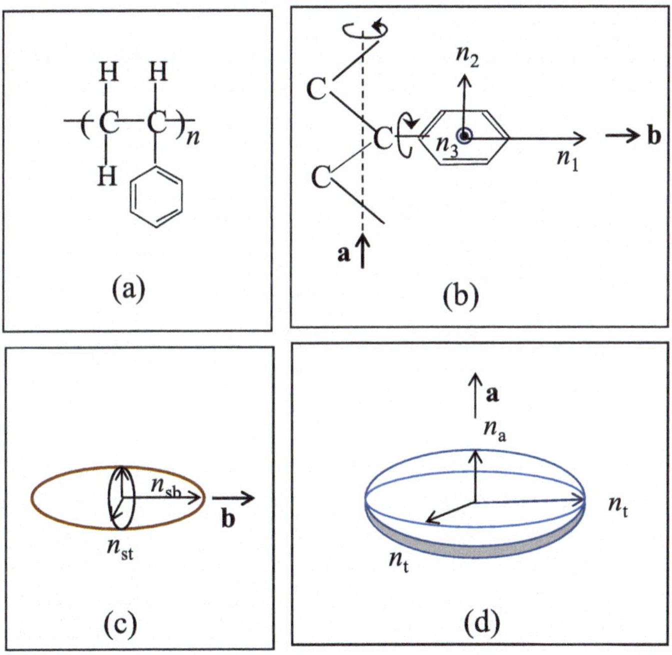

Fig. 5.34 A monomeric unit of the PS chain (part **a**), local structure of the PS segment oriented along unit vector **a** parallel to Oz direction (FD) (part **b**), refractive-index ellipsoid of the side group of the segment having rotational freedom around unit vector **b** as shown in part (**b**) (part **c**), and the refractive-index ellipsoid of the whole segment having rotational freedom around **a** as shown in part (**b**) (part **d**). n_1, n_2, and n_3 are the refractive indices of benzene ring along the three orthogonal directions where n_3 is the refractive index normal to the plane of benzene ring (part **b**)

Figure 5.34 schematically presents the monomeric unit of the PS chain (part a), the local chemical structure of the PS segment oriented along the flow direction, Ox-axis parallel to unit vector **a** (part b), the refractive-index ellipsoid of the side group of the PS segment (part c), and the refractive-index ellipsoid of the PS segment containing the side group (part d). In the case where the side group has a rotational freedom around the C-C bond axis, along the unit vector **b,** connecting between the main chain and the side group (part b), the refractive-index ellipsoid of the side group has a uniaxial optical anisotropy with $n_{\mathrm{sb}} > n_{\mathrm{st}}$, where n_{sb} and n_{st} are the refractive index parallel and perpendicular to the principal optical axis **b**, respectively (part c). In the case where the PS segment shown in part (b) further has the rotational freedom around the main-chain axis (parallel to the unit vector **a**) (part b), the refractive-index ellipsoid of the segment as a whole has the uniaxial optical anisotropy with $n_{\mathrm{a}} < n_{\mathrm{t}}$ where n_{a} and n_{t} are the refractive index parallel and perpendicular to the principal optical axis of the PS segment oriented along the unit vector **a** (part d). Thus, the intrinsic birefringence of the segments $\Delta_{\mathrm{PS}}{}^{\mathrm{o}}$ defined by the equation given below becomes negative,

$$\Delta_{\mathrm{PS}}{}^{\mathrm{o}} \equiv n_a - n_t < 0.$$

The large difference between n_t and n_a contributes to the negative intrinsic birefringence of the PS segment. This is because the contribution of the negative birefringence of the side group unit with its principal optic axis along the unit vector **b** dominates the contribution of the positive birefringence of the main-chain unit with its principal optic axis along the unit vector **a** (part b). Thus, the increased chain orientation from the random orientation, which changes f_{segment} from zero to unity, brings the change of Δn_{xz} from zero to the limit of the negative birefringence of $\Delta_{\mathrm{PS}}{}^{\mathrm{o}}$.

5.10 Summary

Figure 5.35 summarizes the steady-state or transient dissipative structures formed under the shear flow with the increased shear-rate protocol or with the protocol of the increased time after the step-up shear flow, respectively, found for solutions of entangled *noncrystallizable* polymers [49]. Upon increased $\dot{\gamma}$ through the critical shear rates $\dot{\gamma}_{\mathrm{c},x}$, $\dot{\gamma}_{\mathrm{c},z}$, and $\dot{\gamma}_{\mathrm{a}}$ or increased time t, after the step-up shear from $\dot{\gamma} = 0$ to $\dot{\gamma} > \dot{\gamma}_{\mathrm{a}}$, through the critical times $t_{\mathrm{c},x}$, $t_{\mathrm{c},z}$, and $t_{\mathrm{c,a}}$ as shown in the top row (A), the shear-SALS patterns (schematics), the contrast-enhanced transmission shear-OM images, and the rheological properties are found to undergo the characteristic changes as shown in row (B), (C), and (D), respectively. The corresponding changes in the dissipative structures are summarized in row (E). All the patterns shown in row (B) and (C) are taken in Oxz plane where the flow, velocity gradient, and neutral (or vorticity) directions are commonly set parallel to Ox-, Oy-, and Oz-axis, as shown in row (B) and part (c), respectively, where Oy-axis the velocity gradient direction is

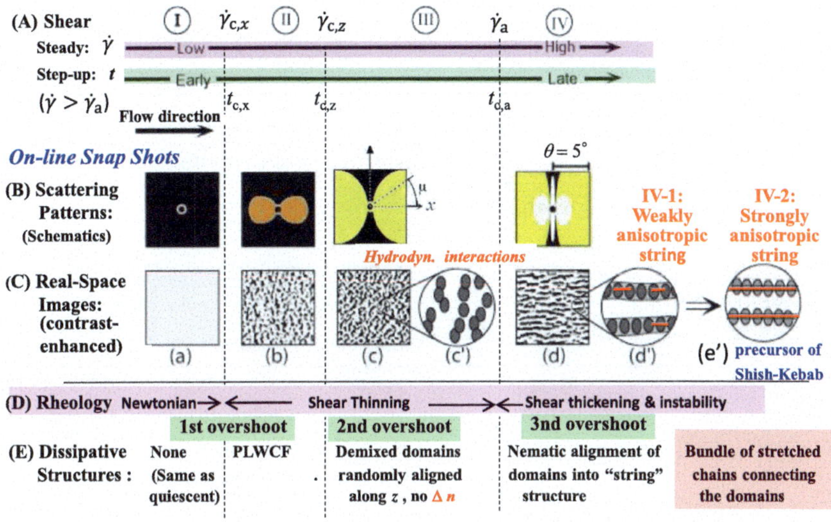

Fig. 5.35 Summary of dissipative structured formed under the steady-state shear flow in the increased shear-rate protocol and the step-up shear flow from $\dot{\gamma} = 0$ to the constant $\dot{\gamma} > \dot{\gamma}_a$ with the increased time after the step-up shear. Based on [50]

set normal to the plane of the paper, and Oz-axis (the neutral direction) is set along the vertical direction of the figure.

The dissipative structures and rheo-optical behaviors under the steady or transient flow are commonly classified into the following four regimes.

5.10.1 Regime I

Here, this Regime exist at $0 < \dot{\gamma} < \dot{\gamma}_{c,x}$ or $t < t_{c,x}$ (corresponding to the terminal rheological relaxation time). The solutions are homogeneous even under the shear flow as in the quiescent solution and in the incubation stage with respect to the shear rate or the time for the shear-induced structure formation.

5.10.2 Regime II

Here, this Regime exists at $\dot{\gamma}_{c,x} < \dot{\gamma} < \dot{\gamma}_{c,z}$ or $t_{c,x} < t < t_{c,z}$. In the experiments in Regime II, any given shear-rates $\dot{\gamma}$'s (in the increased shear-rate protocol) satisfy the shear-rate range $\dot{\gamma}_{c,x} < \dot{\gamma} < \dot{\gamma}_{c,z}$; any given time t's after the step--up shear flow satisfy the time-span $t_{c,x} < t < t_{c,z}$. The flow enhances the plane wave concentration fluctuations (PLWCF) with their wave vectors preferentially oriented along Ox-axis.

However, the flow does not affect much the fluctuations along Oz-axis, i.e., the Fourier modes of the fluctuations with their wave vectors oriented along Oz-axis are effectively absent, giving rise to the characteristic butterfly patterns with the dark streak or sector along Oz-axis [part (b) in row (B) in Fig. 5.35 in Sect. 5.3] [7, 9]. The flow-induced concentration fluctuations is closely interrelated with the first stress overshoot, both of which occur as a consequence of the dissipation of the energy stored in the solution in Regime II due to the applied flow.

5.10.3 *Regime III*

Here, this Regime exists at $\dot{\gamma}_{c,z} < \dot{\gamma} < \dot{\gamma}_a$ or $t_{c,z} < t < t_{c,a}$: $t_{c,a}$ and $\dot{\gamma}_a$ correspond to τ_e and τ_e^{-1}, respectively: τ_e is the rheological relaxation time for the chain stretching (chain retraction time): $\dot{\gamma}_a = 1/\tau_e$ is the chain retraction rate. The scattering along the neutral axis starts to remarkably increase, which may be a signature of a transformation of the dissipative structure evolved under the flow from PLWCF to shear-induced phase-separated domains [12, 35]. The transformation causes the dissipation of the energy stored in the solution in Regime III and the second stress overshoot. Judging from the shape of the butterfly pattern, the domains extend along Oz-axis but compressed along Ox-axis as schematically shown in part (c′) in row (C) in Fig. 5.35 or part (a) and (c) in Fig. 5.32; the uniaxially compressed domains may reflect the memory of the PLWCF existed prior to the transformation in part (b) in row (C) in Fig. 5.35. The centers of the mass of the domains are randomly aligned in space along Oz-axis as illustrated in part (c′) and as actually shown by the OM image in part (c) in row (C) in Fig. 5.35 [12, 38]: They tend to align along the flow direction driven by the hydrodynamic interactions, causing the broad third stress overshoot, in the latter part of Regime III. The rheology exhibits the shear thinning in Regime II and III [9, 12, 35, 36]. The sheared solutions show almost no birefringence up to Regime III, indicating that polymer chains in the domains are essentially in relaxed random coils.

5.10.4 *Regime IV*

Here, this Regime exists at $\dot{\gamma} > \dot{\gamma}_a$ or $t > t_{c,a}$. The SALS exhibits the sharp streak-like scattering pattern with strong intensity centered at $q_x = 0$ and extended along Oz-axis [part (d) in row (B) in Fig. 5.35] [9, 12, 38]. The streak-like pattern is superposed on the butterfly pattern as experimentally presented in part (d) in row (B) in Fig. 5.35. This implies that the phase-separated domains align into string-like structures oriented parallel to the flow as schematically illustrated in part (d′) in row (C) in Fig. 5.35. The string as a whole gives rise to the streak-like SALS pattern, while the individual domain structures within the strings [as schematically shown in part (d′) in row (C) in Fig. 5.35] give rise to the butterfly pattern. The domains structures within the strings could not be clearly resolved by the shear OM in the case of the UHMWaPS solutions as shown in part (d) in row (C) in Fig. 5.35, although they were identified in the case of the UHMWPE solutions [22, 23] to be described in Sects. 6.4 and 6.5. The nematic alignment of the phase-separated domains into the strings may be due to hydrodynamic effects. The strings have only weak birefringence in the early stage in Regime IV (Regime IV-1) as shown in part (d) row (C) in Fig. 5.35. Upon further increasing $\dot{\gamma}$ or t, i.e., in the late stage of Regime IV (Regime IV-2), the negative birefringence suddenly starts to increase, indicating formation of the bundles of stretched chains connecting the nematically aligned domains as schematically shown by the red lines in part (e′) in row (C) in Fig. 5.35. This transformation is accompanied by the shear thickening and instabilities in the shear stress and the first normal stress difference [9, 12, 36, 38]. The large fluctuation (oscillation) of the birefringence with time is observed, under the step wisely increased shear protocol as indicated by Fig. 4.2 in Sect. 4.32, before reaching the steady state shear flow at any step belonging to any given shear rate in Regime IV-2. This fluctuation may reflect the stress instability caused by the repeated stretching and relaxation of chains under the flow, although the relevant data are not presented here (to be presented in Fig. 6.13 in Chap. 6). The domains in Regime IV are predicted to be oblate ellipsoids with their major axes oriented along Ox-axis as illustrated in Fig. 5.32a, c, or as illustrated in the model (d′) and (e′) in row (C) in Fig. 5.7.35, which is validated by the compressed butterfly pattern along Ox-axis (Figs. 5.31a, b and 5.32f in Sect. 5.9).

5.10.5 Remarks on Sect. 5.10

Table 5.3 presents the relationship between the two kinds of the classifications on the flow-induced dissipative structures. The classification of Regime [I] to Regime [IV] (composed of Regime [IV-1] and [IV-2]) specified in the top row (a) is the classification based on the dissipative structures self-organized in the solution under the flow with the increased shear rate or the protocol of the increased time after the step-up shear flow. The classification of Regime [$\dot{\gamma}_{\mathrm{I}}$] to Regime [$\dot{\gamma}_{\mathrm{III}}$] specified in the bottom row (b) is concerned with the classification based on the characteristic shear-rate range predicted in conjunction with Fig. 1.7 and 1.8 in Sect. 1.4. In the dynamically asymmetric systems, the dissipative structures in Regime [I] and Regime [II] to Regime [IV-1] are self-organized under the shear-rate Regime [$\dot{\gamma}_{\mathrm{I}}$] satisfying $\dot{\gamma} < \Gamma_{\mathrm{conc}} \equiv \dot{\gamma}_{\mathrm{c},x}$, and the shear-rate Regime [$\dot{\gamma}_{\mathrm{II}}$] satisfying $\Gamma_{\mathrm{conc}} < \dot{\gamma} < \Gamma_{\mathrm{MD}}$, respectively, whereas the dissipative structures in Regime [IV-2] are self-organized under the shear-rate Regime [$\dot{\gamma}_{\mathrm{III}}$] satisfying $\dot{\gamma} > \Gamma_{\mathrm{MD}}$ (Figs. 1.7 and 1.8 in Sect. 1.4). These classifications described above for the solutions open for the imposed applied shear flow are based on the fundamental general principle proposed in Fig. 1.7 and Fig. 1.8 in Sect. 1.4 for any dynamically asymmetric solutions subjected to the externally applied flow.

Table 5.3 (a) Classification of Regime [I] to [IV] based on shear-induced structures (top row) and (b) classification of Regime $[\dot{\gamma}_{\text{I}}]$ to $[\dot{\gamma}_{\text{III}}]$ based on effects of shear on CF and MD (bottom row) for dynamically asymmetric systems

(b) I	II	III	IV	
			IV-1	IV-2
(c) $[\dot{\gamma}_{\text{I}}]$	$[\dot{\gamma}_{\text{II}}]$			$[\dot{\gamma}_{\text{III}}]$

CF: Concentration fluctuations
MD: Molecular deformation

5.11 Further Remarks

Before closing Sect. 5.5, the following brief remarks are added, regarding the important topics that have not been covered up to this section in this book:

(i) The effects of large amplitude oscillatory shear flow (LAOS) on the self-organization of the dissipative structures, as observed by the combined SALS and SANS [40, 51],
(ii) A possible difference in the dissipative structures self-organized under the continuous shear flow and LAOS [52],
(iii) Shear-induced phase separation in the nonentangled, dynamically asymmetric systems [53].
(iv) Shear-induced structures formed in the phase-separated semidilute solutions in quiescent state [54].

As for the research **item (i),** an oscillatory shear strain γ

$$\gamma = \gamma_0 \exp(\mathrm{i}\ \omega\ t) \tag{5.11.1}$$

was imposed to the semidilute solutions, and the SALS profiles as well as the SANS profiles along q_x and q_z were investigated as a function of the strain amplitude γ_0, the angular frequency ω, the strain phase $\phi \equiv \omega t$. The dynamic response of the SALS and the SANS in the LAOS reaches the steady state with the time after applying the LAOS at the given set of γ_0 and ω, e.g., at 20 min (or 120 cycles) at $\gamma_0 = 4.8$ and $\omega = 0.6133$ rad/s [40, 51]. We discuss below the steady-state dynamic response of the SALS, SANS, and relevant dissipative structures evolved in the solutions subjected to the LAOS flow. the LAOS.

Using the steady-state SALS under the LAOS flow constructed in the parameter space of γ_0 and ω at any given temperature where the quiescent solution is in the single-phase state for the PS548/DOP 6.0 wt.% at 27 °C. At given γ_0's, greater than the critical value of γ_0, defined as $\gamma_{0,\text{c}}$, where $\gamma_{0,\text{c}} \cong 2$ for this solution, the shear-induced structure was observed in the angular frequency range ω given by

$$\omega_{\text{l}} < \omega < \omega_{\text{u}}, \tag{5.11.2}$$

where $\omega_l \sim \dot{\gamma}_{c,x}$ and $\omega_u \sim \Gamma_{squeeze}$, where $\Gamma_{squeeze}$ is the rate of the solvent squeeze from the deformed entangled polymer chains, mediated by the elastic effects and stress-diffusion coupling effects, to the matrix solution. The frequency range $\Delta\omega = \omega_u - \omega_l$ where the shear-induced SALS-butterfly patterns are observed tends to increase with γ_0. At $\omega > \omega_u$, there is no time available for the solvent squeeze during LAOS. Thus, the entangled polymer chains will be deformed and relaxed uniformly without the solvent-squeezing process, so that the solution is kept homogeneous under the flow. At $\omega < \omega_l$, deformed entangled polymer chains can be relaxed under the flow via disentanglements of the chains in the any given matrix solutions under the given LAOS flow. Thereby, the solutions are kept homogeneous.

The steady-state SALS-butterfly pattern along q_z under the LAOS flow elucidated that the characteristic length of the shear-induced dissipative structure (PLWCF) decreases with the increased ω for the any given $\gamma_0' s$: The higher the ω is, the smaller is the length scale of the dissipative structure developed (see Figs. 7, 9, and 11 in Ref. [40]).

The steady-state SANS profiles $I(q_x)$ along Ox axis under LAOS flow also show the crossover in the power-law behavior with the increased q_x, as given by Eq. 5.5.1 Sect. 5.5, from $p_x > 2$ to $p_x = 2$ relevant to the OZ scattering from the concentration fluctuations of the entangled polymer chains in the matrix solution, as shown in Fig. 5.36a [51, 52]. The power-law exponent p_x increases with the increased ω from 2.5 and 3.4 at $q_x < q_{x,c} = 0.1\,\text{nm}^{-1}$, the value of which is larger than $q_{x,c} = 0.06$ for the case of the continuous shear flow (Fig. 5.15a). In the case of the continuous shear flow, p_x at $q_x < q_{x,c} = 0.06$ increases from ~ 1.6 to ~ 3 with the increased $\dot{\gamma}$ from 0.2 to 2.6 s^{-1}[41, 52] as presented in Fig. 5.15a. It should be important to note that the LAOS flow specified by the given set of the parameters (γ_0, ω, T) in Fig. 5.36 corresponds to the shear-rate Regime $[\dot{\gamma}_{II}]$ and evolves the phase-separated domains in Regime [III] at any given strain phases ϕ_s's [51]. This is evidenced by the existence of $I(q_z)$ excess to the OZ scattering [51]. After the subtraction of the OZ contribution from $I(q_x)$ in part (a), $I_d(q_x)$ is obtained in part (b): $I_d(q_x)$ tends to exhibit the power-law scattering with the exponent of 4 at any given ω values as shown in part (b), indicative of the domain structures with the sharp interface boundary. On the other hand, in the case of the continuous shear flow, the power-law of $I_d(q_x)$ obtained after the correction for the OZ contribution exhibits the increased exponent from 2.4 to 3 with increased $\dot{\gamma}$, never reaching 4 (see Fig. 5.15) [41] and Fig. 1 in ref. [52]. It is worth noting that the crossover q value, $q_{xc}(\sim 0.1\,\text{nm}^{-1})$ in $I(q_x)$ in LAOS flow is higher than that in the continuous shear flow ($q_{xc} = 0.06\,\text{nm}^{-1}$).

A comparison of the two pieces of evidence described above, regarding the shear-induced butterfly-SANS patterns observed under the continuous shear flow and LAOS flow, addresses a fundamental question raised by the **item (ii)** in the beginning of this section: LAOS can drive the semidilute solutions into the phase-separated state with the domains having the sharp interfacial boundaries, whereas the continuous shear flow does not. For the continuous shear flow, interfacial boundaries remain less well defined or rough at all shear rates covered in this experiment. This important experimental discovery was qualitatively interpreted in the context of the HFMO theory [32–34].

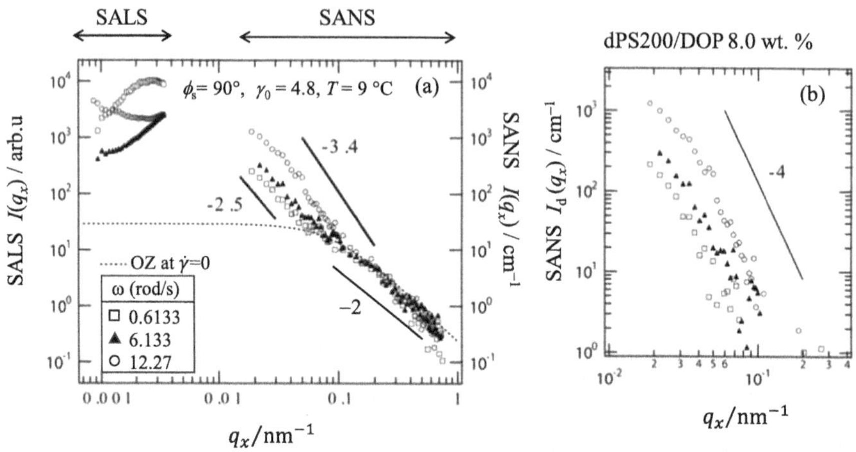

Fig. 5.36 Part **a**: The steady-state scattering profiles $I(q_x)$ parallel to the flow direction obtained under the large amplitude oscillatory shear flow at the angular frequency $\omega = 0.6133$, 6.133 and 12.27 rad/s with strain amplitude $\gamma_0 = 4.8$ and strain phase $\phi_s = 90\,°\text{C}$ for the dPS200/DOP 8.0 wt.% at 9 °C. The SALS profiles were equally multiplied by the shift factor to be smoothly connected with the SANS profiles with the absolute scattered intensity units (cm^{-1}). The dotted line is the OZ function at $\dot{\gamma} = 0$ obtained by the best fitting with the experimental scattering profilesarising from the parasitic scattering at the high q range. The deviation of the experimental scattering from the OZ at the high q tails greater than ca. 0.6 nm^{-1} is due to the artifacts at the detector edge. Part **b**: $I_d(q_x)$ evaluated by subtracting the OZ contribution from the SANS scattering profiles $I(q_x)$. Based on [20]

Below, we shall give a very brief explanation about the important difference described above. For the details, readers should refer to [52]. Under the continuous shear flow, the wave vector for q-Fourier modes of the concentration fluctuations are convected (i.e., rotated and stretched or compressed depending on their orientation) (see Eq. 5.7.8b). They are enhanced for the growth in the first and third quadrants because the wave vector is convected toward smaller q and suppressed in the second and fourth quadrants in the q_x-q_y plane because the wave vector is convected toward larger q; this acts as a stabilizing mechanism of the solution against indefinite growth of the fluctuations with the shear rate. More specifically, the thermodynamic driving force for the growth of the Fourier modes of the concentration fluctuation, $F_{\text{growth}}(q)$, is given by Eq. 5.7.21 which predicts $F_{\text{growth}}(q)$ is maximum at $q = 0$ and parabollically decreases with q. This further predicts that the q-Fourier modes of the fluctuations convected and distorted toward smaller q and larger q under the continuous shear flow are subjected to enhanced and suppressed growth, respectively. Therefore, one can predict that the growth of the q-Fourier modes are enhanced in the first and third quadrant but suppressed in the second and fourth quadrant in Oq_xq_y plane.

However, under the oscillatory shear flow the wave vectors for all the Fourier modes of the fluctuations oscillate. The wave vector of the Fourier mode of the fluctuations which is initially oriented along q_x axis will oscillate around q_x axis and hence will repeatedly pass the q_x axis where the enhancement of fluctuations by the normal stress is strongest (see Fig. 3 in ref [52]). Thus, the convective stabilizing mechanism is absent, and hence, LAOS flow can drive the system all the way to form phase-separated domains with sharp interfacial boundaries, consistent with experimental observation. All other Fourier modes, which will not pass q_x axis in the oscillating shear flow, are subjected to shear forces which alternately enhance and suppress the growth of the fluctuations. Thus, fluctuations with these modes should not be enhanced as much by the shear stress under the oscillatory flow as under the continuous flow. This is why the butterfly patterns in Oq_xq_z plane are most intense along q_x direction.

So far, the shear-induced structures have been discussed for dynamically asymmetric systems with entanglements. The **item (iii)** concerns the shear-induced structures for a system without entanglements of polymer chains but having the dynamical asymmetry due to a disparity of the glass transition temperatures of the two components in the blend. The detailed discussions on this item should be referred to elsewhere [53].

Up to this point, we have discussed shear-induced dissipative structures for a single-phase solutions in quiescent state. The **item (iv)** concerns how the shear flow affects formation of the dissipative structures in the semidilute solutions which are brought into the two-phase region in quiescent state. The details were reported elsewhere [54].

References

1. Ver Strate, G. and Phillipoff, W. (1974) *J. Polm. Sci. Polym. Lett. Ed.*, **12**, 267.
2. Rangel-Nafaile, C., Metzner, A.B., and Wissbrun, K.F. (1984) *Macromolecules*, **17**, 1187.
3. Schmidt, R. and Wolf, B.A. (1979) *Colloid. Polym. Sci.*, **257**, 1187.
4. Kramer, H. and Wolf, B.A. (1985) *Makromol. Chem. Rapid Commun.*, **6**, 21.
5. Hashimoto, T., Takebe, T., and Fujioka, K. (1990) *Phase Transition and Self-Assembling Structures of Polymer Mixtures Under Shear Flow*. In *Dynamics and Patterns in Complex Fluids*, Onuki, A. and Kawasaki, K. (eds), Springer, Berlin, pp. 86–99.
6. Wu,X-L., Pine,D.J., and Dixon, P.K. (1991) *Phys. Rev. Lett.*, **66**, 2408.
7. Hashimoto, T. and Fujioka, K. (1991) *J. Phys. Soc. Jpn.*, **60**, 356–359.
8. Yanase, H., Moldenaers, P., Mewis, J., Abetz, V., van Egmond J. W., and Fuller, G.G. (1991) *Rheol. Acta.*, **30**, 89.
9. Hashimoto, T. and Kume, T. (1992) *J. Phys. Soc. Jpn.*, **61**, 1839–1843.
10. van Egmond, J. W., Werner, D. F., and Fuller, G. (1992) *J. Chem. Phys.*, **96**, 7742.
11. van Egmond, J. W. and Fuller, G. G. (1993) *Macromolecules*, **26**, 7182.
12. Kume, T., Hashimoto, T., Takahashi, T., and Fuller, G.G. (1997) *Macromolecules*, **30**, 7232–7236.
13. Matsuzaka, K. and Hashimoto, T. (1999) Rev. *Sci. Instr.*, **70**, 2387–2397.
14. Boué, F., and Lindner, P. (1994) *Europhys. Lett.*, **25**, 421.
15. Morfin, I., Lindner, P., and Boue, F. (1999) *Macromolecules*, **32**, 7208.
16. Kume, T., Asakawa, K., Moses, E., Matsuzaka, K., and Hashimoto, T. (1995) *Acta Polym.*, **46**, 79–85.
17. Moses, E., Kume, T., and Hashimoto, T. (1994) *Phys. Rev. Lett.*,**72**, 2037–2040.
18. Murase, E., Kume, T., and Hashimoto, T. (1996) *Macromol Symp.*, **104**, 159–182.
19. Murase, H., Kume, T., Hashimoto, T., Ohta, Y., and Mizukami, T. (1995) *Macromolecules*, **28**, 7724–7729.

20. Hashimoto, T., (2008), *Light Scattering from Multicomponent Polymer Systems in Shear Fields: Real-time, In Situ Studies of Dissipative Structures in Open Nonequilibrium Systems.* In *Soft Mater Characterization*, Borsali, R. and Pecora, R. (eds.), Springer New York, Chap. 8, pp. 378–462.
21. Endo, M., Saito, S., and Hashimoto, T. (2002) *Macromolecule,* **35**, 7692–7699.
22. Murase, H., Kume, T., Hashimoto, T., and Ohta, Y. (2005) *Macromolecules*, **38**, 6656–6665.
23. Murase, H., Kume, T., Hashimoto, T., and Ohta, Y. (2005) *Macromolecules,* **38**, 8719–8728.
24. deGroot, Jr. J.V., Macosko, C.W., Kume, T., and Hashimoto, T. (1994) *J. Colloid, and Interface Sci.*, **166**, 404–413.
25. Mani, S., Malone, M.F., Winter, H.H., Halary, J.L., and Monnerie, L. (1991) *Macromolecules*, **24**, 5451.
26. Gerard, H., Higgins, J. S., and Clarke, N. (1999) *Macromolecules*, **32**, 5411.
27. Hindawi, I.A., Higgins, J.S., and Weiss, R.A. (1992) *Polymer*, **33**, 2522.
28. Gerard, H. and Higgins, J.S. (1991) *Phys. Chem. Chem. Phys.*, **1**, 3059.
29. Gerard, H., Cabral, J.T., and Higgins, J.S. (2003) *Phil. Trans. Soc. Lond. A*, **361**, 767.
30. Chopra, D., Vlassopoulos, D., and Hatzikiriakos, S.G. (1998) *J. Rheol.*, **42**, 1227.
31. Saito, S., Takenaka, M., Toyoda, N., and Hashimoto, T. (2001) *Macromolecules*, **34**, 6461–6473.
32. Helfand E. and Fredrickson, G. H. (1989) *Phys. Rev. Lett.*, **62**, 2468.
33. Milner, S. T. (1993) *Phys. Rev. E.*, **48**, 3674.
34. Onuki, A, (1997) *J. Phys. Condens. Matter.*, **9**, 6119.
35. Saito, S. and Hashimoto, T. (2001) *J. Chem. Phys.*, **114**, 10531–10543.
36. Endo, M.K., Takenaka, M., Inoue, T., Watanabe, H., and Hashimoto. T (2008) *J. Chem. Phys.*, **128**, 164911-1 - 164911-12.
37. Bird, R.B., Armstrong, R.C., Hassager, O. (1977) *Dynamics of Polymeric Liquids, Wiley*, New York.
38. Kume, T., Hattori, T., and Hashimoto, T. (1997) *Macromolecules*, **30**, 427–434.
39. Osaki, K., Inoue, T., Uematsu, T., and Isomura, T. (2008) *J. Polym. Sci., Part B: Polym. Phys.*, **38**, 1917.
40. Saito, S., Matsuzaka, K. and Hashimoto, T. (1999) *Macromolecules*, **32**, 4879–4888.
41. Saito, S., Hashimoto, T., Morfin, L., Lindner, P., and Boué, F. (2002) *Macromolecules,* **35**, 445–459.
42. Hashimoto, T. (2005) *Bull. Chem. Soc., Jpn.*, **78**, 1–39.
43. de Gennes, P.G. (1979) *Scaling Concept in Polymer Physics*, Cornell University, Ithaca.
44. Zernike, F. (1992) *Z. Tech. Phys.*, **16**, 454.
45. NIH Image is the public domain Image program on a Macintosh computer written by Wayne Rasband at the U.S. National Institute of Health and available from the Internet by anonymous ftp from zippy. nimh.nih.gov or floppy disk from NTIS, 5825 Port Royal Rd., Springfield, VA 22161, part number PB93–504868.
46. Link, A. and Springer, J. (1993) *Macromolecules*, **26**, 4648.
47. Kume, T. and Hashimoto, T. (1995) *Chapter 3 String Phase in Semidilute Polystyrene Solutions under Steady Shear Flow*: In *Flow-Induced Structure in Polymers, ACS Symp. Series*, **597**, 35–47, Nakatani, A.I. and Dadman, MD. (eds.).
48. Hashimoto, T. (2022) *Principles and Applications of X-ray, Light and Neutron Scattering*, Springer.
49. Murase, H., Ohta, Y., and Hashimoto, T. (2009) *Polymer*, 50, 4727–4736.
50. Hashimoto, T., Murase, H., and Ohta, Y. (2009) *Macromol. Symp.*, **279**, 88–95.
51. Saito, S., Koizumi, S., Matsuzaka, K., Suehiro, S., and Hashimoto, T. (2000) *Macromolecules*, **33**, 2153–2162.
52. Saito, S., Hashimoto, T., Morfin, L., Lindner, P., Boue, F., and Pine, D.J. (2003) *Macromolecules*, **36**, 3745–3748.
53. Takenaka, M., Miyazawa, M., Nishitsuji, S., and Hashimoto, T. (2004) *J. Chem. Phys.*, **121**, 7501–7504.
54. Endo, M., Takenaka, M., and Hashimoto, T. (2006) *Polymer* **47**, 7271–7281.

Chapter 6
Flow-Induced Ordering of Crystallizable Polymer Solutions into Shish-Kebab and Extended Chain Crystal Texture through Cascade Evolution of Dissipative Structures

Abstract This chapter will discuss the self-organization of ultrahigh molecular weight polyethylene (UHMWPE) in decline or paraffin as solvent as typical dynamically asymmetric crystallizable solutions under flow. The self-organization of solutions of crystallizable UHMWPE is discovered to be essentially identical to that of dynamically asymmetric noncrystallizable UHMWa-PS solutions under flow as discussed in Chap. 5, regarding the evolution of the series of the dissipative structures in Regime I–IV via flow-induced liquid–liquid phase transition (L–L) except one important difference regarding presence or absence of the flow-induced liquid–solid phase transition (L–S) (crystallization). The L–S is discovered to interplay with the L–L in the late stage of the self-organization in Regime IV, resulting in the self-organization into shish-kebab crystalline texture (SK) by means of the online shear rheo-optical method. The morphological details of the cascade evolution of the series of the dissipative structures are confirmed by means of the observation of the off-line transmission electron microscope images, taken by the special method as described in the text, along the fiber spinning line as a function of the distance along the spinneret. The results disclose the detailed pathway regarding the self-organization of SK through the series of various dissipative structures and general principles of the cascade pattern formation in nonequilibrium systems under external flow fields: both shear flow fields for UHMWPE and UHMWa-PS solutions universally evolve the same dissipative structures controlled by the work done and dissipated by the mechanical energy imposed to the systems. This chapter will also discuss the self-organization of extended chain crystal texture through dissipative structures under the hot drawing of the SK textures at temperatures close to but below their melting temperatures where the hot-drawn-induced sliding diffusion of chains occur along the chain axis in the shish and kebab crystals.

T. Hashimoto, *Polymer Phasing*, https://doi.org/10.1007/978-981-97-9648-9_6

6.1 Introduction: Basic Concept Regarding Interplay of L-L and L-S Phase Transition

Up to this chapter, formation of the flow-induced structures is discussed for semidilute noncrystallizable polymer solutions only. What would happen if polymers were crystallizable? This case may possibly involve an intriguing interplay of the two kinds of phase transitions on formation of flow-induced dissipative structures: **flow-induced liquid–liquid (L–L) phase transition** (between a single-phase solution and a phase-separated solution) and the **flow-induced liquid–solid (L–S) phase transition** (crystallization and melting). The interplay of the two phase transitions in the crystallizable polymer solutions generally depend on the following two characteristic temperatures, the critical temperature for the phase separation $T_c(\dot{\gamma} = 0)$ and the melting temperature $T_m(\dot{\gamma} = 0)$ for the quiescent semidilute solutions. The interplay depends also on the shear rate $\dot{\gamma}$, because $\dot{\gamma}$ increases the characteristic temperatures $T_c(\dot{\gamma} = 0)$ and $T_m(\dot{\gamma} = 0)$ into $T_c(\dot{\gamma})$ and $T_m(\dot{\gamma})$, respectively, for the dynamically asymmetric solutions. $T_c(\dot{\gamma} = 0)$ and $T_m(\dot{\gamma} = 0)$ are designated hereafter as $T_c(0)$ and $T_m(0)$, respectively, for the sake of brevity. Table 6.1 presents a list of frequently used abbreviations in this chapter.

Table 6.1 Glossary of frequently used abbreviations

BD	Bridging domains	NG	Nucleation and growth
BOSC	Bundle of Stretched Chains	ND	Neutral direction
CF	Concentration Fluctuations	OM	Optical microscope
CSCT	Coil-to-Stretched-Chain Transition	PLWCF	Plane wave concentration fluctuations
ECC	Extended Chain Crystal Texture	SALS	Small-angle light scattering
FD	Flow Direction	SD	Spinodal decomposition
L–L	Liquid–Liquid	SK	Shish-kebab crystalline texture
L–S	Liquid–Solid	UHMWPE	Ultrahigh molecular weight polyethylene
		UHMWaPS	Ultrahigh molecular weight atactic polystyrene
$T_m(0)$	Melting temperature at $\dot{\gamma} = 0$ (in quiescent state)	τ_e	Chain retraction time $\tau_e = 2\tau_R$
$T_m(\dot{\gamma})$	Melting temperature under the shear flow at a given $\dot{\gamma}$	τ_m	The longest rheological relaxation time, disentanglement time
T_{obs}	Working temperature for observation of the solution under shear flow and in quiescent sate	τ_R	The longest Rouse relaxation time of polymer chains

Figure 6.1 presents a scheme of the interplay between the L–L and L–S phase transition. Part **a** presents the L–S phase transition. In the quiescent solution ($\dot{\gamma} = 0$), part (a-1), the L–S phase transition occurs at $T = T_{\mathrm{m}}(0)$. The solution is (amorphous) liquid at $T > T_{\mathrm{m}}(0)$ but solid at $T < T_{\mathrm{m}}(0)$ where the crystallization occurs even without the shear flow.

In the solutions under the flow, part (a-2), the shear-induced phase separation may occur in addition to crystallization in the case where the working temperature T, defined here as T_{obs}, is in between $T_{\mathrm{m}}(\dot{\gamma} = 0)$ and $T_{\mathrm{m}}(\dot{\gamma})$, because $T_{\mathrm{m}}(\dot{\gamma} = 0)$ may increase to $T_{\mathrm{m}}(\dot{\gamma})$ with increased shear rate, so that the solution phase becomes the solid phase after the uprise of the melting temperature at the given $\dot{\gamma}$ above any given working temperatures T_{obs}'s. Thus, the dissipative structure formation becomes quite complex. We shall discus below only the case where any given experimental working temperatures T's satisfy the condition of $T > T_{\mathrm{m}}(0)$, so that the crystallization occurs only under the shear flow. The melting temperature of the solution under the flow increases from $T_{\mathrm{m}}(0)$ to $T_{\mathrm{m}}(\dot{\gamma})$ with the increased shear rate as shown in part (a-2), so that the shear-induced crystallization occurs at T satisfying $T_{\mathrm{m}}(0) < T < T_{\mathrm{m}}(\dot{\gamma})$.

The L–S phase transition under the flow occurs at any given working temperature T above $T_{\mathrm{m}}(0)$, defined as T_{obs}, when $T_{\mathrm{m}}(\dot{\gamma})$ reaches or becomes higher than T_{obs}, i.e., when $T_{\mathrm{obs}} \leq T_{\mathrm{m}}(\dot{\gamma})$ with the increased $\dot{\gamma}$ (part a-2 in Fig. 6.1a). The critical

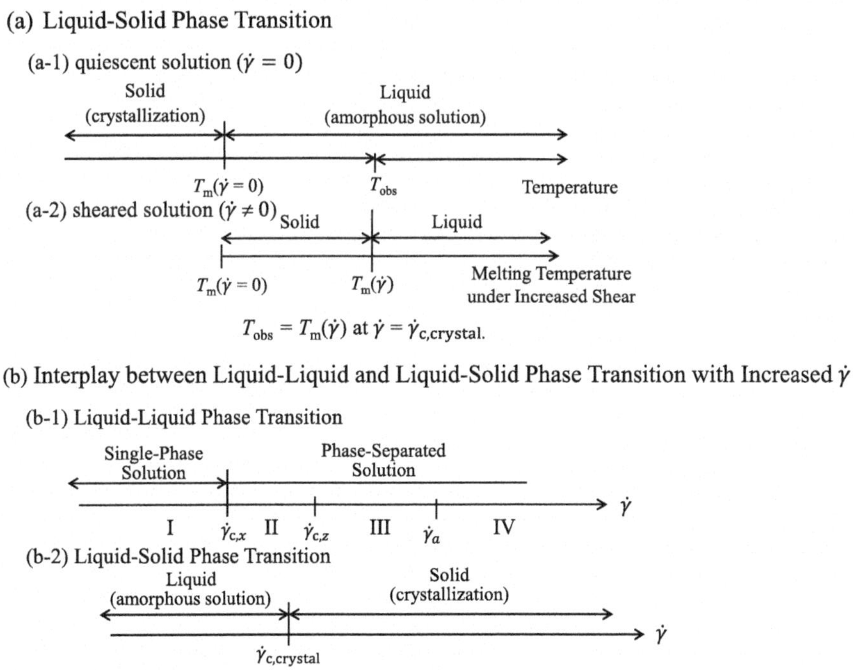

Fig. 6.1 The liquid-solid phase transition (part **a**) and the interplay between the liquid–liquid phase transition and the liquid–solid phase transition (part **b**) for the crystallizable polymer solution with the increased shear rate protocol

shear rate which satisfies this condition is defined as $\dot{\gamma}_{\mathrm{c,cryst}}$, i.e.,

$$T_{\mathrm{obs}} = T_{\mathrm{m}}(\dot{\gamma}) \text{ at } \dot{\gamma} = \dot{\gamma}_{\mathrm{c,cryst}}. \tag{6.1.1}$$

The solution under the flow has the L–S phase transition at T_{m} ($\dot{\gamma}$): it is (the amorphous) liquid at $T_{\mathrm{obs}} > T_{\mathrm{m}}$ ($\dot{\gamma}$) but solid (crystallized) at $T_{\mathrm{obs}} < T_{\mathrm{m}}$ ($\dot{\gamma}$) as shown in part (a-2) in Fig. 6-1a.

Figure 6.1b presents the interplay between the L–L (b-1) and L–S phase transition (b-2) under the flow with the increased shear-rate protocol. The L–L phase transition of the solution under the shear flow (part b-1) occurs at the critical shear rate $\dot{\gamma}_{\mathrm{c},x}$, above which the shear-induced concentration fluctuations are built up along the flow direction, due to the **spinodal decomposition** (**SD**) driven by the thermodynamic instability of the solution under the shear flow along flow direction (**FD**). The SD first builds up the "**plane wave concentration fluctuations**" (**PLWCF**) with its wave vector oriented along FD at the early stage of the SD in Regime II at $\dot{\gamma}$ satisfying $\dot{\gamma}_{\mathrm{c},x} < \dot{\gamma} < \dot{\gamma}_{\mathrm{c},z}$ and subsequently "**phase-separated domain structures**" with their principal axes oriented also along FD at the late stage of the SD in Regime III at $\dot{\gamma}$ satisfying $\dot{\gamma}_{\mathrm{c},z} < \dot{\gamma} < \dot{\gamma}_{\mathrm{a}}$.

Thus, the "**SALS- butterfly scattering pattern**" appears first in the early stage SD and subsequently the "**SALS domain scattering pattern**" in the late stage SD. At $\dot{\gamma} < \dot{\gamma}_{\mathrm{c},x}$ (in Regime I), the solution under the shear flow forms the homogeneous solution in the single phase state. At $\dot{\gamma} > \dot{\gamma}_{\mathrm{c},x}$, the solution under the shear flow first builds up the dissipative structures of the so-called PLWCF which is a single-phase solution with a large amplitude periodic concentration fluctuations (in Regime II) and subsequently at $\dot{\gamma} > \dot{\gamma}_{\mathrm{c},z}$ the dissipative structures composed of the two-phase solution (phase-separated domains in the matrix solution) with the increased shear-rate protocol (part b-1) in Fig. 6.1b. The two-phase solutions under the flow at $\dot{\gamma} > \dot{\gamma}_{\mathrm{c},z}$ evolve the dissipative structures, the evolution pathway of which can be classified into Regime III to Regime IV with the increased shear-rate above $\dot{\gamma}_{\mathrm{a}}$ as shown in part (b-1) in Fig. 6.1b: Here, $\dot{\gamma}_{\mathrm{c},z}$ is the critical shear rate where the phase-separated domains with well-defined interface are formed along the Oz axis (neutral direction ND), and $\dot{\gamma}_{\mathrm{a}}$ is the critical shear rate above which the optically anisotropic fibrillar (string-like) structures are formed. The solutions at any given $\dot{\gamma}$'s above $\dot{\gamma}_{\mathrm{a}}$ exhibit the scattering pattern composed of the butterfly pattern with its two lobes oriented along q_x plus the strong streak-like pattern centered at $q_x = 0$ and extended along the q_z axis as presented in Fig. 5.31 and 5.32 in Sect. 5.9 for the a-PS/DOP solutions (of non-crystallizable polymers).

The structures and properties of the dynamic asymmetric crystallizable polymer solutions under flow in Regime I to IV will be shown later in this chapter to be identical to those of dynamically asymmetric noncrystallizable polymer solutions under the flow in the Regime I to IV as presented in Figs. 5.7, 5.12, 5.13, and 5.35 in Chap. 5.

The critical shear rate $\dot{\gamma}_{\mathrm{c},x}$ and $\dot{\gamma}_{\mathrm{a}}$ are generally functions of T. If $T_{\mathrm{obs}} > T_{\mathrm{m}}(0)$, as in the case at $T_{\mathrm{obs}} = 150\,^{\circ}\mathrm{C}$ for the 5.0 wt.% solution of ultrahigh molecular weight polyethylene with $M_{\mathrm{w}} = 2.0 \times 10^6$ in n-paraffin with $T_{\mathrm{m}}(0) = 110\,^{\circ}\mathrm{C}$, the critical shear rate for the formation of the string-like structures $\dot{\gamma}_a$ exceeds the shear rate range covered by the present experiments ($\dot{\gamma} < 1000\,\mathrm{s}^{-1}$), so that only the phase-separated structures Regime II and III are observed above $\dot{\gamma}_{c,x}$ [1]. However, if T_{obs} is sufficiently low, as in the case of $T_{obs} = 124\,°\mathrm{C}$, the systems first evolve the flow-induced concentration fluctuation in Regime II, then phase-separated domains in Regime III, and finally anisotropic fibrillar (string-like) structures in Regime IV with the increased $\dot{\gamma}$ [2] as shown in Fig. 6.1 (part b-1).

The crystallizable polymer solution can undergo the L–S phase transition at $\dot{\gamma} = \dot{\gamma}_{c,cryst}$ with the increased shear rate in a suitable temperature range of T_{obs}, as shown in part (b-2), as $\dot{\gamma}_{c,cryst}$ also depends on temperature T_{obs}. The interplay of these two kinds of the phase transitions depend on the value of $\dot{\gamma}_{c,cryst}$ relative to the value of $\dot{\gamma}_{c,x}$, $\dot{\gamma}_{c,z}$, and $\dot{\gamma}_a$, which could bring a rich variety of the interplay and thereby a rich variety of the dissipative structures.

In this chapter, we discuss the unique pattern formation process of **shish-kebab (SK)** and **extended chain crystal (ECC)** texture via the crystallization of the crystallizable polymer solution subjected to external flow fields. The SK and the ECC texture to be discussed here involves an extremely large conformational entropy loss of the polymer chains for their crystallization processes. Thereby, the associated free energy barrier can be extremely large and hardly be overcome thermally without the external fields. The external fields may effectively reduce the barrier step by step in a "**cascade**" manner, enabling to form various dissipative structures which are initially amorphous. These amorphous dissipative structures evolved under the applied flow field may provide excellent fields or "**templates**" for a birth of the unique crystalline textures even in the case where the bulk stress and the bulk chain orientation induced by the applied flow are relatively low. This is because the local stress or orientation built-up in the templates can be large enough due to the local stress concentration, which causes stress-induced or orientation-induced crystallization even under the relatively low bulk stress or orientation. In this way, the external fields, in the form of shear flow, extensional flow, hot drawing involving deformation and flow, etc., effectively reduce entropy of the systems via the cascade evolution of the dissipative structures to bring about ordered structures through fluctuations [3].

The "**shish-kebab**" self-assembly is a most intriguing but fundamental crystalline superstructure formed in polymer solutions and melts under the various flow fields. Since the first discovery of the shish-kebab structure [4,5] about a half century ago, a great number of researchers have been attracted to this flow-induced evolution of this superstructure and tried to elucidate its formation mechanism based on the **flow-induced crystallization** phenomenon [6–19]. Nevertheless, the detailed kinetic pathways leading to the shish-kebab formation have not yet been fully clarified. Furthermore, its investigation has now been one of the extremely hot research themes in polymer physical science. One may investigate the kinetic pathway as one of general research themes which belong to open nonequilibrium phenomena, regarding the self-organization of non-equilibrium polymer systems open for the externally applied energy flows [20,21].

Investigation of the early-stage pathway may be important for the self-organization of the SK. This is because our systems involve in general two kinds of the flow-induced phase transitions: the phase separation (L–L phase transition) and crystallization (L–S phase transition). Hence, a coupling of them seems to be crucial to unveil the pathway from entangled coils swollen with solvents to the SK. If the flow-induced crystallization occurs prior to the flow-induced phase separation, so that the effects of the phase separation do not utterly play roles on the pathway, the traditional approach would suffice to solve the problem.

However, if the opposite is the case, the traditional approach would not suffice; we must account for roles of the phase separation on the crystallization as an early-stage pathway leading to the shish-kebab structure. Surprisingly, *the coupling effects* have been left almost unexplored up to now in this research field, except for the work by McHugh and coworkers [22] and by us [2, 23–26]. This may be because the flow-induced phase separation, which could occur prior to the crystallization (in the early-stage kinetic pathway leading to SK), can be observed only in situ at real time, as they decay rapidly after the cessation of the flow, and hence it is difficult to be observed ex-situ after the cessation of the flow. It is also because signatures of the flow-induced phase separation do not cause the remarkable increase of birefringence Δn and are discerned at the length scale much larger than that relevant to SAXS, SANS, and WAXD which are conventionally employed for the studies of the oriented crystallization of SK.

6.2 Background

6.2.1 Crystallization of Homogeneous Solutions into Shish-Kebab Crystalline Texture and Extended Chain Crystal Texture

Polyethylene (PE) is used as a model of a crystallizable polymer having the simplest chemical structure unit $-(CH_2-CH_2)_n-$ as shown in part (a) in Fig. 6.2. Each chemical bond is linked with a fixed valence angle θ_b and a varying rotational angle ϕ_b as shown in part (b): the potential energy for the bond rotation has minima at the rotational angle $\phi_b = 0°$ and $\pm$ 120 °, designated, respectively, as t (trans), g^+ and g^- (gauche) as shown in part (c). If an ultrahigh molecular weight PE ("UHMWPE") has molecular weight of 2×10^6, its number of bonds $\cong 1.4 \times 10^5 \equiv$ 140K. If each bond adopts the three rotational states (t, g^+, g^-), the number of possible states per chain $\Omega \sim 3^{140K}$. Thus, each chain has an extremely large conformational entropy; the chain has a random-coil conformation in amorphous melts and solutions, thereby each chain is thermodynamically stabilized by the large entropy [27]. A snapshot of a chain is shown in part (d). Part (e) presents the skeletal model (e-1) and the space-filling model (e-2) for a fully stretched chain with all bonds having t (trans) conformation. The length of the fully stretched chain is calculated to be ~ 50 μm for the UHMWPE

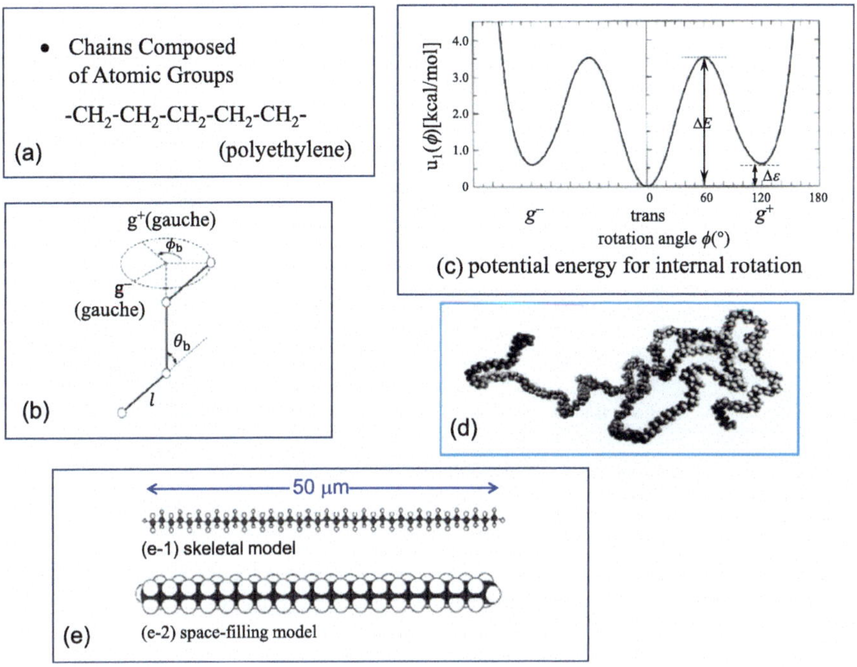

Fig. 6.2 Polyethylene chain (part **a** and **b**) with internal rotation around C–C bonds (part **b** and **c**), a snapshot of random-coil configuration (part **d**), and the skeletal and space-filling model for a stretched chain (part **e**)

with molecular weight of $\sim 2 \times 10^6$ and number of bonds $\sim 1.4 \times 10^5 \equiv 140\text{K}$. The probability having such fully extended chain $\sim (1/3)^{140\text{K}} \sim 0$.

When the chains are crystallized, they are incorporated in the crystal lattice of orthorhombic unit cell with all bonds in the trans conformation as shown in Fig. 6.3a. In the limit where all the chains are stretched out and packed in the orthorhombic crystal lattice, we could attain the **extended chain crystal ("ECC")** as schematically shown in Fig. 6.3b: its length parallel to *c*-axis is ~ 50 μm for the given molecular weight as indicated in part (e-1) in Fig. 6.2e. Here, the cost of the large free energy due to the large entropy loss is outweighed by the large free energy gain attained by the large enthalpy release due to inter-chain associations in the crystal. However, the ECC has never been obtained by crystallization from the homogeneous solutions of random coils in the absence of the applied fields, because of infinitely large conformational entropy loss and thereby infinitely large free energy barrier encountered by the crystallization process.

The most fundamental crystalline texture attained by crystallization in flexible polymers in the absence of the applied fields is the chain-folded lamella [28, 29], as shown in Fig. 6.3c as a consequence of a balance of the two opposing physical factors encountered by the crystallization process: the cost of conformational entropy loss and the gain of bulk free energy due to the interchain associations. In this article, we

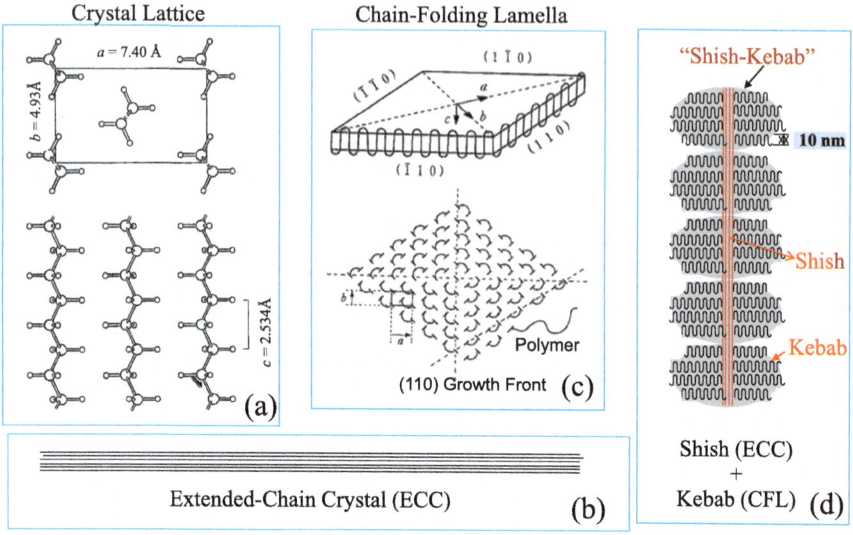

Fig. 6.3 Polyethylene chain in the crystal lattice (part **a**), extended chain crystal (part **b**), chain-folded lamellar crystal (part **c**), and shish-kebab crystalline texture (part **d**)

define the **ECC** as **extended chain crystal "texture"**, because it may contain some defects such as chain ends, chain-folded parts, etc. Figure 6.3d presents schematically the SK model. The SK is a most intriguing but fundamental crystalline texture in polymers, as it has a mixed feature of the two important crystal elements: the ECC (the part shown in Fig. 6.3b) which constitutes the so-called shish at the central core of the SK and the chain-folded lamellae (the part shown in Fig. 6.3c) which constitutes the so-called kebabs overgrown from the as-grown shish.

About a half century ago, the **shish-kebab crystalline texture ("SK")** was discovered in stirred polymer solutions [4, 5]. Figure 6.4 presents a stirrer made of a concentric cylinder to enclose and stir a temperature-controlled polymer solution (part a), where the SK structure was discovered, a transmission electron micrograph (TEM) of the SK (part b), and the SK model obtained from the TEM (part c) reported by Pennings et al. [30]. The SK is composed of the so-called "shish" crystal as a central core and the so-called "kebab" crystal (part c). The shish crystal is the ECC, while the kebab crystal is the chain-folded lamella epitaxially overgrown on the surface of as-grown shishs from random coils coexisting with the shishs. Since the discovery of the SK, a great number of researchers have been trying to clarify its formation mechanism and kinetic pathway. Nevertheless, the mechanism and pathway are not fully understood yet up to date (see review articles by Keller and Kolnaar [6] and Somani et al. [9]).

It is most important to understand the mechanism which enables the reduction of extremely large conformation entropy of macromolecules for the crystallization into the ECC and the SK. Let us consider an ordering into a given ordered structure [**F**] from an initially homogeneous structure [**I**] under the condition as shown in Fig. 6.5

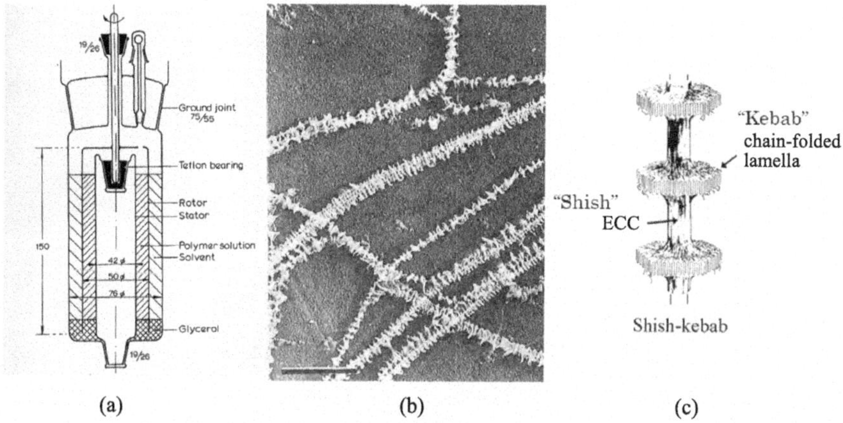

Fig. 6.4 Stirrer made of a concentric cylinder to enclose a temperature-controlled polymer solution (part **a**), a transmission electron micrograph (TEM) (part **b**), and the SK model obtained from the TEM (part **c**)

[31], where the ordering is thermodynamically favored. Even under this condition, the ordering from [**I**] to [**F**] would never occur in the absence of the field applied to the system in the case where the corresponding free energy barrier [**A**] is exceedingly high as illustrated in Fig. 6.5. This is because the ECC is composed of essentially fully extended chains, and SK is composed of a significant (or substantial) amount of fully extended chains.

However, the applied field may lower the free energy barrier as illustrated in the curve [**B**] via the cascade evolution of a series of the dissipative structures ① to ④, for example, which is anticipated to occur as a consequence of the cascade dissipation of the mechanical energy imposed to the solution by the applied field. The evolution of each structure involves a small free energy barrier as illustrated by the anticipated free energy landscape [**B**], which may lead to a novel and rational kinetic pathway for the ordering.

How do the applied fields assist the crystallization of the crystallizable polymer solution into the SK and the ECC, through the cascade formation of what kind of dissipative structures and through what kind of mechanisms? We may be able to clarify these fundamental questions based on online, in situ explorations of time-evolutions of the dissipative structures with a rheo-optical method [21, 32, 33] involving simultaneous measurements of **the stress (σ)**, the **birefringence (Δn)**, **the optical microscope images ("OM")**, **the light transmittance**, and the **small-angle light scattering** ("**SALS**") as described in Chap. 3. Fine details of the dissipative structures captured by the online investigation may be further explored by off-line ex-situ observations of the dissipative structures with TEM.

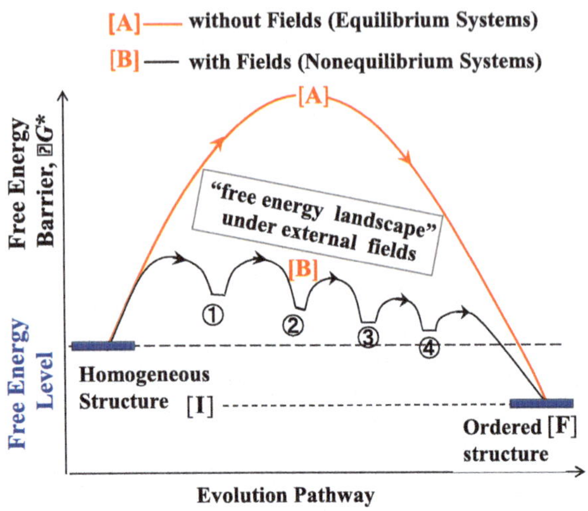

Fig. 6.5 Concept of the flow-induced reduction of a free energy barrier from **[A]** to **[B]** via the cascade evolutions of a series of dissipative structures ① to ④ for the ordering from an initially homogeneous structure **[I]** into an ordered structure **[F]**. External fields assist to suppress an excessively large free energy barrier **[A]** in the absence of the fields into a free energy landscape **[B]**, which enables the ordering from **{I]** to **[F]** that can be hardly attained in the absence of the fields. Based on [31]

6.2.2 Previous Studies of Shish-Kebab Formation

Keller pointed out that the coil-to-stretched chain transition (CSCT) [6], which was theoretically predicted by de Gennes [34], plays an important role on the flow-induced crystallization of polymers into the shish-kebabs. He revealed that an abrupt increase in Δn above a critical strain rate is indicative of an appearance of the fully extended chains in the dilute solution under the extensional flow. Stimulated by his work, many experimental [7–19] and simulation works [35–37] have been carried out to clarify the CSCT and subsequent shish-kebab formation. Many researchers have investigated polymer melts, especially binary melt blends of the same polymer species having high and low molecular weight [7, 9, 11, 12] to elucidate the CSCT occurring in the sections of the chains between the entanglements. Here, we would like to address the following question: whether do the CSCT and subsequent formation of bundles of stretched chains occur in uniform solutions and melts or in the phase-separated solutions and melts under the flow fields. In the case of the binary melt blends, the high molecular weight species may act as "polymer", while the low molecular weight species may act as "solvent" for the flow-induced phase separation in the context of the dynamic asymmetry [20, 21, 24].

6.3 Experimental Systems and Methods

6.3.1 *Materials*

The polymer used is an ultrahigh molecular weight linear polyethylene (UHMWPE, Hizex 240 M, Mitsui Chemicals, Tokyo, Japan) having the weight-average molecular weight $M_\mathrm{w} = 2.0 \times 10^6$, the weight-average degree of polymerization $N_w \cong 1.7 \times 10^5$, and the heterogeneity index $M_\mathrm{w}/M_\mathrm{n} = 12$, where M_n is the number-average molecular weight. The polymer solutions studied are the solutions with the polymer concentration $c = 10.0$ wt.% ($c/c^* = 20$) in decalin as a solvent, where c^* is the overlap concentration, and the solutions with the polymer concentration $c = 5.0$ wt.% ($c/c^* = 11$) in the paraffin wax with $M_\mathrm{n} \sim 50$ as a solvent at temperature higher than its melting point 69 °C [1, 2, 25]. The melt of the paraffin wax forms athermal solutions with polyethylene (having the thermodynamic segmental interaction $\chi = 0$). The two solutions with the different polymer concentration have $T_\mathrm{m}(0) = 118$ °C and $T_\mathrm{c}(0)$ below 69 °C.

Purification of the solutions . Purification of the solutions is extremely important and indispensable for the SALS experiments, because the refractive index increment of the semidilute polyethylene/paraffin solutions $(\partial n/\partial c)_\mathrm{T} = -0.001$ ml/g is very small compared with that of the semidilute solutions of PS in DOP (0.119 ml/g). The concentrated UHMWPE solutions were prepared by mixing UHMWPE, decalin or paraffin, and an antioxidant agent (2,6-di-tert-butyl-p-cresol) by an amount of 1 wt.% of the total solution, at 210 °C by using a twin-screw extruder, and small impurities insoluble to the solutions were filtered off by a mesh filter (400 lines/inch) as illustrated in Fig. 6.6a. Figure 6.6b, c present the SALS patterns from the UHMWPE/paraffin 3.0 wt.% solution before and after the filtration, respectively, observed at $\dot{\gamma} = 46\,\mathrm{s}^{-1}$ and at 150 °C [1]. The true scattering (the butterfly-SALS pattern) in part (c) is demonstrated to be completely masked by the relatively strong parasitic scattering from the impurities in part (b).

6.3.2 *Online Rheo-Optics*

The solution under shear flow is investigated by the shear-rheo-optical method which measures in situ, at real time and simultaneously the SALS, transmission OM, birefringence Δn, normal stress, and shear stress. Figure 6.7 illustrates the principle of the method. The details of the method should be referred to Chap. 3.

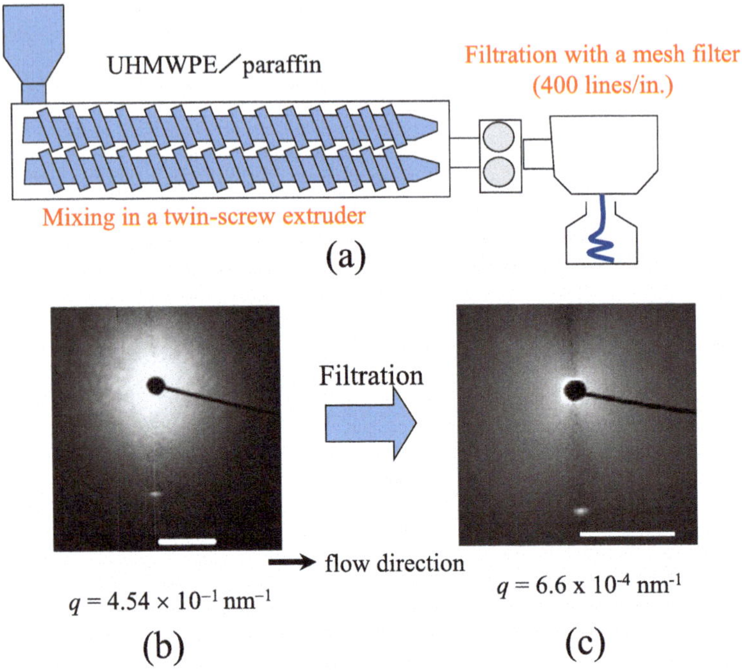

Fig. 6.6 Filtration of impurities in UHMWPE/paraffin 3.0 wt.% solution (part **a**) and its effect on the flow-induced SALS patterns (part **b** and part **c**: before and after the filtration, respectively) taken in situ at $\dot{\gamma} = 46\,\text{s}^{-1}$ and at 150 °C

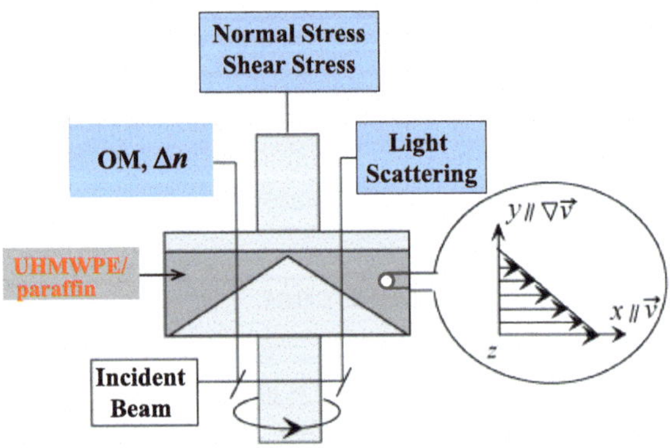

Fig. 6.7 Shear-rheo-optical method: the simultaneous measurements of the normal stress and the shear stress, the small-angle light scattering and the transmitted light intensity, the transmission optical microscope image (OM) and birefringence (Δn) at real time and in situ under the shear flow

6.3.3 *Off-Line TEM Investigations*

Morphological details of a series of the dissipative structures evolved along the fiber spinning line are investigated with an aim to disclose the secrets of the self-organization of solution under the flow-field into the SK and to discover general principles of the cascade pattern formation in the nonequilibrium systems.

(A) **Fiber spinning**. The off-line observations of the dissipative structures developed in the UHMWPE/decalin 10 wt.% solution along the fiber spinning lines were conducted with transmission electron microscopy (TEM). The fiber spinning process involves an extrusion of the purified, homogeneous decalin solutions, which are prepared by a twin-screw extruder, through a spinneret with the throughput rate of 1 g/min as shown in Fig. 6.8 [31]. The spinneret has an orifice diameter of 0.8 mm and is controlled at 170 ± 1 °C. The extrudates were drawn into fibers with a take-up velocity of 60 m/min, while they were cooled down by the nitrogen gas at room temperature as schematically illustrated in Fig. 6.8.

(B) **Sampling of specimens for TEM observations along fiber spinning line**. To explore the flow-induced structures along the fiber spinning line at positions

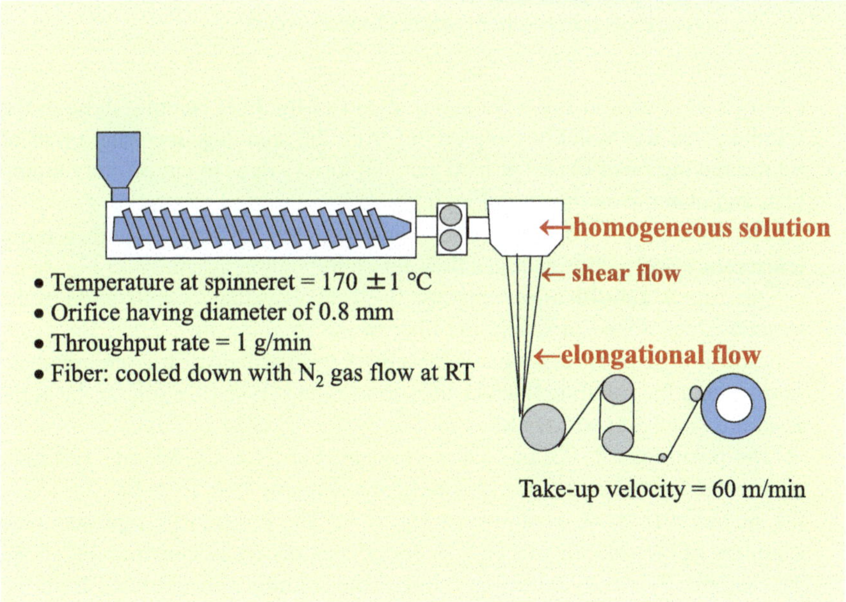

Fig. 6.8 Schematic illustration of the fiber spinning process of the purified, homogeneous UHMWPE/decalin 10 wt.% solutions prepared with a twin extruder. The extruded solution at 170 °C with the throughput rate = 1 g/min was drawn into a fiber with the take-up velocity of 60 m/min. The fiber running through the spinning line was cooled down with the nitrogen gas at room temperature. Based on [31]

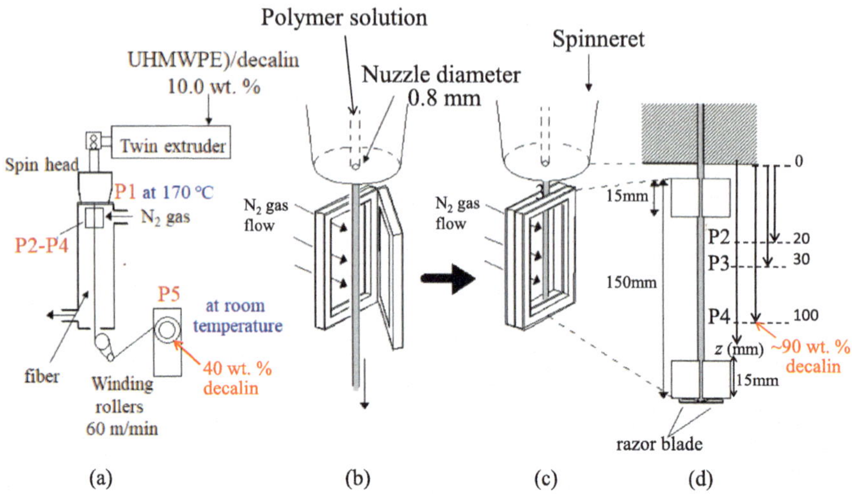

Fig. 6.9 Part **a**: a Schematic illustration of the fiber spinning experiment. Part **b**: Part of the fiber running through the spinning line was quickly sampled out from the spinning line with a pair of frames having razor blades at its upper and lower edges. Part **c**: The specimen fixed by the frame was quickly cooled to room temperature with a nitrogen gas flow for the solidification. The structures in the solidified specimens at the positions P2–P4 shown in part **d** as well as at the positions P1 and P5 specified in part **a** were investigated under TEM after the fixation process illustrated in Fig. 6.10. Based on [38], reprinted with permission, American Chemical Society

P2 to P5 as shown in Fig. 6.9a and d, a part of the fiber running through the spinning line was quickly sampled out from the spinning line with a pair of frames having razor blades at its upper and lower edges to cut off the running fiber and clamp it as shown in Fig. 6.9b, c [38].

The specimen fixed by the clamping frame was quickly cooled to room temperature with a nitrogen gas flow for solidification as shown in Fig. 6.9b, c. This “quick **solidification**” process itself involves the crystallization in the structures developed in situ in the fiber in the spinning line. Thus, the quick solidification process serves to effectively fix the dissipative structures evolved in situ along the spinning line via diffusion-limited crystallization for the high-resolution analyses with TEM.

The positions P2, P3, and P4 were located at $z = 20$, 30, and 100 mm, where z is defined as the distance along the spinning line, from the surface of the spinneret ($z = 0$) as shown in Fig. 6.9d. The position P1 represents the sampling of the solution at 170 °C before the extrusion as shown in Fig. 6.9a. The sample P1 was obtained as follows: the solution was extruded from the orifice at the same temperature (170 °C), with the same rate of throughput (1 g/min), and allowed to be freely fallen under the gravitational force on a flat plate set at 50 mm below the orifice. The solution received on the flat plate was cooled down to room temperature by nitrogen gas flow at room temperature and solidified (crystallized) on the plate without an excess deformation. The samples obtained from the positions P1 to P4 were expected to retain the original structures evolved in the spinning line through the solidification process via diffusion-limited crystallization.

The structures observed in the solidified specimens, especially those at the length scale greater than the lamellar crystals evolved in the solidification process, are anticipated to reflect online (in situ) structures evolved along the spinning line. This is simply because the large-length scale structures relax slower than the solidification rate (the **diffusion-limited solidification** criterion). The structure developed at the end of the spinning line was sampled at position P5 at a winding roller (Fig. 6.9a).

(C) **Fixation of internal structures of fiber specimens swollen with solvent.** The solidified specimens taken out from the spinning line at the positions P1 to P4 are gel-like and swollen with the solvent where the solvent concentration is close to but less than 90 wt.%, and the specimen taken out at the position P5 contains about 30–50 wt.% of the solvent. The solvent must be completely removed before TEM observations without changing the structures developed in the as-sampled spun fibers.

UHMWPE/decalin solution. In the case of the UHMWPE/decalin solution, the solvent decalin was first exchanged to acetone at room temperature by repeating a stepwise soaking of the specimen into an acetone/decalin mixture, in which the composition of acetone/decalin was stepwise changed from 50/50 to 100/0 wt.%/wt.%, as shown in part (a_1) to part (a_4) in Fig. 6.10A [38]. Acetone was then exchanged to epoxy monomer by using the same solvent exchange procedure as described above, as shown in part (a_5) to part (a_7) in Fig. 6.10A. The epoxy monomer was then cured into epoxy resin at 60 °C for 8 h (part a_8 in Fig. 6.10 A). We define this process including both the solvent exchange and the curing as the "**fixation process**". This process should be distinguished from the process defined as the "solidification" in Sect. 6.3.3 (item (B)). The fixation process was confirmed to hardly affect the structures developed in the solidified fibers as will be shown later in Sect. 6.6.2 (item (A)).

UHMWPE/Paraffin solution. The UHMWPE/paraffin solution after the cessation of the imposed shear at $\dot{\gamma} = 2.9\,\text{s}^{-1}$ for 130 s at 124 °C in the cone-and-plate fixture, under the condition of which the SK is expected to be self-organized in the solution, was rapidly cooled down below T_m of the paraffin (69 °C) in the time scale of roughly about 10 s. The rapid cooling was attained by soaking the whole assembly composed of the cone-and-plate fixture made of copper and the specimen in between the fixture, as shown in Fig. 6.11a, into liquid nitrogen in order to solidify the specimen by means of crystallization of paraffin as a solvent [23].

For investigation of the structures developed in the solution under the transmission electron microscopy (TEM), a small piece of the sample was cut out from the solidified solution with a recognition of $Oxyz$ coordinate with respect to FD (Ox direction), the velocity gradient direction (Oy direction), and ND (Oz direction) based on the sample shape and the cutting directions as shown in Fig. 6.11b and as further detailed later in Fig. 6.12. The sample was then immersed in the liquid paraffin at 79 °C

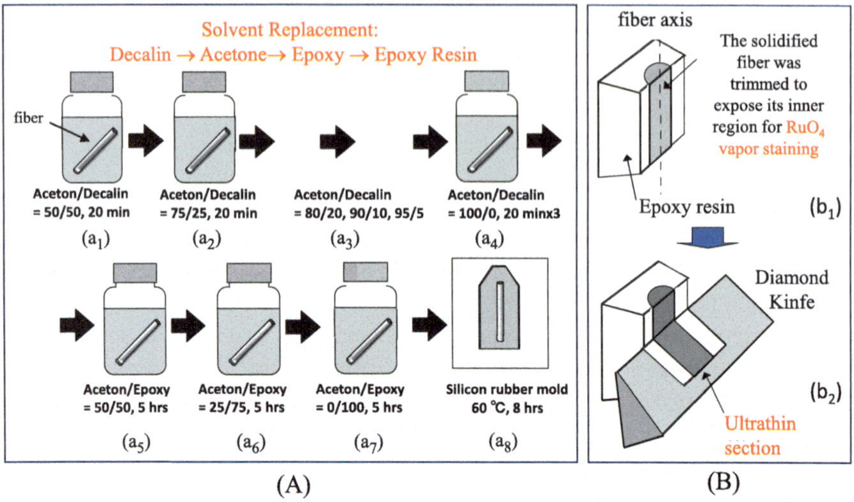

Fig. 6.10 Part **A** : Fixation process of the solidified specimens swollen with the solvent involves first the stepwise solvent replacement from decalin to acetone (parts a_1 to a_4), then acetone to epoxy monomer (part a_4 to a_7), and finally the epoxy monomer into the epoxy resin at 60 °C for ∼ 8 h. (part a_8). Part **B:** The block of the epoxy resin was trimmed to expose the inner region of the fiber for staining with RuO_4 vapor for 10 h. at room temperature (part b_1). The stained specimen was cut into ultrathin sections of approximately 80 nm in thickness parallel to the fiber axis with a diamond knife (part b_2). The ultrathin sections had the size of ∼ 2 mm along the fiber axis and contained a whole fiber in the direction perpendicular to the fiber axis

above the T_m of the paraffin used as a solvent under a gentle agitation for 1 h, as shown in Fig. 6.11c, in order to exchange the paraffin to the noncrystallizable liquid paraffin. The liquid paraffin was then replaced stepwise by propylene oxide from liquid paraffin/propylene oxide with 100/0 wt.%/wt.% (part c) to 50/50 wt.%/wt.% for 30 min (part d), 25/75 wt.%/wt.% for 15 h (part e) and 0/100 wt.%/wt.% for 1 h: the last step was repeated five times (part f). Then propylene oxide also was stepwise replaced by epoxy monomer from propylene oxide/epoxy monomer with 100/0 wt.%/wt.% (part f) to 50/50, 25/75, and 0/100 wt.%/wt.% with each step for 15 h (part g): the last step was repeated three times. The specimen replaced by epoxy monomer was solidified by curing the monomer into the resin at 60 °C for 15 h (part h).

(D) **TEM observations.** A block of the epoxy resin, in which the solidified and fixed fiber specimen was embedded, as described in Sect. 6.3.3 (item (C)), was trimmed to expose the inner region of the fiber for staining with RuO_4 vapor for 10 h at room temperature as illustrated in Fig. 6.10 in part (b_1) in part (B). The stained specimen with the length of ∼ 2 mm along the fiber axis, which includes the whole fiber along its width direction, was cut into ultrathin sections of approximately 70 nm in thickness parallel to the fiber axis with a diamond knife mounted on the ultramicrotome (Ultratome V, LKB Co.

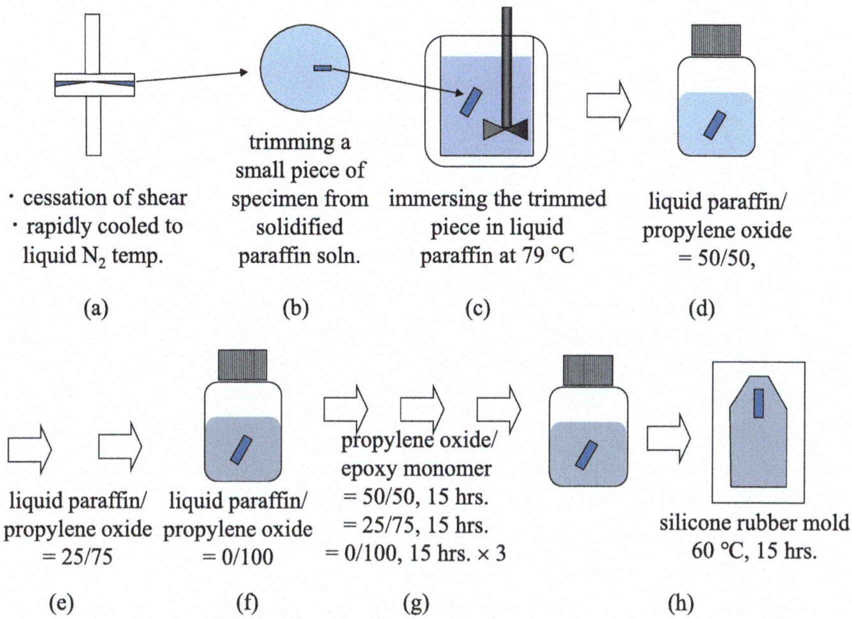

Fig. 6.11 Fixation of the solidified specimen of the sheared UHMWPE/paraffin solution in a cone-plate fixture (part **a** and **b**) by the successive solvent replacement from paraffin, liquid paraffin, propylene oxide to epoxy monomer (part **c** to **g**), and subsequently by embedding the specimen in the cured epoxy resin for the ultramicrotomy

Ltd.) at room temperature as illustrated in Fig. 6.10 in part (b_2) in part (B). The sliced sections were prepared in such a way that they contain the central part of the fiber in the middle of the sliced sections and mounted on carbon-coated poly(vinyl formal) film on copper grids for the TEM observations with JEM-200CX (JEOL, Tokyo, Japan) operated at 100 kV.

The RuO_4 vapor selectively stains the polymer-rich phase but not its matrix phase of the epoxy resin, by which the solvent phase is replaced. The staining is a "negative staining" which involves the selective deposition of RuO_4 nanoparticles on the surfaces of the chain-folded lamellae of the UHMWPE in the polymer-rich phase which is crystallized during the solidification process [39]. Thus, polymer-rich regions appear dark and the solvent-rich regions replaced by the unstained epoxy resin appear bright in the TEM micrograph.

Figure 6.12 presents the ultramicrotomy of the sheared solution subjected to the solidification and fixation as shown in Fig. 6.11 for the TEM observation [23]. A small piece of the specimen(part b) is cut from the solidified UHMWPE/paraffin solution sheared under the cone-plate fixture (part a) with a recognition of the *OXYZ* coordinate from the shape of the trimmed sample specimen and cutting direction as shown in part (a) and (b). After the solvent replacement and fixation of the specimen with the epoxy resin as shown in Fig. 6.11, the specimen is cut into the ultrathin

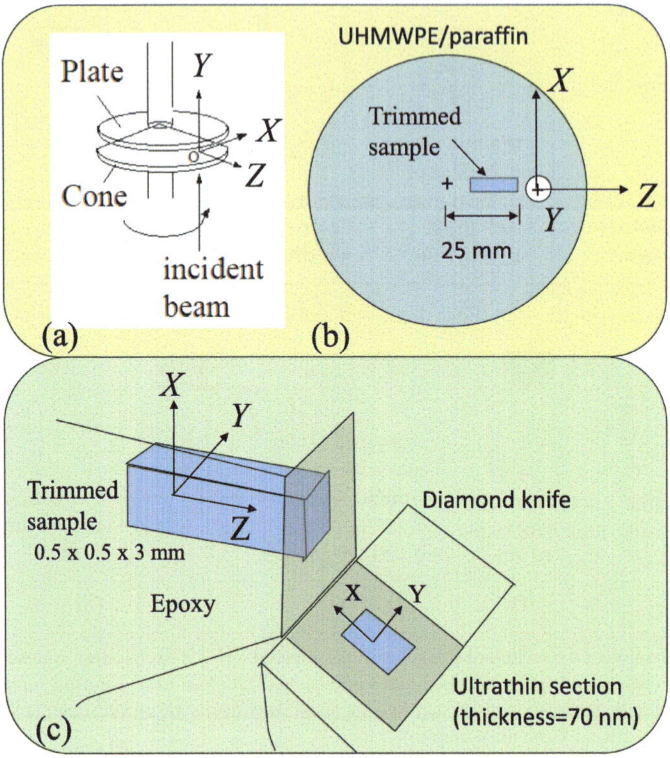

Fig. 6.12 The ultramicrotomy of the UHMWPE/paraffin solution sheared under the cone-plate fixture. Part **a**: The sheared solution and the Cartesian coordinate *OXYZ*. Part **b**: The trimmed small piece of the solidified sample specimen with the recognition of the *OXYZ* coordinate from the shape of the specimen and cutting direction. Part **c**: The ultramicrotomy of the specimen embedded in the epoxy resin, which is obtained after the fixation shown in Fig. 6.11, with the recognition of the *OXYZ* coordinate

section with the thickness of 70 nm with a diamond knife, with the recognition of *OX* axis (FD), *OY* axis (the velocity gradient direction), and *OZ* axis (the ND), along the thickness direction of the ultrathin section as shown in part (c).

6.4 Self-Organization of Polymer Solutions into Shish-Kebab Texture under Flow: Online Rheo-Optical Investigation

The time evolution of a series of the transient dissipative structures is explored at real time and in situ for the UHMWPE/paraffin 5.0 wt.% solution at 124 °C after the step up $\dot{\gamma}$ from 0 to 2.9 s^{-1}, higher than the critical shear rate $\dot{\gamma}_a = 1.2\,s^{-1} \sim {\tau_e}^{-1}$ [23,

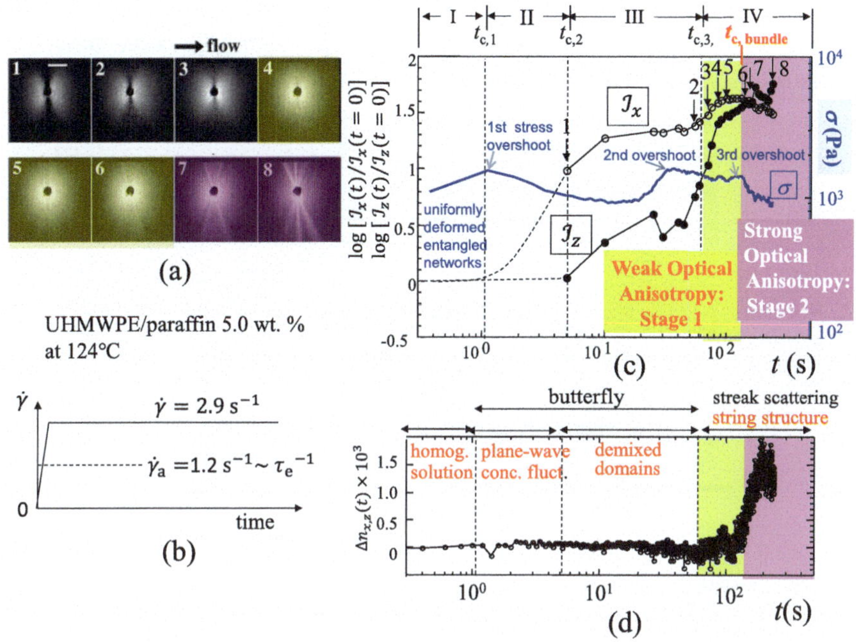

Fig. 6.13 The time evolution of the SALS patterns (part a), the integrated SALS intensities parallel ($\mathcal{J}_x$) and perpendicular to the FD ($\mathcal{J}_z$), and the shear stress σ (part c), and the birefringence $\Delta n_{x,z}(t)$ (part d) simultaneously measured after the step-up shear rate $\dot{\gamma}$ from 0 to 2.9 s^{-1} (part b). $\mathcal{J}_x$ and $\mathcal{J}_z$ taken at the given times t after the step-up shear rate are normalized with $\mathcal{J}_x(t)$ and $\mathcal{J}_z(t)$ at $t = 0$ s, respectively. $\Delta n_{x,z} \equiv n_x - n_z$ where $n_k (k = x$ or $z)$ is the refractive index along the k axis. The data based on [31]

24, 31], as presented in Fig. 6.13b, with the shear-rheo-optical method described in Chap. 3. Here, $\tau_e = 2\tau_R$ where τ_e and τ_R are the chain retraction time required for the stretched chain being relaxed with the Rouse mode and the longest Rouse time, respectively. Figure 6.13 summarizes the online snapshots regarding the time evolution of the shear-SALS patterns #1 to #8 (part a) taken at the particular times after the step up shear as indicated by the arrows with the numbers on the data of $\mathcal{J}_x(t)$ in part (c), the time evolution of the shear stress $\sigma(t)$ and the integrated SALS intensity parallel and perpendicular to the shear flow normalized with those at $t = 0$, $\mathcal{J}_x(t)/\mathcal{J}_x(t = 0)$ and $\mathcal{J}_z(t)/\mathcal{J}_z(t = 0)$, respectively, (part c), and the time evolution of the $\Delta n_{x,z}$ (part d) [31].

Figure 6.14 summarizes the online snapshots of the shear-OM images (a) to (e) together with the corresponding sketches of the dissipative structures (g) to (k) [24, 31]. The OM image (f) was taken in situ at 124 °C after the shear cessation from stage 2 in Regime IV. The TEM image (l) was obtained after cooling the structure shown in image (f) [23, 24]. All the results shown in Figs. 6.13 and 6.14 were simultaneously measured at real time except for the TEM image part (l).

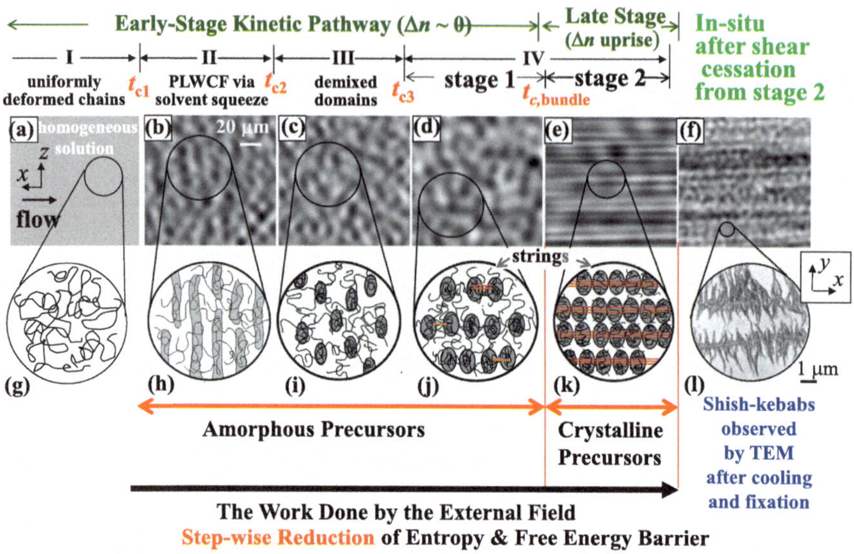

Fig. 6.14 The online snapshots of the OM images **a**–**e** simultaneously measured with the results shown in Fig. 6.13 and their corresponding sketches **g**–**k**. The OM image **f** is taken in situ at 124 °C immediately after the shear cessation from the stage 2 in Regime IV. The TEM image **l** is taken after cooling the structure shown in the image **f**. Each snapshot of the OM images was taken at a shutter speed of 10^{-4} s. In the sketch **g**–**k**, thin lines schematically represent the entangled polymer chains. The time regions I to IV are common with those defined in Fig. 6.13c. The data based on [31]

All those results shown in Figs. 6.13 and 6.14 indicate that the entire time-evolution process in all these experiments, covered up to ~ 300 s at $\dot{\gamma} = 2.9\,\text{s}^{-1}$, can be classified into the four fundamental processes occurring in the time span of Regime I to IV as specified in Figs. 6.13 and 6.14. The results reveal the cascade time evolution of the dissipative structures. It is worth noting that the entire time-span required for the homogeneous solution to self-organize into the shish-kebab texture is ~ 300 s under the step-up shear flow at the constant shar rate of 2.9 s^{-1}. Physical significance found in each regime I to IV is summarized below, separately, one after the other.

6.4.1 Regime I, $t < t_{c1}$: Incubation Period

Here, t_{c1} defines the "solvent-squeezing time". In this regime, the OM image is featureless, not showing any distinct patterns, as shown in Fig. 6.14a, and the SALS pattern could not be discerned as revealed by $\mathcal{J}_x(t)/\mathcal{J}_x(t=0) = \mathcal{J}_z(t)/\mathcal{J}_z(t=0) = 1$ in Fig. 6.13c. These results reveal that: the sheared solution in this regime remains homogeneous as sketched in Fig. 6.14g; the entangled chains schematically represented by thin lines are uniformly deformed, thereby causing the stress uprise up to $t = 1$ s as shown in Fig. 6.13c.

6.4.2 Regime II, $t_{c1} < t < t_{c2}$: PLWCF (Plane Wave Concentration Fluctuations)

Here, t_{c2} defines the characteristic time for the formation of the demixed (phase-separated) domains. In this regime (Regime II in Fig. 6.13c) $\mathcal{J}_z(t)/\mathcal{J}_z(t = 0) = 1$, but $\mathcal{J}_x(t)$ starts to increases with t from $\mathcal{J}_x(t)/\mathcal{J}_x(t = 0)$ as shown in Fig. 6.13c, which is consistent with the fact that the butterfly SALS pattern starts to evolve at 1 s after the step-up shear flow and subsequently grows into the butterfly- SALS pattern #1 in Fig. 6.13a at 5 s (corresponding to the time designated $t_{c,2}$) as marked by point 1 on the curve $\mathcal{J}_x(t)$ in Fig. 6.13c This fact is attributed to occur as a consequence of the building up of the **PLWCF** along **FD** (flow direction) due to the solvent squeeze as validated for the PS/DOP solution discussed in Sects. 5.3 and 5.6. This conclusion is confirmed by the real-space OM image shown in Fig. 6.14b and its sketch in Fig. 6.14h. The experimental result presented in Fig. 6.14b and its sketch shown in Fig. 6.14h are further validated quantitatively by the result to be presented below by Fig. 6.15d. In Fig. 6.14h the dark shaded regions represent those rich in polymers, and the bright unshaded regions represent those poor in polymers. The conclusion is supported also by the stress decay from the maximum stress at $t \cong 1$ s observed in the first stress overshoot shown in Fig. 6.13c, because the solvent squeeze causes the stress relaxation into a lower stress level, i.e., the dissipation of the energy imposed to and stored in the solution by the applied flow, as discussed in Sect. 5.4 in conjunction with Figs. 5.17, 5.18, 5.19, 5.20 and 5.21. In Fig. 6.13c, $\mathcal{J}_x(t)/\mathcal{J}_x(t = 0)$ is presented to start increasing from 1 at $t = 1$ s to the given value at $t = 5$ s, as shown by the dotted line. However, the data from $t = 1$ to 5 s are not actually obtained in this experiment and thereby only signify a visual aid. Nevertheless, we believe this trend is well supported by the time evolution of the shear stress in this period and at $t < 1$ s.

Figure 6.15 presents a close-up of the SALS pattern (part a and b) and the OM image (part c and d) in Regime II which were simultaneously measured for the UHMWPE/paraffin 5.0 wt.% solution at 150 °C and $\dot{\gamma} = 4.6\ s^{-1}$ [1]. The butterfly pattern and its contour pattern composed of the iso-scattered-intensity lines, as shown in part (a) and part (b), respectively, are the typica butterfly-SALS pattern observed for the solution in Regime II. The POM image (c) shows the filtered image which enhances the low frequency Fourier modes relevant to the butterfly scattered pattern [1]. The filtered image was taken according to the method described in conjunction with Fig. 5.24 in Sect. 5.6.3. The contour map of the POM image shows the iso-intensity lines in the regions rich in the UHMWPE having brightness lower than

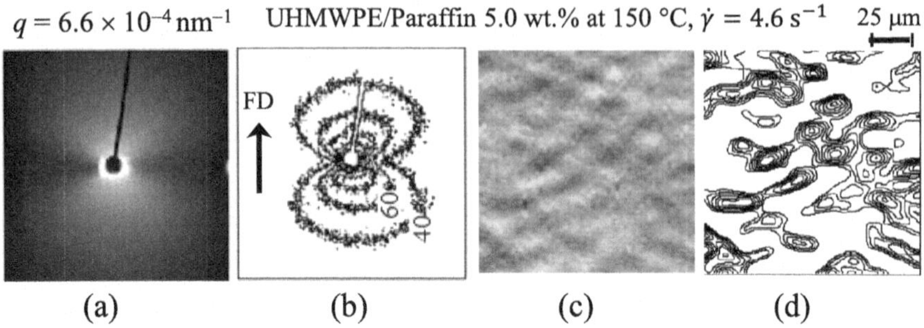

Fig. 6.15 Close-up of the SALS pattern (part **a** and **b**) and the OM image (part **c** and **d**) in Regime II which were simultaneously measured for the UHMWPE/paraffin 5.0 wt.% solution at 150 °C and $\dot{\gamma} = 4.6\ \mathrm{s}^{-1}$

average. The image may be well reflecting the superposition of the PLWCF whose wave vectors preferentially orient along the FD.

6.4.3 Regime III, $t_{c2} < t < t_{c3}$: Demixed Domains

Here, t_{c3} defines the characteristic time for the string formation. In this region, $\mathcal{J}_z$ starts to increase with t, and hence the "dark sector" of the butterfly pattern observed along **ND** (neutral direction) becomes less clear with t as shown by the SALS patterns #2 and #3 in Fig. 6.13a. The increase of $\mathcal{J}_z$ with time reflects the nonlinear effect driven by the flow as discussed in Sect. 5.7.3 in conjunction with Eqs. (5.7.37) and (5.7.38) and can be interpreted as a signature of a transformation of the dissipative structures from the PLWCF to the demixed domains with well-defined interface along ND as discussed in Sect. 5.8.

The physical origin of this transformation is considered as follows. As time elapses in this regime, the amplitude of the PLWCF increases because of the progressive solvent-squeeze process with time, which causes (i) the PLWCF to be unstable, due to the increased hydrodynamic interactions between solvent flow and the PLWCF, as supported by (ii) the stress uprise at $t > 20$ s to bring about the second stress overshoot as shown in Fig. 6.13c. Thus, the PLWCF will be eventually broken into the demixed domains, which (iii) creates the domain interfaces along ND also (as illustrated in part (b) in Fig. 5.30), hence giving rise to the increased $\mathcal{J}_z$. (iv) The evolution of the domain interface causes the second stress overshoot to decay from the maximum value at $t \gtrsim 30$ s. The dissipation of the stored energy in the solution imposed by the flow, accompanied by the transformation of the dissipative structure from the PLWCF to the demixed domains, gives rise to the second stress overshoot centered around $t \sim 40$ s as presented in Fig. 6.13c.

The OM image in Fig. 6.14c suggests formation of the demixed domains, whose mass centers are randomly placed in space as sketched in Fig. 6.14 (i), consistently with the SALS and stress behaviors. The demixed domains appear as dark objects in the transmission OM images shown in Fig. 6.14c, d. This is because the transmitted light intensity through the objects is suppressed by the SALS scattering from them.

6.4.4 Regime IV, Stage 1, $t_{c3} < t < t_{c,bundle}$: Strings with Weak Optical Anisotropy

Here, $t_{c,bundle}$ defines the critical time ($\sim$ 125 s) at the boundary between yellow and red region in Regime IV in Fig. 6.13c, d) for formation of **bundles of stretched chains ("BOSC")** interconnecting the demixed domains within the strings, as will be clarified below. In Regime IV, the randomly aligned demixed domains in space (in Regime III) tend to be aligned into "string-like assemblies" oriented parallel to FD (Ox-axis) driven by the hydrodynamic interactions between the domains and solvent flow, so that the centers of the domains within the strings were nematically aligned parallel to FD, as schematically illustrated in Fig. 5.30c and as shown in the OM images in Fig. 6.14d, e and their corresponding sketches (j) and (k), respectively.

The number density of the domains in the strings increases with time in Regime IV, because the number of the off-aligned domains to be incorporated in the strings increases with time at any given $\dot{\gamma}$'s in Regime IV: that is, this increased number density of the domains in the strings brings about a densification of the domains in the strings with time. The strings can be recognized by the appearance of the "streak-like pattern" centered at $q_x = 0$ and oriented and extended along ND (the q_z-axis, Oz-axis) but with a small width of the intensity distribution along q_x (FD) as shown in the butterfly-SALS patterns #6 to #8 in Fig. 6.13a: the strings can be recognized also by the large integrated intensity $\mathcal{J}_z$ in Regime IV, while the increased number of the individual demixed domains inside the string can be discerned by the increased integrated intensity $\mathcal{J}_x$ with time due to the diffuse butterfly scattering along FD.

It is striking to note that: (i) the $\Delta n_{x,z}$ dramatically increased at $t \sim t_{c,bundle} \sim$ 125 s, at the boarder of the two regions marked yellow and red in Fig. 6.13c: (ii) the $\Delta n_{x,z}$ oscillates with time around the average value which increases with time. These two regions in region IV are defined as stage 1 and stage 2, respectively. In stage I, the birefringence is still small, revealing that the strings have weak optical anisotropy. On the other hands, in stage 2, the birefringence reaches a large maximum value, revealing that the strings have the strong optical anisotropy.

6.4.5 Regime IV, Stage 2, $t > t_{c,bundle}$: Strings with Strong Optical Anisotropy

The sharp crossover from the weak to strong optically anisotropic strings at $\boldsymbol{t_{c,bundle}}$ is the most intriguing and fundamental physical process to be further studied, because it must imply a **coupling between the L–L phase transition** and **L–S phase transition (crystallization)** at this stage. In the stage 2, $\mathcal{J}_z$ finally becomes larger than $\mathcal{J}_x$, indicating that the streak-like butterfly SALS pattern arising from the string as a whole becomes stronger than the diffuse butterfly scattering arising from the individual domains aligned and densified inside the strings. This piece of evidence further implies (i) an increased integrity of the strings with increased number density of the domains within the strings as supported by the further increase of $\mathcal{J}_z$, and (ii) this enriched domain density may **trigger formation of the BOSCs which interconnect the domains at their boundaries and traverse through them** as supported by the $\Delta n_{x,z}$ uprise, i. e., and by the strings having strong optical anisotropy in Fig. 6.13d.

Correspondingly sketch (k) in Fig. 6.14 schematically illustrates formation of the BOSCs traversing through the domain in the strings by the red lines along FD. The formation of the BOSCs must be intricately related to the third stress overshoot centered around $t \sim t_{c,bundle} \sim 125$ s as shown in Fig. 6.13c. It is important to note that (iii) the **BOSCs** may be formed via **coil-to-stretched-chain transition ("CSCT") of polymer chains in the local regions of the strings** rather than the **CSCT induced in the medium of homogeneous solutions** free from the phase separation under extensional flow [6, 34, 35]

(A) **Evidence for crystallization in the strings.** Figure 6.16 presents supporting evidence for the crystallization which locally occurs in the strings in Regime IV-1 [25]. Part (A) presents the SALS patterns (upper half) and their contour patterns composed of the iso-intensity lines (lower half). Part (B) shows the normalized integrated scattered intensity along FD $\mathcal{J}_x(t)/\mathcal{J}_x(t=0)$ and ND $\mathcal{J}_z(t)/\mathcal{J}_z(t=0)$ as a function of time t under the shear protocol indicated in part (C) for the UHMWPE/paraffin 5.0 wt.% solution at 124 °C [25]. The solution is subjected to the step-up shear from $\dot{\gamma} = 0$ to $2.9\ \mathrm{s}^{-1} (> \dot{\gamma}_a = 1.2\ \mathrm{s}^{-1})$ (Fig.6.16 (C)) and then the shear is ceased at 95 s. This shear protocol brings the solution into the stage 1 in Regime IV. The SALS patterns (a) to (d) in part (A) are taken at the times as indicated by the point a to point d on the data of $\mathcal{J}_x$ in part (B) and on the shear protocol in part (C), respectively. The SALS pattern (a) in part (A) and the integrated intensity $\mathcal{J}_x$ and $\mathcal{J}_z$ at point a in part (B) are measured at 90 s after the onset of the step-up shear (but before the shear cessation as shown by the red broken line): they are the same as the pattern 5 in part (a) and the integrated intensity at point 5 in part (c) in Fig. 6.13 or at point a in Fig. 6.16 (A),(B),(C), respectively.

After the cessation of the shear, the scattered intensity almost instantaneously decays as shown by the change in the SALS pattern from (a) to (b) (part A) and in the integrated intensity from the point a to b (part B). The integrated intensity along FD

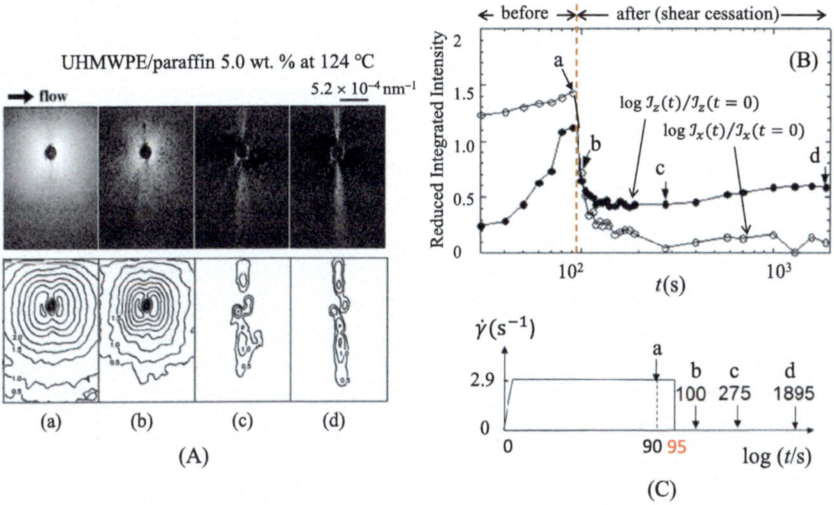

Fig. 6.16 Part **(A)** presents the SALS patterns (upper half) and their contour patterns composed of the iso-intensity lines (lower half), and part **(B)** shows the normalized integrated intensity along FD ($\mathcal{J}_x$) and ND ($\mathcal{J}_z$) as a function of time t after the step-up shear flow under the shear protocol indicated in part **(C)** for the UHMWPE/paraffin 5.0 wt.% solution at 124 °C. The solution is subjected to the step-up shear rate from $\dot{\gamma} = 0$ to $2.9\ s^{-1}$ ($> \dot{\gamma}_a = 1.2\ s^{-1}$) in the stage 1 in Regime IV (Fig. 6.13c)

($\mathcal{J}_x$) continues to decay slowly over the time span of 175 s (from point b to c as presented by the symbol of the open circles in part B) to the value, $\mathcal{J}_x(t = 0)$, in the quiescent solution at $t \geq 275\ s^{-1}$ (point c). On the other hand, the integrated intensity along ND ($\mathcal{J}_z$) instantaneously reaches the constant value of $\sim 3 \times \mathcal{J}_z(t = 0)$ (point c as presented by the symbol of the filled circles) and then slowly increases to reach the constant value of $\sim 4 \times \mathcal{J}_z(t = 0)$ at $t = 1895$ s (point d). Thus, at 275s and 1895 s after the cessation, only the streak-like SALS pattern along ND remains (part c and d in Fig. 6.16A, respectively), and its intensity increases with time as shown in the change of the SALS pattern from (c) to (d) (part A) and the integrated intensity $\mathcal{J}_z(t)$ from point c to point d (part B).

The result reveals that (1) the flow-induced crystallization already occurs locally within the strings at 90 s after the step-up shear flow (point a before the cessation of the flow in part C): this evolution of the flow-induced crystallization may be well revealed by the enhanced strong SALS intensity along ND in the butterfly-type SALS patterns (as presented part a and b in Fig. 6.16A). The enhanced strong SALS intensity along ND conceals their dark sectors along ND (see Fig.5.3 in Sect 5.3). The crystallization keeps integrity of the string structure even after the cessation of the flow and brings about the weak optical anisotropy with the small but sizable birefringence under the flow in the stage 1.

This result together with the results in the stage 2 shown in Figs. 6.13 and 6.14 may predict that (2) the crystallization into extended chain crystals (ECC) may be nucleated from the BOSCs developed in situ in the stage 1 under the flow: the nuclei may grow along FD. The nuclei of the ECC grown under the flow may provide a seed for further crystallization into the ECC even after the cessation of the flow, mediated by the promoted growth of the BOSCs in the tip of the front face of the growing ECC, driven by the mechanism proposed by Peterman and Keller and their coworkers, i.e., the autocatalytic orientation of the chain molecules in front of the rapidly growing ECC nuclei [40].

(B) **Splitting of the streak-like SALS pattern**. The streak-like pattern extended along ND (Oz-axis) appears at the center of the SALS pattern at $q_x = q_y = 0$, as shown in the SALS pattern #4 and #5 in Fig. 6.13a, which emerges in the stage 1 of Regime IV, although the streak pattern is not clearly observed in the 2D patterns, because it is superposed and concealed by the butterfly pattern with much stronger intensity than the streak pattern in stage 1. It starts to split into multiple streaks with the common center (at $q_x = q_y = 0$) but different orientation around ND as shown in the SALS patterns #6 to #8 in stage 2 of Regime IV (Fig. 6.13a). The multiple streaks in the SALS patterns have the following characteristics.

(1) the orientation of the streak pattern observed in situ and at real time is fluctuating with time under the flow, giving rise to such a momentary variation as the multiple streaks or the single streak.
(2) the orientation of the streak pattern changes with time around the ND by the angular range of $\sim \pm 20°$ at maximum.

The result implies that the string structures developed under the shear flow exist under such a dynamic equilibrium that the structures may be laterally bent and straightened or broken and reformed with time. The string-like dissipative structures developed may have a stiff solid-like integrity due to some crystallinity developed along the string after the local CSCT-induced crystallization; the structures may be subjected to lateral forces originating from the normal stress difference. The lateral forces may account for the orientation fluctuations of the strings with respect to FD and thereby the orientation fluctuations of the streak-like SALS pattern with respect to ND. The fluctuations of the streak-like pattern were never observed in Regime IV for the noncrystallizable polymer solutions, the PS/DOP solutions, discussed in Chap. 5, probably because the strings in this case are soft but not solid like because of the absence of crystallinity in the strings in PS/DOP systems. Thus, the dynamic orientation fluctuations of the streaks in the UHMWPE/paraffin solutions and thereby the strings in those solutions are predicted to cause the large fluctuation of the birefringence (Fig. 6.13d) and shear stress (Fig. 6.13c) in the stage 2 of Regime IV.

6.5 Summary on Online Rheo-Optical Studies

In addition to the important pieces of information concerning the cascade pattern formation revealed in Regimes (II) to (IV) as described above, we further described below important physical factors discovered from the above experiments in Fig. 6.13 and 6.14 for the UHMWPE/paraffin 5.0 wt. % at 124°C as follows.

(1) The early-stage kinetic pathway from Regime I to the stage 1 in Regime IV (specifically in the early part of the stage 1 before 90 s, i.e., before the up-rise of the birefringence presented in Fig. 6.13d) evolves a series of the amorphous dissipative strictures as schematically shown in part (h) to part (j) in Fig. 6.14. Especially, the dissipative structures evolved in the stage 1 in Regime IV are amorphous precursors of the SK having $\Delta n_{x,z} \sim 0$.
(2) If the shear is ceased at any times in the early-stage kinetic pathway as described above in (1), the structures evolved disappear, and the solution is brought back into the homogeneous solution accompanied by the stress relaxation, i.e., the dissipation of the energy stored in the solution.
(3) In the later part of the stage 1 and in the stage 2 in Regime IV (specifically after 90 s, where the birefringence starts uprising as presented in Fig. 6.13d), the crystallization into shish crystals is predicted to evolve within the string structures via the nucleation and growth of the ECC from the BOSCs developed locally in the strings. This predicts the formation of the crystalline precursors for the SK in these stages in Regime IV as indicated and predicted by Fig. 6.14e, k, respectively.
(4) Strikingly, the OM image exhibiting the strong optical anisotropy in the late stage 2, as shown in Fig. 6.14e, does not disappear even after the shear cessation at 124 °C as shown in Fig. 6.14f. This implies that the dissipative structures evolved in the solutions under the flow at 124°C and at any given $\dot{\gamma}$'s in the stage 2 in Regime IV do not disappear but rather remain even after the shear cessation. In relation to this implication, the intensity of the streak-like SALS pattern after the shear cessation further increased with t, as validated by the increased scattered intensity from the pattern (c) to (d) in Fig. 6.16(A) and by the increased reduced integrated scattered intensity from point c to d in Fig. 6.16 and (B) [25, 26], and the solution after cooling to room temperature showed the shish kebabs as shown in Fig. 6.14 (l) under the TEM observations [23] utilizing the method as detailed in Sect. 6.3.3.
(5) Those results described in the item (4) suggest that: the crystallization into shishs occurred in situ under the flow at the any given t's satisfying $t > t_{\mathrm{c,bundle}}$; this crystallization not only prevents the relaxation of the structures evolved in situ at 124 °C under the shear after the shear cessation but also promotes further the crystallization into the shish-kebabs after the shear cessation at 124 °C, according to the mechanism proposed by Peterman and co-workers [40], although the shish-kebabs shown in Fig. 6.14 (l) might grow also during the cooling process (i.e., the solidification process) of the solution for the TEM observations.

(6) Strikingly enough, the evolution of the series of the dissipative structures found for the UHMWPE solution is essentially the same as that found for the UHMWaPS solution described in Chap. 5 [26, 33, 41], except for the fact that the optically anisotropic strings developed in situ in the stage 2 cannot be crystallized for the amorphous UHMWaPS solutions under the flow. Thus, the BOSCs developed in the UHMWaPS are relaxed back into the random-coil chains in the homogeneous UHMWaPS solutions, unless the solutions are rapidly vitrified below their glass transition temperatures before the relaxation of the optically anisotropic strings.

6.5.1 Remarks on the Models

The work done to the solution of volume V by the flow, W, is given by,

$$W/V = \sigma\gamma = \eta\dot{\gamma}^2 t, \tag{6.5.1}$$

where σ, γ, and η are the shear stress, the shear strain, and the viscosity of the solutions under the flow at any given $\dot{\gamma}$'s and t's, respectively. The energy density W/V imposed to the solution increases with t, after the step-up shear flow at any given $\dot{\gamma}$'s which step by step reduces the entropy and free energy barriers for the SK formation via the cascade time evolution of the amorphous dissipative structures driven by the flow-induced L–L phase separation: the cascade evolution of the dissipative structures is accompanied by cascade dissipation of the energy imposed to the solution. The dissipative structures (precursor) developed in the stage 1 of Regime IV (the strings) provides the special templates which efficiently promote the crystallization into the SK. This is because the nucleation and growth of the ECC may occur at the boundaries of the domains in the strings, and the growth of the ECC along the string axis may provide the crystal surface for the epitaxial nucleation of the chain-folded lamellae and the growth of the lamellae from the random coils existing in the reservoir of the demixed domains in the strings.

The pattern formation in Regime II is predicted by the HFMO theory given by Eq. 5.7.1. The theory is a generalized version of the Ginzburg–Landau (GL) theory which accounts for the effect of shear flow on the dynamic concentration fluctuations by incorporating the coupling between the stress and diffusion into the GL theory. The linearized version of the theory predicts the shear-rate dependent thermodynamic driving force, $F_{\mathrm{growth}}(q; \dot{\gamma})$ given by Eq. 5.7.21, for the growth of PLWCF driven by the SD (the L–L phase separation) at the any shear rates $\dot{\gamma}$'s larger than the critical shear rates $\dot{\gamma}_{c,x}$'s (see Figs. 5.7, 5.12, 5.17, 5.19, 5.20, and 5.24). The theory predicts that the Fourier modes of the structure elements having a smaller wave number q and hence a larger length l are subjected to a larger driving force for the growth. In other words, the large structure of the PLWCF evolves first, as presented in Fig. 6.14b, h, under a low bulk shear stress $\sigma \sim O(10^3\ \mathrm{Pa})$, as shown in Fig. 6.13c.

The evolution of the large length-scale structure in Regime II is followed by the evolution of the increasingly smaller length-scale structures in the order of Regime III and Regime IV with the increasingly lager expense of the work done by the applied flow. In Regime IV-2 the bundles of stretched chains are formed within the strings triggered by the stress-induced CSCT in the local regions of the strings, specifically at the boundaries of the impinged domains in the strings as will be demonstrated later in conjunction with Fig. 6.25 in Sect. 6.5.2(E). Although the bulk stress σ involved in Regime IV-2 is as low as in Regime II, the local stress σ_l in the local regions of the strings is predicted to be much higher than σ, as will be clarified in the next section.

In other words, the weak bulk stress σ imposed to the strings may exert a high local stress σ_l to the boundaries of the interconnected domains within the strings through the so-called **stress concentration** due to the cross-sectional area of the boundaries being smaller than the average cross-sectional area of the strings. The amorphous precursors for the SK (the strings in stage 1 of Regime IV shown in Fig. 6.14 part (j)) with the lateral width of the order of several μm along ND cannot be observed by SAXS and SANS and by the conventional birefringence experiments also as $\Delta n_{x,z} \sim 0$, so that their existence has been overlooked for a long time since the discovery of the shish-kebabs.

The predicted model for the SK formation described here in this Chapt. is different from and/or more complicated than the conventional model for the polymer solutions or melts under extensional flow. The previous model is based exclusively on the flow-induced crystallization (liquid–solid phase transition) in the homogeneous solutions and melts (free from phase-separated structures) under the extensional flow at the flow rate higher than the critical flow rate. The role of the liquid–liquid phase transition (the phase separation) has not been explicitly taken into account in the concept of the flow-induced crystallization.

(A) **Previous model: polymer solutions**. The flow-induced crystallization into the ECC in polymer solutions involves formation of the bundles of the fully stretched chains (Fig. 6.17 a) in the homogeneous solution via the CSCT [6] mediated by the hydrodynamic interactions between the solvent flow and individual chains in the random-coil conformation as shown later in Fig. 6.18.

(B) **Previous model: polymer melts**. The flow-induced crystallization of polymer melts into the ECC involves formation of the bundles of the fully stretched chains with some entanglement coupling points in the homogeneous melts of entangled chains [9] (Fig. 6.17b).

The solutions and melts universally form the bundles of stretched chains (BOSCs) with the growth rate of G_{Bundle} in which the nucleation and growth of the ECC crystals occur with the growth rate of G_{ECC} along the axis of the bundles to result in formation of the ECC (Fig. 6.17 c). The necessary and sufficient condition for the ECC formation may be as follows:

The necessary condition: formation of the bundles of stretched chains,

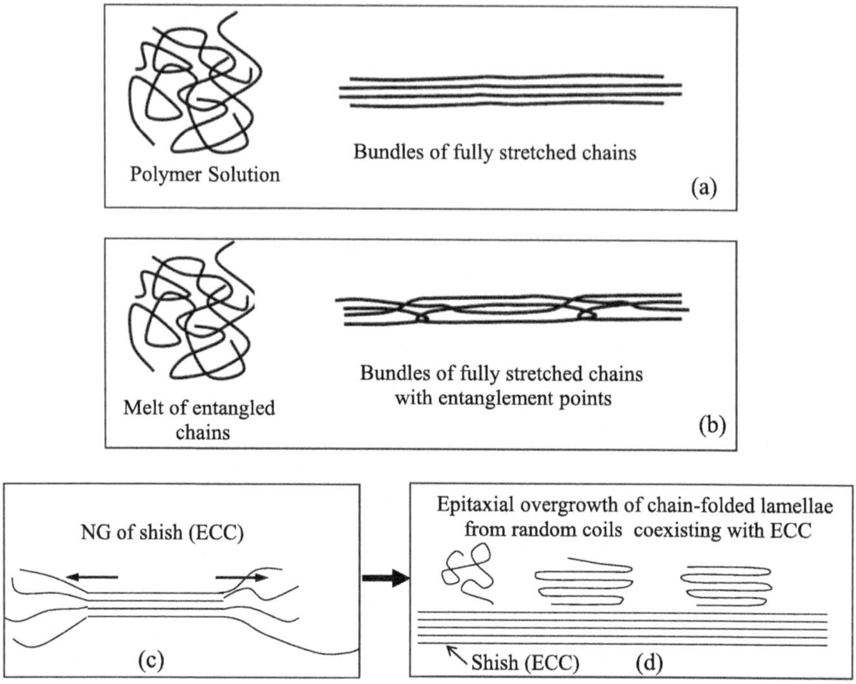

Fig. 6.17 Model for flow-induced crystallization into the ECC in the homogeneous solutions (part **a**) and melts (part **b**). The nucleation growth of the ECC from the bundle of the stretched chain (part **c**) and the epitaxial overgrowth of the kebab lamellae from the ECC (part **d**).

The sufficient condition:

$$G_{\text{Bundle}}, G_{\text{ECC}} \gg G_{\text{Lamella}}, \tag{6.5.2}$$

where G_{Lamella} is the growth rate of the chain folding lamellar crystal from the BOSCs.

The random coils, coexisting with the growing ECC, are crystallized into the chain-folded lamellae mediated by the epitaxial growth of the lamellae onto the ECC to result in formation of the SK where the ECC and the overgrown lamellae from the ECC correspond to "Shishs" and "Kebabs", respectively, as schematically illustrated in Fig. 6.17d.

Figure 6.18 presents a schematic illustration of the CSCT of a single chain in solution under the elongational flow. A random coil nondraining with solvent flow (part a) is deformed under the flow (part b). When the flow rate reaches the critical flow rate, the deformed coil reaches the unstable state: the unstable random coil is stabilized by the transition into a stretched chain (part c) under the flow at the flow rate higher than the critical flow rate [34]. The stretched single chains are further stabilized by inter-chain associations of the neighboring stretched chains to form the bundles of stretched chains (BOSC), the process of which may be hypothesized to cause L–L

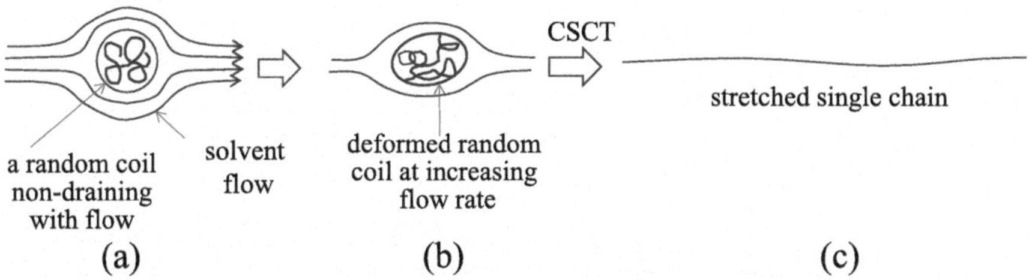

Fig. 6.18 Model for the random-coil chain (part **a**) to the stretched chain (part **c**) through the CSCT transition (CSCT) via the deformed random-coil chain (part **b**)

phase separation between the stretched chains and matrix solution of random-coil chains; the rate of the phase separation should be faster than the orientation relaxation rate of individuals stretched chains to form the BOSC in homogeneous solutions.

6.6 Self-Organization of Polymer Solutions into Shish-Kebab Texture under Flow: Cascade Evolution of Dissipative Structures along Fiber Spinning Line

6.6.1 Objectives

Figure 6.19 highlights the objective in this section (row A) in conjunction with summaries of the results obtained by the online investigations in the preceding Sect. 6.4 regarding the kinetic pathway of the shish-kebab formation under the shear flow (rows B, C, and D) [23, 38]. This section is devoted to explore structure evolution along the spinning line of fibers spun from the homogeneous, entangled UHMWPE/decalin solutions to clarify the kinetic pathways leading to the shish-kebab structure along the spinning line (Fig. 6.19A). In the spinning process, a homogeneous solution extruded from a spinneret is subjected to a complex flow field with the decreased temperature and the **increased total strain** or the **work done by the flow** imposed to the solution **along the spinning line**.

The objective in this work is to clarify the following two problems: (i) how do the flow-induced structures evolve along the fiber spinning line (Fig. 6.19A)?; (ii) are the fundamental base regarding evolution of the flow-induced structures appeared along the spinning line essentially similar to the time evolution of the shear-induced structures observed after the step-up shear from $\dot{\gamma} = 0$ to $\dot{\gamma} > {\tau_e}^{-1}$ at the constant temperature T higher than the nominal melting temperature $T_{m,n}$ of the quiescent solutions, as summarized in parts (a) to (e) in Fig. 6.19B –D ? In the latter case also, the **total strain and work done by the flow imposed to the solution increase with time** after the step-up shear flow.

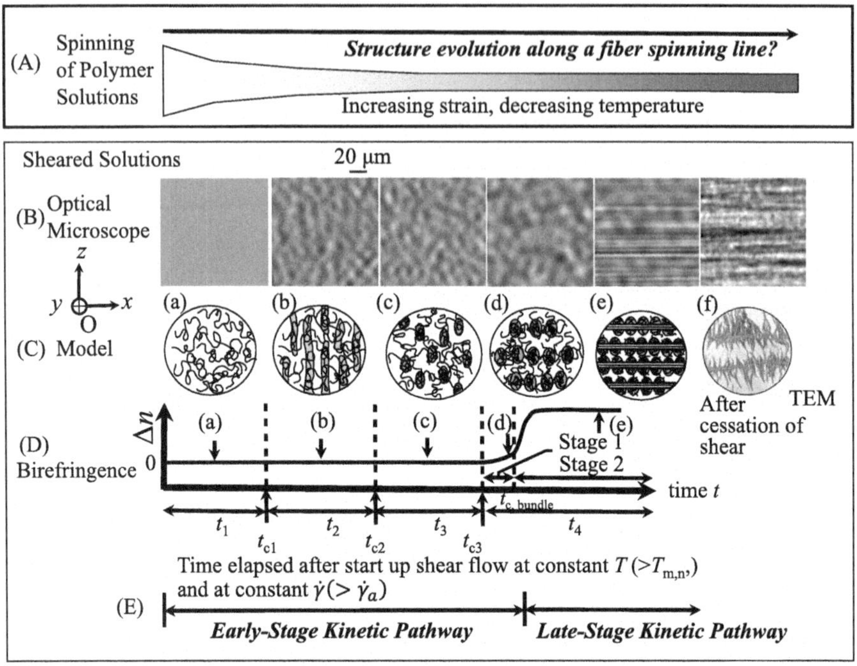

Fig. 6.19 Highlight of the objective in this section (part **A**) in conjunction with the summaries of the results obtained in the preceding Sect. 6.4 regarding the kinetic pathway of the shish-kebab formation under the shear flow (row **B** , **C** , and **D**). From [38], reprinted with permission, American Chemical Society

If there are similarities in the two systems regarding the self-organization of dissipative structures into the shish-kebab texture, the real-time, in situ investigation of the latter system is reinforced by the off-line, ex-situ investigation of the former system with the high-resolution TEM images along the fiber spinning line, and vice versa. Figures 6.19B, C present the time evolution of the transmission optical micrographs (OM) and their sketches, respectively, while Fig. 6.19D presents the time evolution of birefringence Δn observed in the Oxz plane, where the Ox-, Oy-, and Oz-axis are taken parallel to the flow direction, the velocity gradient direction parallel to the propagation direction of the incident light beam, and the neutral (or vorticity) direction, respectively.

The entire time-evolution process of the self-organized structures after imposing the step-up shear flow to the homogeneous solutions is classified into the four time-regions t_1 to t_4 as detailed previously [24] and summarized in Fig. 6.19(B)-(D). In the region t_1, where the time t satisfies $t < t_{c1}$, t_{c1} being the characteristic time for the solvent squeeze, the entangled polymer networks swollen with solvent are uniformly deformed, and hence the sheared solution remains essentially homogeneous, so that the OM image is featureless, showing no remarkable structures as shown in part (a) in Fig. 6.19B.

In the region t_2 where the time t satisfies $t_{c1} \leq t < t_{c2}$, t_{c2} being the characteristic time for formation of the demixed domains, we discerned the OM image in part (b) in Fig. 6.19 B which essentially represents the plane-wave type concentration fluctuations with their wave vectors oriented along the flow direction as presented in Fig. 5.19 in Sect. 5.6 and indicated by the sketch in part (b) in Fig. 6.19 C, where the shaded and unshaded regions represent those rich and poor in polymers, respectively.

In the region t_3 where the time t satisfies $t_{c2} \leq t < t_{c3}$, t_{c3} being the characteristic time for the string formation, the OM image in part (c) in Fig. 6.19 B is discerned, which suggests formation of the demixed domains whose centers of mass are randomly distributed in space, as sketched in part(c) in Fig. 6.19 C.

In region t_4 where the time t satisfies $t \geq t_{c3}$, the demixed domains are aligned into the string-like assemblies oriented parallel to the flow direction, as shown in the part (e) in Fig. 6.19 B and in the sketch in part (e) Fig. 6.19 C. In the string phase, the strings are aligned side by side laterally (along the direction perpendicular to the flow direction), so that the centers of mass of the demixed domains are nematically aligned parallel to the flow direction.

It should be stressed that the Δn is very small up to the time t_{c3} compared with that in stage 2 in the time region t_4. It is still small in stage 1 in the time region t_4, at any given time t_4 satisfying $t_{c3} < t_4 < t_{c,\text{bundle}}$, as schematically shown in part (d) in Fig. 6.19D; thus, polymer chains themselves are essentially in random coils up to this stage (the stage 1 in the time region t_4) even under the flow, as schematically illustrated by the chains in part (d) in Fig. 6.19C.

The observed drastic increase in Δn by a factor of about three times in the time region t_4 above $t_{c,\text{bundle}}$, the characteristic time for the formation of the bundles of the stretched chains, as illustrated in part (e) in Fig. 6.19C, leads us to further classify this time region t_4 into two stages: (i) stage 1 where Δn becomes slightly more positive than that in the time regions t_1 to t_3, so that the strings have only a weak Δn and (ii) the stage 2 where Δn sharply increases with time, which reveals that the bundles of the stretched chains interconnecting the demixed domains start to be formed parallel to the string as illustrated by the straight lines in part(e) in Fig. 6.19C, thereby the string starts to have a strong optical anisotropy, as shown in part (e) in Fig. 16.19D. The OM image in part(f) in Fig. 6.19 B represents the OM image taken in situ, at real time and at the given temperature after cessation of the shear flow. The OM image is proved to reflect the shish-kebab structure by applying the ex-situ TEM observation as presented in part (f) in Fig. 6.19C with the methods as detailed in Sect. 6.3.3.

In this section we present a full account of the structure formation along the fiber spinning line to elucidate the kinetic pathway and mechanism of the shish-kebab formation, including the CSCT. We especially aim to clarify the early-stage kinetic pathway as defined in the bottom of Fig. 6.19E and the detailed structures evolved therein.

6.6.2 *Off-Line TEM Observations of Structure Evolutions along the Fiber Spinning Line with the Solidified Specimens*

Here, we investigate the cascade evolution of the dissipative structures along the fiber spinning line by applying the off-line TEM method. The method of the off-line TEM investigation is described in Sect. 6.3.3.

(A) **Conservation of the solidified structures before and after the fixation**. The fiber specimen which was sampled at a given position in the fiber spinning line by using the sampling method described in Sect. 6.3.3 (B) was solidified at room temperature. The solidified specimen contains the solvent (decalin) by 90 wt.% at the position P4 in the spinning line (Fig. 6.9d): thereby the solidified fiber is called as the gel fiber. The solvent in the gel fiber was replaced by epoxy monomer, and the monomer was further cured to obtain the solidified specimen embedded in the epoxy resin using the fixation method described in Sect. 6.3.3 (C) (Fig. 6.10) for the observation under TEM as described in Sect. 6.3.3 (D).

To confirm the conservation of the internal structures of the fiber during the fixation process, the small-angle light scattering (SALS) and OM experiments were conducted before and after the fixation to compare the results before and after. For this purpose, however, we did not use the fiber, because a round shape of the fiber is not suitable for the SALS experiment. Instead, as a model system of the fiber, we used a sample prepared by the same UHMWPE/decalin 10 wt.% solution subjected to the specified shear flow with a parallel-plate fixture at a gap distance of 30 μm by using the shear apparatus described in Sect. 3.1.

The homogeneous solution used for the shear experiment was prepared by keeping first the mixture of the UHMWPE and decalin at 160 °C for 10 min to assure the complete homogenization of the solutions and subsequently by cooling it down to the working temperature 118 °C($> T_{m,n} = 115 - 119$ °C). The step-up shear flow was imposed to the homogeneous solution at 118 °C for 240 s and then the flow was ceased. After the cessation of the shear, the solution was cooled down to room temperature for the solidification of the shear-induced structures: the gel-like sample thus obtained was first carefully removed from the shear cell to avoid any excess deformation and then soaked in decalin to prevent drying. A small piece of the specimen ($\sim$ 2.5 mm^2) was cut out from the sample with the thickness of 30 μm at the specified position from the center of the sample where the shear rate is equal to the specified shear rate, $\dot{\gamma} = 9.3\,\text{s}^{-1}$, using a sharp blade. The specimen was embedded in the epoxy resin after the "fixation" process as described in Sect. 6.3.3 (C).

In Fig. 6.20, the upper and lower rows of part (a) and (b) demonstrate the SALS patterns and the OM images, respectively, taken for the solidified specimens before (a) and after (b) the fixation process, respectively [38]. The arrow marked FD indicates the flow direction common for all the SALS patterns and OM images. The scales inserted in the SALS pattern and the OM images in part (b) are also common to those

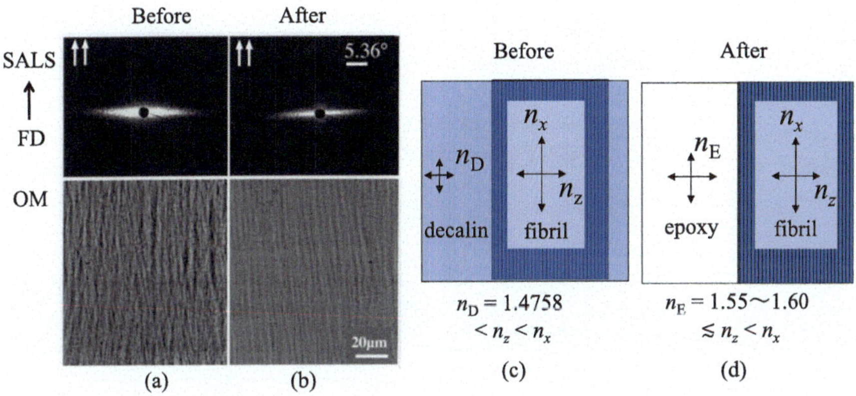

Fig. 6.20 Upper and lower row presents the SALS patterns and the OM images, respectively. Part (a) and (b) are the SALS pattern and OM image taken for the solidified specimens before and after the fixation process (b), respectively. Parts **c** and **d** present the refractive indices of the solidified specimens before and after the fixation, respectively

in part (a). The optical axes of both the polarizer and the analyzer for the SALS and OM are set parallel to FD, defined as V_v polarization, as indicated by the two arrows in part (a) and (b).

The OM images in part (a) and (b) taken before and after the fixation are essentially identical to the OM image shown in Fig. 6.14 f in the stage 2 of Regime IV. The SALS patterns and the OM images before (part a) and after the fixation (part b) are confirmed to be essentially identical except (i) a slight difference in the orientation of the streak-like SALS patterns and the string structure in the OM images. However, there are (ii) the large difference in both the scattering intensity of the SALS pattern and the contrast of the OM image before and after the fixation. The former difference (i) is an artifact due to a slight mismatch between the orientation of the reference axis (FD) of the specimen before and after the fixation. On the other hand, the latter difference (ii) is due to the difference in the refractive indices between the internal structure of the specimen and its surrounding medium under the V_v polarization before and after the fixation.

If the refractive indices of the internal structure (the fibril) of the specimen (n_x, n_z) do not change before and after the fixation just as anticipated, and only that of the surrounding medium changes from n_D (decalin) to n_E (epoxy resin) before and after the fixation as shown in part (c) and (d), respectively, the refractive index difference under the V_v polarization decreases after the fixation, because $(n_x - n_E) < (n_x - n_D)$, since $n_E > n_D$, which accounts for the latter difference (ii).

It is important to note that the SALS patterns and the OM images observed before and after the fixation are identical, except the natural difference in the scattered intensity and the image contrast. These results assure that there are no changes in the solidified structures throughout the fixation processes.

Fig. 6.21 Typical appearance of the fibers at positions P2–P4 (part **a**) and P4–P5 (part **b**) after the solidification and the fixation under the crossed polarizers with the polarization directions of the polarizer and analyzer shown by the arrows. From [38], reprinted with permission, American Chemical Society

(B) **Variation of the birefringence along the spinning line**. Birefringence of the solution extruded from the nozzle must be nearly zero at the position P1 at 170 °C (Fig. 6.9a), and it would increase along the spinning line due to the evolution of the structures oriented along the fiber axis in the fiber spinning process [42, 43]. In this work, the off-line measurement of birefringence $\Delta n(z)$ was conducted for the solidified fiber at room temperature as a function of the distance z from the spinneret. We may conjecture that the $\Delta n(z)$ found for the solidified fiber qualitatively reflects a relative change in $\Delta n(z)$ with z for the fiber in situ in the spinning line.

Figure 6.21 presents the optical micrographs of the fiber after the solidification and the fixation under the crossed polarizers with the polarization direction of the polarizer and analyzer as indicated by the arrows [38]. Parts (a) and (b) represent the typical optical micrographs of the fiber sampled at the position from P2 to P4-1 ($z < 100$ mm) and from P4-2 to P5 ($z > 100$ mm), respectively. The points P4-1 and P4-2 are defined as the point P4 slightly upstream and downstream of the spinning line at $z = 100$ mm, respectively (see Fig. 6.24 and 6.25 below). The brightness of the fiber in part (b) is found to be much higher than that of part (a).

It is worthy of noting that the birefringence $\Delta n(z)$ may increase due to the post crystallization induced by the solidification process of the spun fiber also in addition to the birefringence inherent to the fiber in the spinning line before the solidification. The small $\Delta n(z)$ observed for the solidified fiber in part (a) thereby may suggest that the $\Delta n(z)$ observed in situ along the spinning line, designated as the "online $\Delta n(z)$", at positions from P2 to P4-1 must be even smaller than that expected from part (a).

On the contrary to part (a), the higher brightness observed in part (b) clearly suggests a lager $\Delta n(z)$ developed in situ in the fiber in the spinning line from P4-2 to P5. Thus, the results reveal that the online $\Delta n(z) \sim 0$ at $z < 100$ mm (defined as the stage 1) and the online $\Delta n(z) > 0$ at $z > 100$ mm (defined as the stage 2). The value of the $\Delta n(z)$ in stage 2 is larger than the value of the $\Delta n(z)$ in the stage 1 at least by about a factor of 3.

(C) **Position P2: the plane wave concentration fluctuations (PLWCF)**. Figure 6.22 presents a typical TEM (part b) taken for the specimen sampled at P2 located at $z = 20$ mm downstream from the nozzle (part a) [38]. The method for the TEM observation is detailed in Sect. 6.3.3 (D). The RuO_4 vapor selectively stains the polymer-rich phase but not its matrix phase of the epoxy resin which replaces the solvent phase via the fixation process. Thus, the polymer-rich regions as highlighted by the red arrow in the TEM image (b1) in part (b) appear dark, and the solvent-rich regions, as highlighted by the bright arrow, appear bright in the TEM image. The TEM image b1 in part (b) reveals that the polymer-rich regions, which are stained dark, are extended normal to FD and periodically arranged with the average spacing of a few μm along FD.

This TEM image in part (b) reflects the PLWCF with the wave vector parallel to FD and is consistent with the PLWCF predicted by the SALS patterns and the OM images observed at real time and in situ under the flow in Regime II: see Fig. 6.15a, b and 6.15c, d, respectively, at $\dot{\gamma} = 4.6\,s^{-1}$ under the increased shear-rate protocol and part (b) in Fig. 6.14b observed in Regime II after the step-up shear at the shear rate

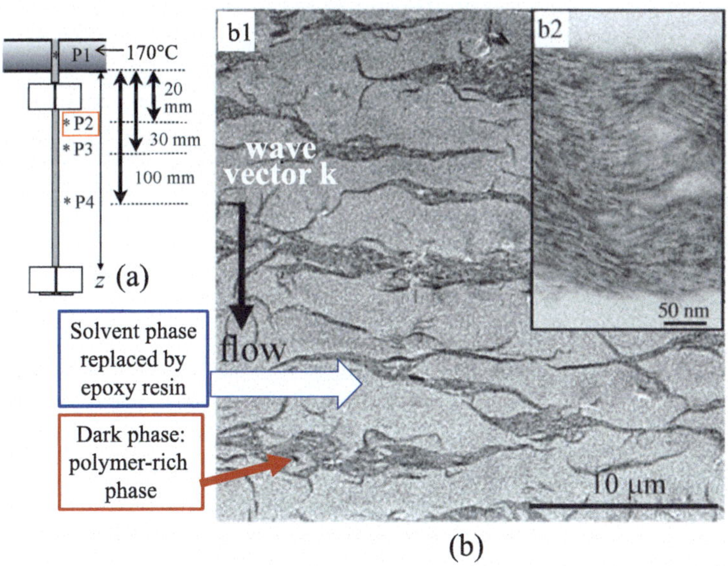

Fig. 6.22 Typical TEM (part **b**) taken for the specimen sampled at P2 located at $z = 20$ mm downstream from the nozzle (part **a**)

of $\dot{\gamma} > \dot{\gamma}_a$ as shown in Fig. 6.13b [6]. Thus, the TEM image presented in Fig. 6.22b, which is taken ex-situ in the running spinning line, is expected to reflect the plane wave concentration fluctuations which exists in situ at position P2 in the spinning line.

The large length-scale (global) structure of the PLWCF existed in situ at P2 is found to be effectively conserved in the fiber specimen solidified by cooling the specimen sampled at P2 from the temperature at P2 to room temperature. The cooling effectively fixes the *global structure* of the PLWCF via **diffusion-limited crystallization** of the UHMWPE in the PLWCF. The off-line TEM image shown in part (b1) in Fig. 6.22b reveals the structure at P2 more clearly and with a higher resolution than the online OM image shown in Fig. 6.14b. The diffusion-limited crystallization is the crystallization which occurs locally in the PLWCF without involving a large length-scale diffusion of chains in the process of a rapid solidification of the specimen and thereby without changing the global spatial distribution of chains in PLWCF.

The diffusion-limited crystallization alters only the *local structure* of polymer-rich regions of the PLWCF through formation of chain-folding lamellar crystals. Therefore, the solidification transforms polymer-rich region composed of solution of random-coil chains into the polymer-rich region composed of the stacks of lamellae dispersed in the solution.

A part of the dark regions in part (b1) is enlarged in part (b2): the part (b2) reveals that the dark region in part (b1) is composed of the stacks of lamellae of the thickness $D \sim 5.9$ nm and the interlamellar spacing $L \sim 8.8$ nm. In part (b2), the surfaces of lamellae containing chain-folded loops (defined as **SLCFL**) are standing normal to the surface of the ultrathin section. The SLCFL are stained dark by RuO_4 [39], and the unstained cores of the chain-folded lamellae appear bright. The SLCFL appear to be oriented normal to the FD. It is intriguing to note that this special orientation of the lamellae is developed by crystallization from the solution of unoriented chains (randomly coiled chains) in the polymer-rich domains during cooling the specimen sampled at P2 from the temperature at P2 in the spinning line to room temperature (in the process shown in Fig. 6.9), underlying physics of which is discussed in the item (F) in Sect. 6.6.3 (F) below.

(D) **Position P3: demixed domains**. Figure 6.23a presents a typical TEM obtained from the solidified specimen sampled from the spinning line at P3 located at $z = 30$ mm downstream from the nozzle [38]. In the image, the polymer-rich domains appear dark due to the selective staining with RuO_4, whereas the solvent-rich matrix phase replaced by the epoxy resin appears bright. The TEM image reveals that the polymer-rich domains are extended perpendicular to the flow direction, suggesting that a statistically averaged domain-shape is characterized by an oblate-ellipsoidal domain shape with the axis of revolution oriented parallel to the flow direction and their mass centers are randomly aligned in space. Interestingly the off-line TEM image is consistent with the online OM image shown in part (c) in Fig. 6.19B, C which is identical to part (c) and (i) in Fig. 6.14. Part (b) in Fig. 6.23 presents a zoom-in image of part (a)

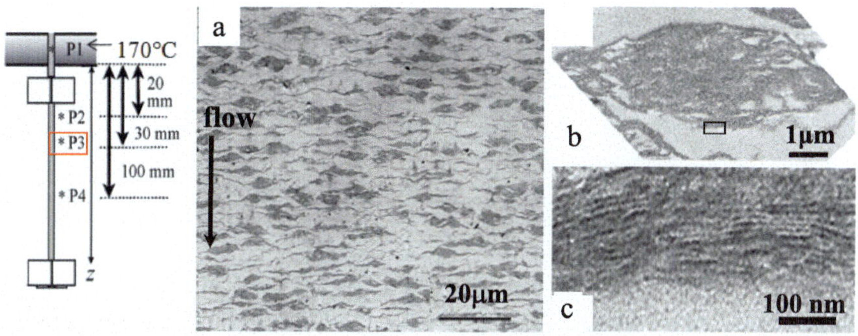

Fig. 6.23 The TEM images of the fiber in the solidified specimen sampled from the spinning line at P3 located at 30 mm downstream from the spinneret: **a** the low magnification TEM image, **b** the zoom-in image of the single demixed domain, **c** the zoom-in image of the part of the demixed domain encompassed with a black rectangle in part **b**. Based on [38], reprinted with permission, American Chemical Society

to a single demixed domain. The image of the single domain elucidates that the domain contains chain-folded lamellar crystals developed by the crystallization during cooling the specimen sampled at P3 to room temperature which serves to the solidification of the specimen. This solidification of the specimen due to the crystallization serves to conserve the demixed domains existed in situ at P3 in the spinning line at room temperature. Part (c) shows the zoom-in image of the boundary region of the demixed domain encompassed by the small solid black rectangle in part (b). The demixed domain, facing the matrix phase replaced by the epoxy resin, is shown to be composed of the stacks of the lamellae with the lamellar thickness $D \sim 5.6$ nm and the interlamellar spacing $L \sim 8.8$ nm. The direction normal to the SLCFL are preferentially aligned parallel to FD and perpendicular to the interface between the demixed domain and the matrix phase. This special, preferential orientation of the lamellae is discussed further in the item (F) in Sect. 6.6.3 (F) below.

(E) **Position P4: Strings composed of aligned demixed domains**. At P4 located at $z \sim 100$ mm from the nozzle, the demixed domains start to align with their centers of mass in rows parallel to the flow-direction, as typically shown in Fig. 6.24a [38]. Part (b) is the same micrograph as part (a), but it presents sketches of the string-like self-assemblies and the demixed domains inside the strings with the broken black lines and the solid black ellipses, respectively. Interestingly, the off-line TEM image is consistent with the online OM image shown in part (d) and (j) in Fig. 6.14 and part (d) in Fig. 6.19C. About three string-like assemblies of the demixed domains appear in Fig. 6.24a as sketched in Fig. 6.24b, but there are still some domains which are off-aligned from the strings. The trend for formation of the well-defined strings composed of one-dimensional assembly of demixed domains along the FD was found in a slightly downstream portion of the spinning line from the position P4, as will be shown

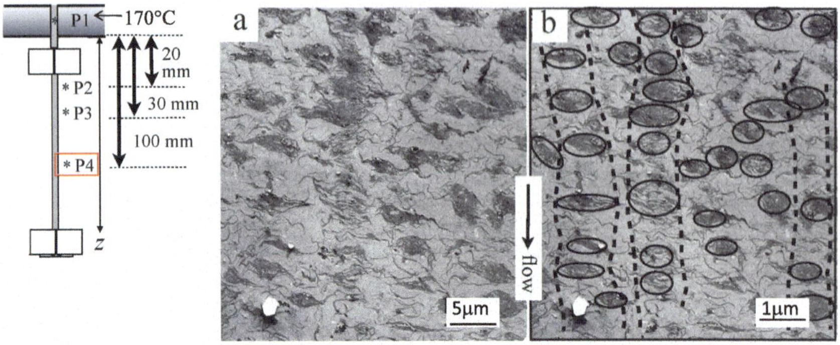

Fig. 6.24 The TEM images of the fiber solidified at P4 located at 100 mm downstream from the spinneret: **a** onset of string formation, **b** the same image as part **a** where the string–like assemblies and demixed domains in the strings are sketched by broken lines and black ellipses, respectively. Based on [38], reprinted with permission, American Chemical Society

in the item (B) in Sect. 6.5.3 (Fig. 6.29) later. The averaged lamellar thickness and long period due to the post crystallization during the solidification process are 5.3 and 8.1 nm, respectively.

(F) **The coil -to-stretched-chain transition and the nucleation of the shishs.** Figure 6.25 presents the TEM image obtained from the specimen sampled at the position slightly downstream from the position P4 at $z = 100$ mm from the nozzle [38] (part a). The position is a critical position where the birefringence $\Delta n(z)$ sharply uprises by about three times with increasing z across the position P4. We designate the regions slightly upstream and downstream side of this point P4 as P4-1 and P4-2, respectively (part a). The result presented in this figure is obtained specifically at P4-2. Part (b) elucidates the string-like assembly of demixed domains A to D which are sketched in part (c). Part (b) reveals formation of shishs in the domains A, B, and C as marked by the two black arrow heads, i.e., the shish traversing from the domain A to domain B as sketched by the line marked X (the arrow head in the upper position) and the shish traversing from the domain B to C as sketched by the line marked Y (the arrow head in the lower position) in part (c). In part (b) and (d) the shishs appear as thin bright lines, seemingly because the RuO_4 vapor does not stain the core of shishs. The shishs are seemingly grown from the boundaries of impinged domains A and B toward their interiors and from the boundaries of impinged domains B and C toward their interiors.

The impingements of the domains at their boundaries may be brought by the flow-induced densification of the domains within the strings and the flow-induced deformation of the densified strings which in turn may bring (i) collisions between the domains within given strings, (ii) resultant partial coalescence and deformation of them at domain boundary regions, and (iii) formation of a narrow and short bridging domain between the neighboring domains. Part (d) shows the zoom-in image of

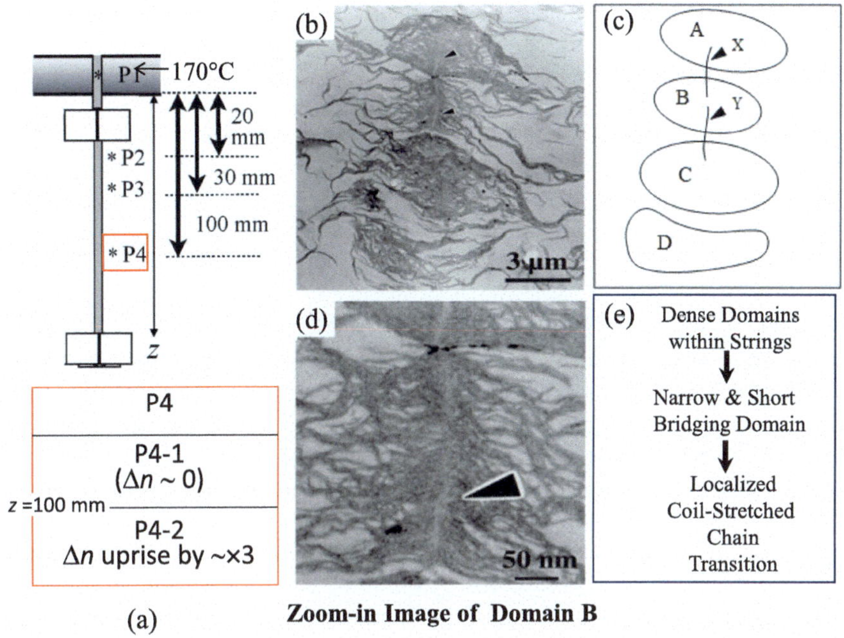

Fig. 6.25 The TEM images of the solidified specimen sampled at P4 (part a) and the assembly of the demixed domains, suggesting the shish formation from the boundaries of the impinged demixed domains (part b). Part (c) presents a sketch of the TEM image shown in part (b): part (c) elucidates the string-like assembly of the demixed domains A to D which suggests formation of shishs, extending from the two black arrow-heads in part **b**, which are grown from the boundary of the impinged domains A and B and from that of the impinged domains of B and C. Part **d** is the zoom-in image of the domain B. (e) Key words for the process leading to the nucleation of the shishs

part (b), The image in part (d) is composed of the bottom part of domain A, the whole domain B, and the top part of domain C and visualizes the two shishs (X) and (Y) connecting the three domains (A),(B),and (C) as illustrated in the sketch (c). The shishs are found to grow from the boundaries of the impinged domains toward their interiors, although the two shishs grown from the respective boundaries of the domain B to the interior do not yet seem to be interconnected within the B domain, as sketched by the two lines marked X and Y in part (c).

We propose that the local stress concentration in the narrow short bridging domains at the impinged domain boundaries is sufficiently large so that it causes the CSCT there to form the bundles of stretched chains where the shish is nucleated and grown. The growth rate of the shishs in the bridge domains may be so high that the growth front of the shishs further propagates into the interior of the amorphous domains composed of unoriented polymer chains through the autocatalytic orientation of randomly coiled chains in front of the rapidly propagating front face of shish [40].

The strings with the interconnected domains are thought to be the precursory structure of shish-kebab because they can form first shishs and subsequently the kebab lamellae overgrown from the shishs. It is of worthy noting that the precursory structure still retains the **memory of the demixed domains** as presented in part (b) and (d) which serve as reservoirs where crystallization of the solution locally occurs to develop the SK. Part (e) presents a list of the keywords which triggers the shish-kebab formation. The flow-induced *densification of the domains within the string* results in formation of the *narrow and short bridging domains* in between neighboring domains where the large stress concentration is built up to bring about the *localized coil-stretched chain transition* and the nucleation growth of the shish.

Figure 6.26a1 at $z \sim 100$ mm, position P4-2, and its sketch (a2) also indicate an example of the string-like assembly of demixed domains which suggests the shish formation from the boundaries between the two impinged domains [38]. The black arrows in parts (a1) and (a2) indicate the boundaries between the two impinged domains which form the short and narrow bridge between them, whereas the yellow arrows in parts (a1) and (a2) indicate the boundaries of the two domains which are flattened due to their collision and which do not yet form the bridges between them. However, they may likely form eventually the bridges through their collisions and elongational deformation involved by the flow. The broken white lines in the sketch (a2) indicate the boundaries of the domains, whereas the solid white lines indicate shishs developed from the boundaries of impinged domains, respectively. One can visualize that three shishs grow from respective boundaries between the impinged domains toward their interiors.

Figure 6.26b presents the SK developed at $z > 100$ mm, sightly downstream of the spinning line at P4-2. The TEM image presented in Fig. 6.26b validates well-developed shishs and kebabs overgrown from shishs [38]. The memory of the demixed domains served as reservoirs for the nucleation growth of the SK is almost relaxed due to the long-range molecular rearrangements of chains through the sliding chain diffusion [44] in shish crystals and kebab crystals.

(G) **The shish-kebab formation: the memory decay of the demixed domains.** Figure 6.27 presents the typical TEM image at the position P5 in the wound fiber shown in Fig. 6.9a and Fig. 6.35a to be shown later [38, 39]. It shows two shish kebabs with interlocked lamellae having average lamellar thickness D and lamellar spacing L of 7.5 and 10 nm, respectively. The memory of the domain boundaries is observed to almost disappear in the shish kebabs. Such memory decay of the demixed domains occurs with an increasing degree in the order of the images shown in Figs. 6.26a1, 6.26b, and 6.27. It is important to note that the whole space of the wound-up fiber at poison P5 is filled with shish-kebabs and the phase composed of solvents molecules replaced by the epoxy resin which is unstained with RuO_4 and thereby appears as the bright phase in the TEM image. The structures such as the demixed domains and the string-like structures having the domain memory are hardly observed. This simply suggests that all the structures observed in Figs. 6.22, 6.23, 6.24, 6.25 and 6.26 represent a series of the transient dissipative structures developed

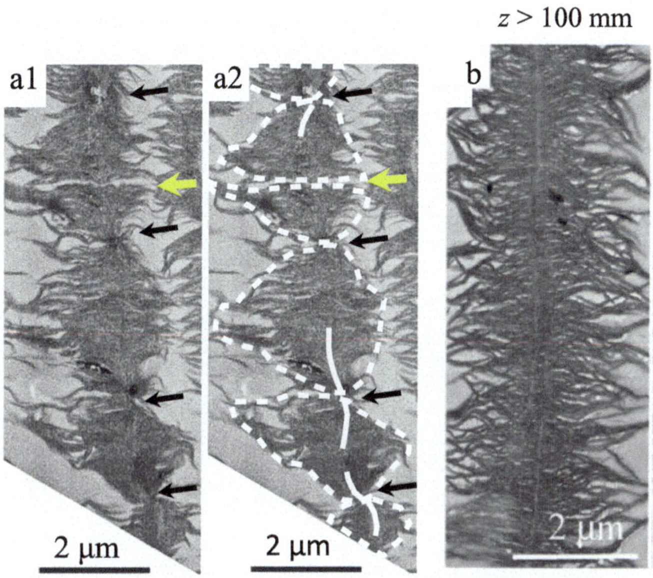

Fig. 6.26 Example of the string-like assembly of the demixed domains suggesting the shish formation from the boundary of the impinged demixed domains [part (**a1**)] and its sketch [part (**a2**)]. Black arrows in parts (**a1**) and (**a2**) indicate the boundaries of the impinged demixed domains, whereas broken white lines and solid white lines in part (**a2**) sketch the boundaries of the domains and the shishs, respectively. Part **b** elucidates the growth of the shish-kebab along the fiber axis together with the decay of the memory of the demixed domains observed at $z > 100$ mm.

in the kinetic pathway from the homogeneous solutions to the shish-kebabs in Fig. 6.27. The close observation of the fine structure of the SK shown in Fig. 6.27 at the position P5 in the spinning line implies existence of "crystalline fibrous bridging structures (FBS)" which bridges neighboring kebabs. This FBS is thoroughly confirmed by means of transmission electron micro-tomography (TEMT) [39]. More specifically, this FBS is predicted to be self-organized in the UHMWPE solution which fills the space between neighboring lamellae through the SD in the late-stage kinetic pathway way shown later in Figs. 6.30b, 6.31f and 6.31g. In this late-stage kinetic pathway way, the temperature of the solution in the spinning line is sufficiently lowered, so that the SD is triggered in the solution to result in the self-organization of the FBS.

6.6.3 *Discussion and Model*

(A) **The shape of the demixed domains**. As shown in Fig. 6.23, the demixed domains have a shape of an oblate ellipsoid with its revolution axis statistically oriented along the flow direction. The shape and the orientation of the domains

Fig. 6.27 The typical TEM image of the shish-kebabs observed in the fiber sampled at P5 shown in Fig. 6.9. From [38], reprinted with permission, American Chemical Society

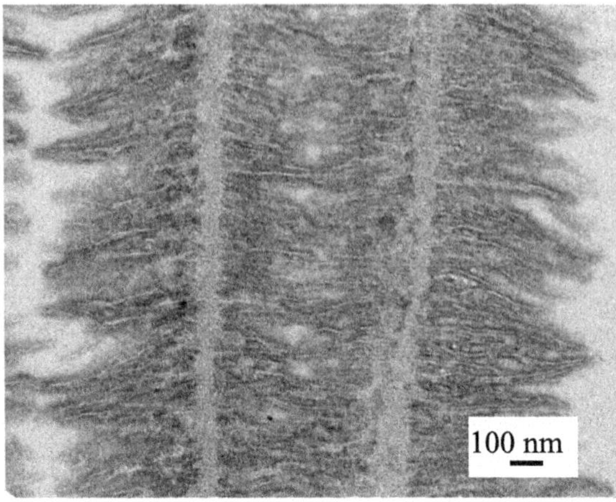

Position P5 along the spinning line

reflect the history regarding the formation of the demixed domain under flow along the spinning line at P3 or after the onset of the step-up shear flow at the constant shear rate higher than $\dot{\gamma}_a$ (Fig. 6.13b).

The thermodynamic instability of the solution brought by the flow first develops the PLWCF as schematically shown in Fig. 5.19. When the amplitude η_k of the PLWCF ($\delta\phi_{\mathbf{k}}$) increases with time (Eq. 5.6.1), the hydrodynamic interactions increase between the solvent flow and polymer chains in the regions rich in polymers (the red regions in Figs. 5.19 and 5.20). Then, the flow destabilizes the PLWCF and breaks them along ND normal to FD, as schematically shown by the structure transformation from part (a) to (b) in Fig. 5.30, at the length scale statistically larger than the wavelength ($\Lambda_{\mathbf{k}}$) of the PLWCF along FD, which results in the formation of the demixed domains with the oblate ellipsoidal shape with its axis of revolution oriented along FD as illustrated by the domain b3 in Fig. 6.28b.

This hypothesis regarding the transformation of the PLWCF into the demixed domains is experimentally supported by the TEM shown in Fig. 6.28a [38]. Figure 6.28a demonstrates the transformation process which generates the partly thickened and thinned PLWCF as shown by the sketch with the white broken lines. The part (b) schematically presents the model for the transformation mechanism. The thinner part marked b2, which is developed in the polymer-rich region marked b1 in the PLWCF driven by the solvent flow, becomes even thinner through the **local capillary flow** (indicated by the thin black arrows) **driven by the capillary pressure**; the thinner part is finally broken, developing the oblate-ellipsoidal domains marked b3. We can predict the reason why the demixed domains have the particular shape of the oblate ellipsoid with its radii along ND (Oz direction) and the velocity gradient direction (Oy direction equal

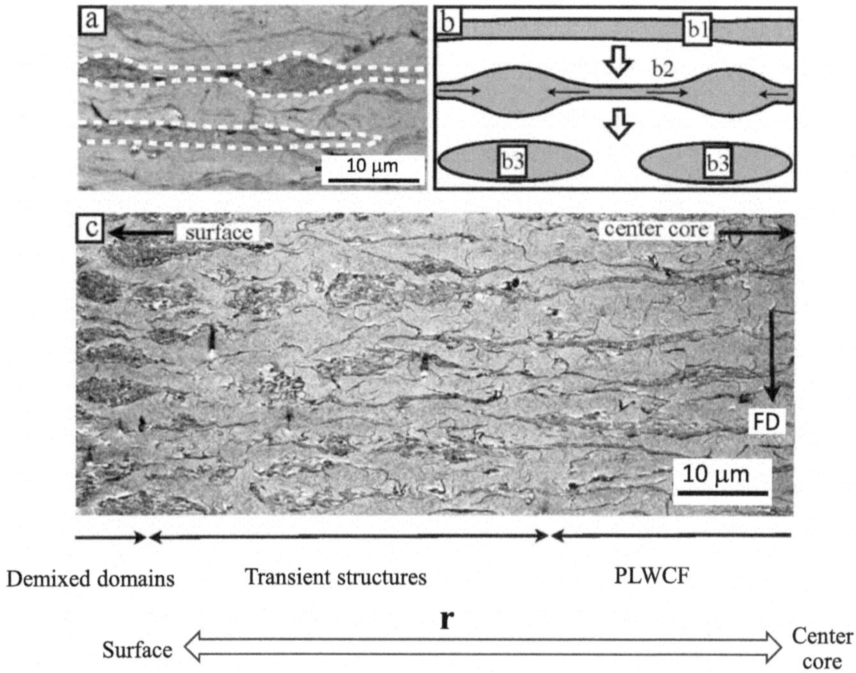

Fig. 6.28 The structural transformation from the plane-wave concentration fluctuations to the demixed domains. Part **a**: TEM image showing the transient structural transformation from the plane-wave concentration fluctuations to the demixed domains; the partly thickened and thinned plane-wave concentration fluctuations are shown by the sketch with the white broken lines. Part **b**: Schematic illustration of the transformation from the plane-wave concentration fluctuations (b1) to the demixed domains (b3) via the transient structure (b2), driven by the hydrodynamic effects. Part **c**: The continuous transition from the plane-wave concentration fluctuations to the demixed domains observed along the lateral direction of the spun fiber from near the center core to near the surface of the fiber. From [38], reprinted with permission, American Chemical Society

to the propagation direction of the incident beam) being larger than the radius along FD (Ox direction).

We can account for the physical origin of the shape described above based on the Ginzburg-Landau theory on thermal concentration fluctuations with the Fourier modes with the wave number q [48]: the smaller the q is, the lower is the free energy. Thus, the Fourier modes with the largest wavelength have the largest thermodynamic driving force. However, these modes have the smallest growth rate. These two opposing physical factors select the modes which have the fastest growth rate. These selected modes become the dominant Fourier modes of the thermal concentration fluctuations to grow along the direction perpendicular to FD in the structure transformation process from the PLWCF (b1) to the two demixed domains (b3) through the transient domain structure (b2) as presented in Fig, 6.28b.

Moreover, this transformation process can be also visualized by investigating the morphological variations of the specimen sampled at the position P2 in the spinning line along the lateral direction **r** from the center core to the skin of the fiber as shown in Fig. 6.28c, where the center of the fiber exists beyond the right-hand edge of the figure as shown by the arrow marked by the "center core", whereas the skin of the fiber exists beyond the left-hand edge of the figure as indicated by the arrow marked by "surface". The figure simultaneously elucidates different morphologies in the three regions along the lateral direction at P2: the PLWCF at the right-hand side part of Fig. 6.28c, the assembly of the densified demixed domains at the left-hand side part of the figure, and the transient structures between the PLWCF and the demixed domains at the middle part. It should be noted that the TEM image presented in Fig. 6.22b is the image taken near the center core region of the fiber in the same ultrathin section at the position P2, where the TEM image in Fig. 6.28c is taken. The unique orientation of the demixed domains under the flow appears to be stabilized by interactions with neighboring domains.

(B) **Densification of the demixed domains in the strings and the densification of the strings in the fiber**. TEM images shown in Fig. 6.29 were obtained from the specimen sampled at P4-1 [38], the downstream of P3 shown in Fig. 6.24. The following two trends are worth noting in Fig. 6.29: (1) a densification of the demixed domains within the strings as shown in the regions encompassed by broken white lines in part (a) and (2) the densification of the strings themselves within the fibers as shown in the regions encompassed by the rectangular broken white lines in part (b). Part (c) in Fig. 6.29 is a micrograph which clearly captures evidence for the former densification process (1); the red painted domains marked by A, B, and C seem to be in the middle of being incorporated into the string, driven by the hydrodynamic interactions of the solvent flow and the domains isolated from the strings.

Because of the hydrodynamic effects on the densification (1), the strings S1 to S4 shown in part (a) have a higher density of the demixed domains than the average density of the demixed domain in part (a). Furthermore, the strings themselves tend to be densified, driven by the hydrodynamic interactions of the solvent flow and the strings; the clusters of the strings in the area encompassed by the rectangular white broken line in Fig. 6.29b are composed of four strings which are both laterally and longitudinally self-assembled closely one another, as shown in the inset to Fig. 6.29b where the four strings are encompassed by the white broken lines. This densification process of strings will be also discussed later in Sect. 6.5.3 (D).

(C) **The model for the self-organization of the dissipative structures along the spinning line: universality found in the self-organized dissipative structures in the dynamically asymmetric systems**. Figure 6.30 summarizes the results described in Sect. 6.5.2 concerning the evolution of the various dissipative structures (the ordered structures developed in the open nonequilibrium systems) in part (b) along the fiber spinning line as schematically shown in part (a) [38]. The kinetic pathway from the solution of the entangled random coils

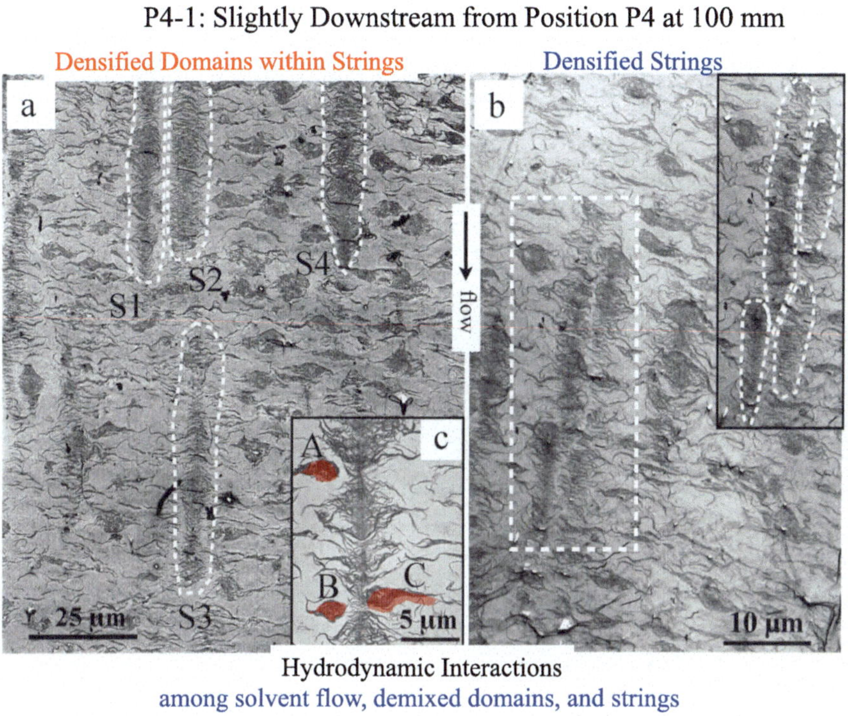

Fig. 6.29 The densification of the demixed domains in the string-like structure (part **a**) and the densification of the string-like structures themselves within the fiber (part **b**). The four strings denoted by S1–S4 in part **a** are composed of the densified demixed domains. In part **b** the clusters of the densified strings encompassed by the white broken rectangle are composed of the four strings, as sketched by the white broken lines in the inset of part **b**. In part **c,** the off-aligned domains marked by A, B, and C tend to be incorporated into a string, which seemingly validates the densification of the string driven by the hydrodynamic interactions among the solvent flow, demixed domains, and the strings. Based on [38], reprinted with permission, American Chemical Society

of the polymers at P1 to the well-developed shish-kebabs swollen with solvent at P5 (Fig. 6.27) comprises the various transient dissipative structures typically found at P2 to P4 along the spinning line, as listed below:

(1) At P2, the plane wave concentration fluctuations are self-organized in situ in the spun solution which is brought into the thermodynamic instability by the imposed flow. Here, the stress-diffusion coupling plays the significant role in the solutions under the flow (see the result shown in Fig. 6.22b).
(2) At P3, the oblate-ellipsoidal demixed domains with their revolution axes oriented parallel to the flow direction with their centers-of-mass aligned randomly in space (Fig. 6.23).

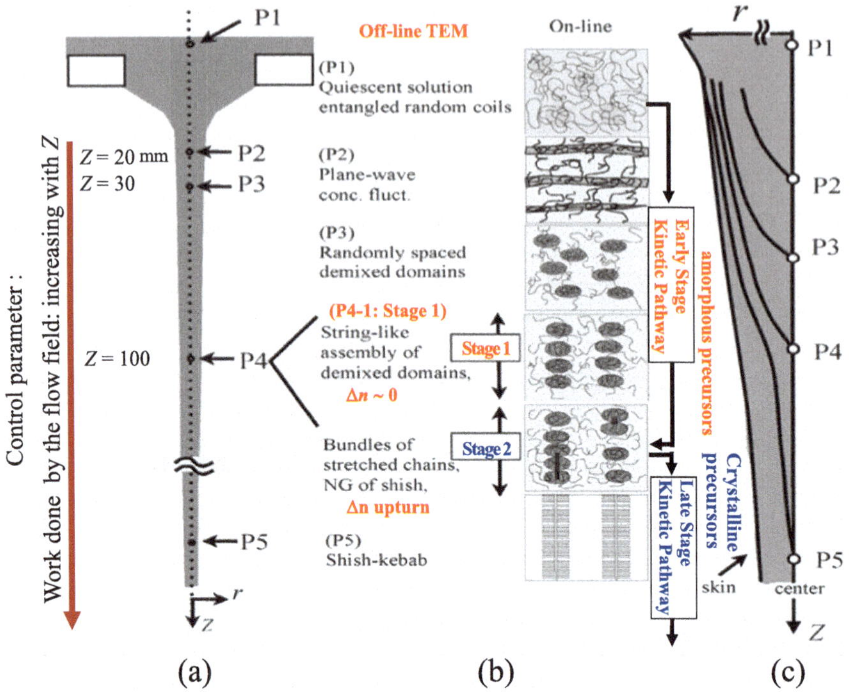

Fig. 6.30 Summary of the dissipative structures evolved along the spinning line of the fiber spun from the UHMWPE solutions: **a** schematic illustration of the fiber spinning line and the sampling positions P1 to P5 for the off-line TEM observations, **b** a proposed kinetic pathway for the solutions (in the spining line) of entangled random coils swollen with solvent at P1 to the shish-kebabs at P5 [the left half of part **b** elucidated based on the off-line TEM observations, and right half of part **b** expected for the on-line structures along the fiber], **c** schematic illustration of the contour lines representing the change in the transient structural transformation found along the fiber axis z and the lateral axis r: the trajectory of each of the characteristic points P2 to P5 shifts upward with the increased r, as schematically shown by the solid lines emanating from P2 to P5. Based on [38], reprinted with permission, American Chemical Society

(3) At P4, the string-like self-assemblies of the demixed domains oriented parallel to the flow direction (Fig. 6.24).

(4) At P4-1, slightly downstream of P4 (Fig. 6.29a), the string-like self-assemblies composed of densified demixed domains.

(5) At P4-2, slightly downstream of P4-1, the bundles of stretched chains parallel to the flow direction are generated via CSCT in the bridging domains formed in between the impinged demixed domains in the assemblies (Figs. 6.25 and 6.26 a1, 6.26 a2). The sharp upturn of the $\Delta n(z)$ by the factor of ~ 3 occurred in the narrow range of z or time span between P4-1 and P4-2 (Fig. 6.25a), corresponding to the crossover from the early-stage (stage 1) to the late-stage (stage 2) kinetic pathway shown in Fig. 6.19E

for the sheared solutions. Up to P4-1, the chains in the demixed domains retain essentially randomly coiled conformation with $\Delta n(z) \sim 0$.

(6) At the position between P4-2 and P5, the nucleation and growth of shishs and the subsequent overgrowth of the kebab lamellae from the shishs occur (Figs. 6.26b and 6.27).

It should be noted here that the positions P4-1 and P4-2 are almost the same, both existing at $z \sim 100$ mm, although P4-1 exists at a slightly upstream side of P4-2. This reveals itself such an important fact that the transformation from the stage 1 to the stage 2 shown in Fig. 6.30b occurs sharply over a small increment of the distance z at a given lateral position r of the spun fiber. The lateral position r will be defined in the last paragraph of this section in conjunction with Fig. 6.30c; effects of a small increase of r at a given longitudinal position z will be also clarified later in this section.

The structure evolution found for the solutions of the entangled UHMWPE/decalin along the fiber spinning line is complex because it involves the variations of temperature, strain rate, and types of the flow. Nevertheless, the structure evolution elucidated via the off-line snapshots of TEMs summarized in Fig. 6.30 for the solutions in the spinning line is strikingly similar to the isothermal time-evolution of the structures, as summarized in Fig. 6.19, after the step-up simple shear flow for the solutions 6.19C, which is elucidated via the online snapshots of the OMs, the Δn, and SALS both for the non-crystalizable the entangled UHMWa PS/DOP solutions and for the crystallizable solutions of the entangled UHMWPE/paraffin under the shear flow fields generated by the cone-and-plate fixture.

It is striking to note a remarkable universality found for those two systems, with respect to the kinetic pathways and the specific self-assembled structures formed therein, by means of the off-line observations shown in Fig. 6.30b and the online observations shown in Fig. 6.19. The universality may well suggest that the model shown in Fig. 6.30b constructed based on the off-line observations essentially provide a good model expected for the dissipative-structure evolution in the online fiber spinning process also. The common important physical factor responsible for the universality is the work done to these different systems by the applied flow fields under a maximum energy dissipation mechanism.

It may be a surprise at first glance to note the similarity of the shear flow effects and the extensional flow effects on these structure evolutions shown in Figs. 6.19 and 6.30, respectively, since these two flows are known to have different effects on the molecular orientation and the stress evolution. However, we should note that the extensional flow field imposed on the fiber in the spinning line is complex, because the extensional strain rate and stress depend on the distance z from the spinneret along the spinning line, as expected from the z-dependence of the Δn, the fiber diameter, the temperature, etc. (see for example [42, 43]). Consequently, it may be plausible that the extensional flow effects are still small, but the shear flow effects are dominant at the positions P2, P3, and P4-1 closer to the spinneret, so that the Δn and hence the chain orientation and stretching are negligibly small. This may explain why the transient structure evolved at P2, P3, and P4-1 are close to those shown in Fig. 6.19 under the step-up simple shear flow at the given temperatures.

Finally, the structure change should be considered along the lateral direction *r* of the fiber, as illustrated in Fig. 6.30c. The structures evolved at P1 to P5 along the center of the fiber are found to shift toward upstream line with shifting the observation-point along the axis *r* from the center to the skin of the fiber: the trajectories of P2 to P5 are schematically presented by the respective solid lines (Fig. 6.30c). This trend was confirmed by scanning the TEM images along the lateral direction *r*, although the TEM images are not included in this book. This shift is mainly due to the temperature gradient with respect to *r* which is brought by the cooling rate at the surface region of the fiber being larger than that at the center region, hence the stress level is higher at the surface than at the center and thereby the work done by the flow increases rapidly with *z* at the surface area.

(D) **Trend toward the flow-induced macrophase separation: flow-induced alignment of demixed domains, densification of demixed domains in the strings, and densification of the strings in the fiber**. Figures 6.24, 6.25, 6.26, and 6.29 clarify the string-like structures appeared as a precursory structure of shish-kebabs at position P4-2. The hydrodynamic interactions between the solvent flow and the demixed domains align the demixed domains into the strings oriented parallel to the flow direction. The strings aligned parallel to the flow direction themselves are aligned side by side laterally (along the direction perpendicular to flow), which gives rise to the nematic alignment of centers of mass of the demixed domains parallel to the flow direction. Moreover, the solvent flow causes off-aligned domains incorporate into the as-grown strings driven by the hydrodynamic interactions described above (Fig. 6.29c), which leads the densification of the demixed domains in the strings, thereby giving rise to the **solvent squeeze** from the strings. The strings themselves also are densified both laterally and longitudinally as shown in Fig. 6.29b due to the hydrodynamic interactions between the solvent flow and the strings, resulting in the solvent being squeezed out from the space in between the string-like assemblies. The two kinds of the densifications, intra-string and inter-string densification mediated by those hydrodynamic interactions, eventually lead to flow-induced growth of phase-separated structures toward macroscopic phase separation between the "polymer-rich phase" (strings-rich phase and/or SKs-rich phase) and the solvent phase under dynamic equilibrium at a given flow field: solvent molecules in one of the macrophase which is rich in solvent molecules are squeezed out from the fiber in the spin line, causing the syneresis and thereby the reduction of fiber diameter in the downstream of the fiber beyond the position P4.

However, as-spun fiber at P5 in the wound bobbin still contains a lot of solvent, 30–50 wt.% solvent, i.e., as-spun fiber at P5 may be considered to be one of the macrophase rich in polymer, whereas the other macrophase composed of solvent is squeezed out from the fiber at P5. The macrophase rich in polymers still contains solvent molecules by 40 wt.% at P5 which appear to form the bright phase rich in solvent molecules in the assembly of SKs as shown in Fig. 6.27.

(E) **Coil-to-stretched chain transition leading to formation of bundles of stretched chains and nucleation-growth of shishs**. The flow-induced phase separation is discovered to provide an excellent and rational field for polymer chains to crystallize into the shish-kebab morphology under the flow. In the proposed scenario for the shish-kebab formation, the demixed domains in the strings play two important roles as follows.

(1) The demixed domains densified within the strings may impinge one another under the flow to form short and narrow bridging domains at their domain boundaries. The large local stress concentration on the bridging domains may develop bundles of stretched chains there via the CSCT, as evidenced by Figs. 6.25 and 6.26, and eventually trigger the nucleation growth of shishs within the bundles.

(2) The demixed domains serve as "reservoirs of random coils" (i) for the formation of shishs and (ii) for the formation of the kebab lamellae. As for (i), the growth front of shishs nucleated in the bridging domains propagates into the reservoirs. As for (ii), the kebab lamellae overgrow epitaxially on the surface of growing shishs from random-coil chains in the reservoirs.

Figure 6.31a–g and Table 6.2a–d summarize the local events involved in the cascade evolution of the dissipative structures in the early-stage and late-stage self-organization process of SK under the flow [31, 38]. FD is set in vertical direction as shown in part (a) in Fig. 6.31. In the early stage, the solutions have effectively no birefringence ($\Delta n \cong 0$), and thereby the polymers are effectively in random-coil conformations (parts a and b) under the flow. On the other hand, in the late stage, the solutions have the positive birefringence ($\Delta n > 0$), and thereby the polymers are partially oriented in parts (c) to (e) and mostly oriented in parts (f) to (g) along FD.

The bulk orientational order parameter f_b of the chains in solutions is defined by the first line in Eq. 6.5.1 below,

$$f_b \equiv \left[3\langle\cos^2\theta_s\rangle_b - 1\right]/2, \quad f_l \equiv \left[3\langle\cos^2\theta_s\rangle_l - 1\right]/2, \tag{6.6.1}$$

Table 6.2 The pattern formation mechanisms in the multilength, heterogeneous, and amorphous precursor: A unique feature found in the particular open non-equilibrium phenomenon involved in the SK formation

Local events	Consequences
(a) Formation of the BDs within the dense strings	$\sigma_l \gg \sigma_b$, leading to the CSCT in the BDs
(b) Formation of the bundles of the stretched chains (BOSCs) in the BDs	$f_l \gg f_b$, $T_{m,h} \gg T_{m,r}$, large $\Delta T_{m,h}$
(c) The NG of the shish starts from the BDs at the very large rate	(d) The growth of the BOSCs and shishs into the domains having $f_l \sim 0$

The local events (a) to (c) as well as the consequence (d) are indicated in Fig. 6.31 also

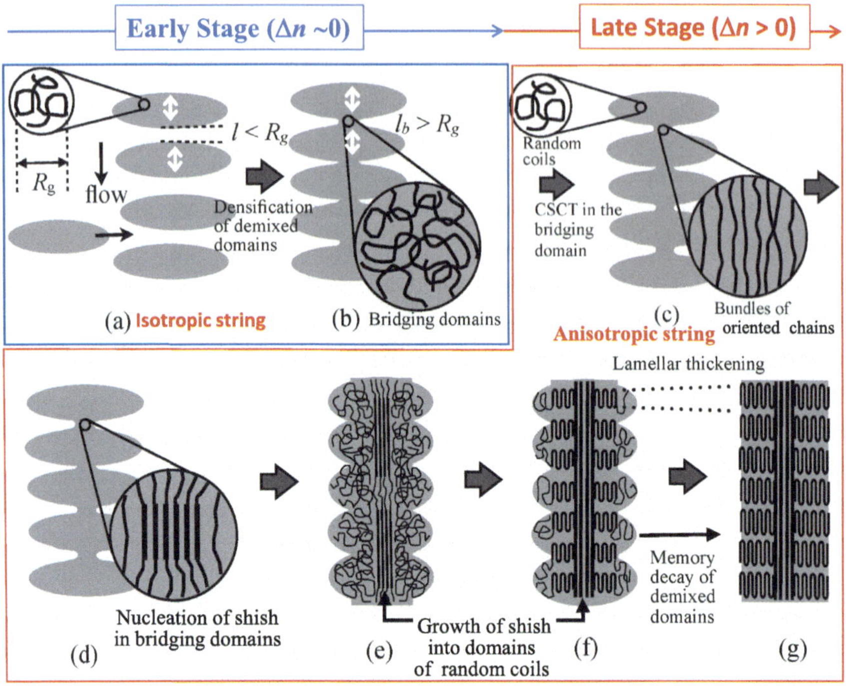

Fig. 6.31 Schematic model showing the formation of the string and the transformation from the string to the shish-kebab: the densification of the demixed domains in the string (part **a**), the formation of the bridging domains or the bridging chains between the domains densified in the strings (part **b**), formation of bundles of stretched chains (part **c**) and nucleation–growth of shishs in the bridging domains at the boundaries of the impinged demixed domains (part **d**), the growth of the bundles of the stretched chains and the shishs from the boundaries toward the interiors of the domains (part **e**), and the epitaxial overgrowth of the kebab lamellae from the random coils in the demixed domains onto the surface of the shish (part **f**), and the long-range rearrangement of the chain molecules within the as-grown shish-kebab which causes memory of the demixed domains within the shish-kebab decay to the more ordered shish-kebab (part **g**). Based on [38], reprinted with permission, American Chemical Society

where θ_s is the polar angle of segments of chains with respect to FD, and $\langle\cos^2\theta_s\rangle_b$ is the average of $\cos^2\theta_s$ over the bulk solutions under the flow. On the other hand, the local orientational order parameter f_l is defined above by the second line in Eq. 6.6.1 where $\langle\cos^2\theta_s\rangle_l$ is the average of $\cos^2\theta_s$ in local regions in the bridging domains (parts b to d in Fig. 6.31) and in the central cores of the string (part e). In the bulk solutions with the dissipative structures shown from parts (a) to (d) in Fig. 6.31, f_b is small, close to zero. On the other hand, the local orientational order parameter f_l is zero in part (b) in the early stage but very large close to 1, in part (c) to (e) in the late stage in Fig. 6.31.

The model shown in Fig. 6.31 schematically illustrates (1) the formation of the strings composed of the demixed domains aligned along FD and the densification of the demixed domains in the string due to the off-aligned domains, with respect to the strings, being incorporated in the strings (part a), (2) the formation of the bridging domains or bridging chains between the neighboring domains in the densified strings; here, the chains in the bridging domains or bridging chains and the demixed domains aligned along FD are clarified to be still in the random-coil state (part b), (3) the formation of bundles of stretched chains (BOSC), due to the CSCT,in the bridging domains (part c); here, the chains in the demixed domains are clarified to be still in the random-coil state, although the BOSC are clarified to be highly stretched chains oriented along FD, (4) the nucleation and growth of shishs in the bridging domains at the boundaries of neighboring demixed domains and further growth of the bundles of the stretched chains in the bridging domains driven by the sufficiently high nucleation-growth rate of the shishs in the bridging domains (the autocatalytic effect on the growth of the BOSCs [40]; here, the bundles of the highly stretched chains oriented along FD have been built up prior to the nucleation growth of the shish (part c and d), and (5) the growth of the shishs nucleated within the bundles of the stretched chains in the bridges from the boundaries to the interiors of the amorphous domains (part e), (6) the crystallization of the random-coil chains in the demixed domains into the kebab lamellae overgrown epitaxially from the shishs which have already been built up within the demixed domains prior to the crystallization of the kebabs (part f), and (7) the long-range translational diffusion of the chain molecules along the chain axis of the crystals within the as-grown shish-kebabs, which causes the memory of the demixed domains within the shish-kebabs (part f) decay to form a more ordered shish-kebabs (part g).

The transformation from part (a) to (b) in Fig. 6.31 indicates the following two local events:

(1) the densification of the demixed domains, which brings about the interstitial space between the neighboring domain interfaces at a distance l apart becomes smaller than R_g, the radius of gyration of chain molecules, assisted by flow-induced dynamical motion of the domains (Fig. 6.31a);

(2) the interstitial space is eventually transformed into the short and narrow domains bridging the neighboring demixed domains with its length l_b along the string axis being grown to become larger than R_g(Fig. 6.31b) so that the bridging domains of the length l_b are composed of some given number of entangled chains.

The bridge may be formed by the diffusion-coalescence and the deformation of the neighboring domains under the flow field. The CSCT in the bridging domains in part (b) mediated by the local stress concentration will cause the formation of the bundles of stretched chains (c). The nucleation and growth of shishs in the BOSC dominantly occurs longitudinally but not laterally as shown in the shish growth from (e) to (f). From (f) to (g), the long-range rearrangements of chains in the kebabs and shish (the translational diffusion of chains along the chain axis of the crystals) leads to (i) lamellar thickening, (ii) the reduction of the lateral size of the shish-kebabs [45], and (iii) the decay of the memory of the demixed domains in the shish-kebabs.

In Fig. 6.31b, c, it may be puzzling why the chains in the bridging domains are stretched from (b) to (c) under the flow and why the chains inside the demixed domains are kept in random coils. We predict that:(i) the demixed domains are only partially or hardly draining with the solvent flow, hence the chains inside the demixed domains being kept essentially in random coils as illustrated in part (a) and (c); (ii) the motions of the demixed domains as a whole mediated by the solvent flow cause the collision and deformation of the demixed domains to result in formation of the chains or domains bridging between the neighboring demixed domains, and the bulk stress σ_b imposed on the demixed domains is highly concentrated locally at the narrow bridging domain so that the local stress σ_l there satisfies $\sigma_l \gg \sigma_b$ (Table 6.2 a); (iii) these bridging chains or bridging domains will be stretched also mediated by the solvent flow, which in turn causes the formation of the bundles of stretched chains within the bridging chains or domains via CSCT (Fig. 6.31c, Table 6.2 b).

The following model is proposed, as one of the plausible models, to explain the growth of the shish toward the interior of the demixed domains which comprise essentially undeformed entangled coils as illustrated schematically in the change from Fig. part (d) to (e) in Fig. 6.31. The growth of the shish at a sufficiently large growth rate from the bridging domains under the high local stress σ_l is predicted to cause the shish propagates toward the interior of the demixed domains composed of unoriented chains with $f_l \cong 0$ (Table 6.2 c and d) according to the mechanism proposed by Petermann and Keller and coworkers [40], which is based on the autocatalytic orientation of relaxed chains existing in front of the rapidly growing shish into the bundles of the stretched chains (BOSC) along FD.

The proposed hypothesis regarding high growth rate of the shishs nucleated in the BOSCs. Why the growth rate of shishs nucleated in the BOSCs is so high that it enables the autocatalytic orientation of the relaxed chains in front of the tips of the growing shish? Figure 6.32 presents the melting temperature, $T_{m,h}$, (part d) of the oriented crystals (part c) crystallized from the highly oriented amorphous chains as in the BOSCs (part a) and the melting temperature, $T_{m,r}$, (part d) of the crystals crystallized from randomly oriented chains (part b). The equilibrium melting temperature of the crystals T_m is given by

$$T_m = \Delta H_m / \Delta S_m, \tag{6.6.2}$$

where ΔH_m and ΔS_m are the enthalpy and entropy change in the melting process.

ΔS_m for the L–S phase transition between (a) and (c) is much less than that between (b) and (c). On the other hand, the ΔH_m between (a) and (c) and the ΔH_m between (b) and (c) are almost the same or the difference in ΔH_m is much less than the difference in ΔS_m. Thus, it is predicted that

$$T_{m,h} \gg T_{m,r}, \tag{6.6.3}$$

as shown in part (d).

The melting temperature for the *bulk solution*, having randomly coiled chains as shown in Fig. 6.32b under the flow at the shear rate $\dot{\gamma} > \dot{\gamma}_a$ is equal to $T_{m,r}$ in the early stage with the $\Delta n \sim 0$ in Fig. 6.31a, b and in the beginning of the late stage prior

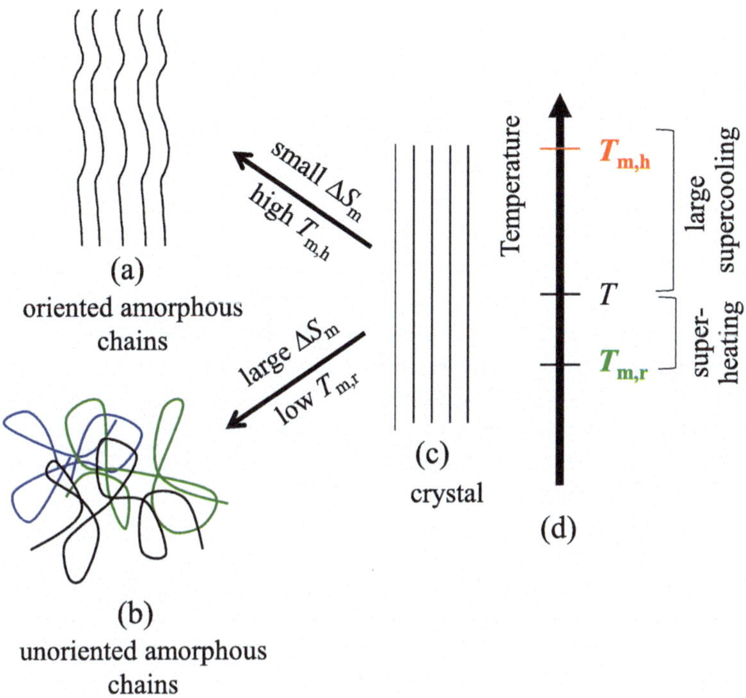

Fig. 6.32 Melting temperature, $T_{m,h}$, of the oriented crystals (part **c** and **d**) crystallized from the highly oriented amorphous chains as in the BOSCs (part **a**) and the melting temperature, $T_{m,r}$, of the crystals crystallized from the randomly oriented chains (part **b** and **d**)

to the stage where the anisotropic strings with the small positive Δn is developed as shown in Fig. 6.31c, because the bulk solution effectively has the random-coil chains. However, the *local melting temperature* in the bridging domains in the late stage (part c and d in Fig. 6.31) is raised to $T_{m,h}$, because of the very high f_l , as shown in Fig. 6.32a i.e., the high degree of the chain orientation brought by the BOSCs developed in the bridging domains driven by the large local stress concentration there. It should be noted that the *local stress* σ_l there is much larger than the *bulk stress* σ_b of the sheared solution (Table 6.2 a). Therefore, the bridging domains with the BOSCs are brought under the extremely large supercooling $\Delta T_{m,h} = T_{m,h} - T$ as shown in Fig. 6.32d at the working temperature T under the shear flow (Table 6.2 b). This may bring about the very high growth rate of the shishs locally nucleated in the bridging domains (Table 6.2c, d).

(F) **The orientation of the lamellae crystallized from the random-coil chains in the "confined space"**. The lamellae formed by the post crystallization during the cooling the specimens sampled at positions P2 and P3 to room temperature is preferentially aligned with their normal to the SLCFL, defined as $\mathbf{n}_L$ hereafter, parallel to FD (Figs. 6.22 and 6.23). It is important to note that the lamellae formed here have the special orientation and a high degree of the orientation with respect to FD, despite the sampled solutions at P2 and P3 are amorphous with random-coil chains and zero birefringence, $\Delta n \sim 0$.

Let us consider the physical factors which cause the intriguing orientation of the lamellae found in the solidified specimen sampled at P2 and P3, is described in Sect. 6.5.2 (C) and (D), respectively, in conjunction with Fig. 6.22b at P2 in Sect. 6.5.2 (C) in conjunction with Fig. 6.23b, c at P3 in Sect. 6.5.2(D). The birefringence Δn at P2 and P3 is almost zero, hence the chains in the plane wave concentration fluctuations at P2 and in the demixed domains at P3 in situ remain essentially in random coils. Therefore, the orientation of lamellae developed during the solidification process, designated as "**oriented crystallization of lamellae**", must not be caused by "conventional oriented crystallization", i.e., the crystallization induced by oriented chains, but be rather caused by the orientation induced by an "**oriented confined space**" where the random-coil chains are enclosed.

The PLWCF and demixed domain provide the respective oriented **confined space**. As for the PLWCF, the concentration gradient of polymer chains exists along FD (Ox direction) i.e., normal to the wave front of the PLWCF (Oyz plane), as presented below in Fig. 6.33a. As for the demixed domain, the concentration gradient which exists along FD normal to Oyz plane is most significant one for the crystallization in the confined space as presented in Figs. 6.23b, c, although concentration gradient exists in all the directions normal to the interfaces. We propose that the preferential orientation of the lamellae must be a consequence of the oriented crystallization induced by the orientation of the PLWCF with their wave vectors preferentially oriented along FD and by the orientation of the oblate-ellipsoidal demixed domains with their revolution axes oriented along FD.

Figure 6.33 presents the k-th Fourier mode of the PLWCFs (part a), $\delta\phi_k$, and the confined space relevant to the "dominant" Fourier modes of the PLWCFs under the flow with the flow direction along FD (Ox direction), i.e., the PLWCF with its wave vector **k** along FD (part b). The k-th Fourier mode has the concentration variation $\delta\phi_k$ with the amplitude η_k, the wave number k, wavelength $\Lambda_k = 2\pi/k$, and the wave vector **k** along Ox axis set parallel to FD as indicated in Fig. 6.33a. The PLWCFs which give rise to the butterfly SALS pattern is given by a superposition of the Fourier modes with varying wave vector **k** where **k** is preferentially oriented with respect to FD.

In part (b) in Fig. 6.33, the confined space relevant to the dominant Fourier mode of the PLWCFs is presented: the confined space has the concentration variation only along FD, keeping a constant concentration perpendicular to FD. The dominant PLWCF has the periodicity of $\Lambda_{\mathrm{m}} = 2\pi/q_{\mathrm{m}}$, where q_{m} is the wave number of the dominant PLWCF or the magnitude of the scattering vector at the maximum intensity of the butterfly pattern, and the persistence length l_x and l_z parallel and perpendicular to FD, respectively. The gradient of the concentration variation, $\nabla\delta\phi_k(\mathbf{r})$, exists only along O$x$ axis, i.e., $\nabla\delta\phi_{\mathrm{k}}(\mathbf{r}) = \nabla\delta\phi_{\mathrm{k}}(x)$. It is predicted that the specific orientation of the lamellar stacks found in the solidified specimen at P2 is due to the $(\nabla\delta\phi_{\mathrm{k}}(x))$-induced oriented crystallization of lamellae in the confined space. In other words, the

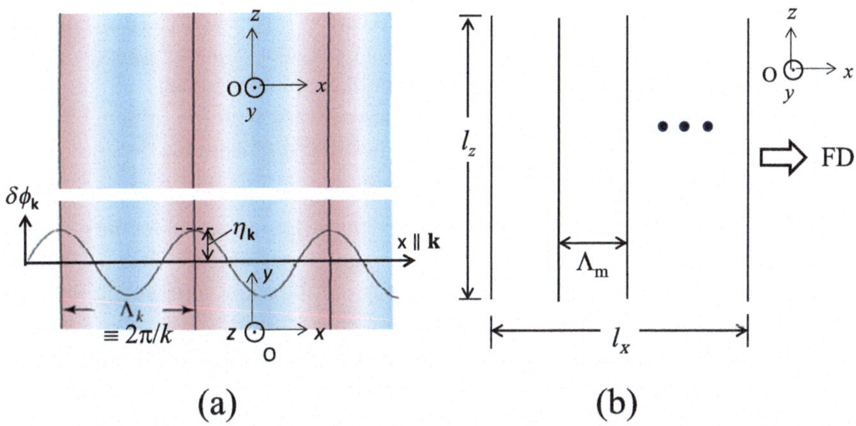

Fig. 6.33 The kth Fourier mode of the PLWCF (part **a**), $\delta\phi_k$, and the confined space relevant to the "dominant" PLWCF (the PLWCF with its wave vector along FD) (part **b**)

$\nabla\delta\phi_k(x)$ induces the oriented crystallization of the lamellae in such a way that the lamellae grow with their $\mathbf{n}_L$ parallel to $\nabla\delta\phi_k(x)$.

Now, we consider the specific orientation of the chain-folded lamellae developed in the demixed domain during the solidification of the spun solution at P3 from the temperature at P3 to room temperature. The lamellae are oriented with their normal vector $\mathbf{n}_L$ parallel to FD as shown in part (b) and (c) in Fig. 6.23. This specific orientation may be interpreted as follows. The oblate ellipsoidal confined space relevant to the demixed domain has the concentration gradient $\nabla\delta\phi_k(\mathbf{r})$ normal to its interface of the oblate ellipsoid with the matrix solution. Among the various directions of the gradient axes, the gradient normal to the interface of the oblate ellipsoid with the least curvature is $\nabla\delta\phi_k(x)$ which exists parallel to the revolution axis of the oblate, parallel to FD (Ox axis); this gradient axis along Ox-axis is the most important axis among others for bringing about the oriented crystallization of the lamella in the demixed domains with the specific orientation. In other words, the growth of the stack of lamellae with its $\boldsymbol{n}_L$ parallel to $\nabla\delta\phi_k(x)$ controls the crystallization of randomly coiled chains in the oriented confined space.

(G) **Remarks on polymer solutions versus polymer melts with respect to the SK formation.** It is concluded that the phase separation mediated by the stress-diffusion coupling inherent to the systems composed of the dynamic asymmetric molecules provides a key for the kinetic pathway leading to the shish-kebab formation under the applied flow. It is also important to note that the nature creates the shish-kebabs from the entangled polymer solutions subjected to the flow via the *complex but rational* kinetic pathway as elucidated in this work. It should be also noted that the observation reported and the model presented here are restricted to the homogeneous solutions of the entangled polymers swollen with the solvents in quiescence state at this stage.

However, the kinetic pathway observed here is proposed to be universal for the homogeneous solutions having different polymer concentrations with the chain entanglements and at some different fiber spinning speeds, although the positions P1 to P4 along the spinning line will vary with the spinning speeds and the polymer concentrations. This is because the positions depend on the work done to the solutions by the flow field.

The generalization of the observed effects and the proposed model for the polymer solutions to (i) the polymer melts composed of the binary blends of the UHMW and low molecular weight (LMW) polymers having the same monomeric units or to (ii) the polymer melts having large MW polydispersity deserves future works. Nevertheless, at this stage the flow-induced phase separation may be proposed to occur and be important in those systems described above which also contain the polymer species having a large dynamic asymmetry. In case (i) described above, the LMW polymer and the UHMW polymer may be regarded as "solvent" and "polymer", respectively. If this analogy works, we anticipate that the melt bends under the flow exhibit cascade evolution of dissipative structures through the phase separation, similar to that found for the polymer solutions. In case (ii) described above, the LMW polymer species and the UHMW polymer species may be treated as "solvent" and "polymer", respectively, and the same analogy as the molten binary blends with a large molecular weight difference may be applied also, to this case. We should note, however, a big difference between the solutions and the binary melt blends or the molten polymers having large MW polydispersity as follows. The "solvent" (the LMW polymers in the binary blends or the low molecular weight species of the polymers having large MW polydispersity) can be in principle co-crystallized with the shish and kebab crystals formed by the "polymers" (the HMW polymers in the blends or the UHMW polymer species in the polymers having the large MW polydispersity), although the solvent in the polymer solutions (liquid paraffin or decalin) cannot be co-crystallized with the polymers whatsoever.

Polymer melts with a narrow MW polydispersities. What would be the necessary and sufficient condition for the polymer melts with a narrow molecular weight polydispersity to crystallize into shish-kebab texture under the flow, free from the phase separation? We may predict the condition as follows based on the concepts of the self-organization of the dissipative structures elucidated in Chaps. 5 and 6. The necessary condition would be the flow-induced evolution of orientation fluctuations of polymer chains with appropriate correlation lengths. The orientation fluctuations may develop local density fluctuations, which in turn may develop local stress fluctuations in the melt. The sufficient condition would be the local stress concentration in the regions with highly correlated chain orientation is large enough, so that the regions further develop the BOSCs where the nucleation of shishs is triggered and the nucleated shishs grow rapidly enough to promote the autocatalytic orientation of amorphous chains in front of the chip of the growing shishs. In other words, formation of the flow-induced BOSCs in the homogeneous melt may require sufficient enough the flow-enhanced local orientation fluctuations of the chains which in turn causes the large enough local density fluctuations, the density-rich regions of which are subjected to large enough stress concentration.

6.7 Self-Organization of Shish-Kebab Texture into Extended Chain Crystal Texture under Hot Drawing

6.7.1 *Introduction*

At equilibrium in the absence of the applied field such as flow, the crystallization of solutions of the random coils to the extended chain crystal (ECC) encounters much larger free energy barrier than the crystallization of solutions of random coil to the shish-kebab texture (SK), because the loss of the conformational entropy of polymer chains is much larger in the former than in the latter, as illustrated by the free energy barrier (A) and (B), respectively, in Fig. 6.34 [31]. The free energy barrier is also very large for the transformation from the SK to the ECC at equilibrium, as illustrated by the barrier (C) in Fig. 6.34. Thus, the crystallization from the random coil to the ECC and the transformation from the SK to the ECC are hardly achieved at equilibrium.

The flow is proposed to changes the free energy barrier at equilibrium. In the case of the crystallization of the SK from the random-coils solutions, the free energy barrier (B) at equilibrium is effectively reduced under the flow and replaced by the free energy land scape (D1) with a stepwise reduction of the barrier accompanied by the cascade evolution of the dissipative structures ①–④, as shown in the left half of Fig. 6.34, and as clarified in Sects. 6.4 and 6.5. The large free energy barrier (C) involved by the transformation from the SK to the ECC at equilibrium may be reduced in cascade as shown by a free energy landscape (D2) in the right half of Fig. 6.34 under hot drawings of the SK into the ECC texture through the cascade evolution of the dissipative structures ⑤ and ⑥. The dissipative structures are formed accompanied by the dissipation of the energy imposed to the systems by the applied hot drawing which involves deformation and flow of the SK.

The entire self-organization process of the ECC from the homogeneous polymer solutions under the applied field is summarized in Scheme 6.1 which is comprised of the flowing two stages.

The stage 1: The flow of the solutions in the fiber spinning line induces first the phase separation and subsequently the crystallization leading to the formation of the SK texture dispersed in the matrix of the solvent in the as-spun gel fiber. The as-spun gel fibers, which are subjected to the process in the stage 2, contain the volume filling SK texture swollen by 40% decalin as a solvent.

The stage 2: The hot drawing close to the melting temperature of the crystals developed in the fibers induces the phase separation of the SK textures against the solvent (syneresis of the solvent) and the deformation and the flow of the shish crystals and

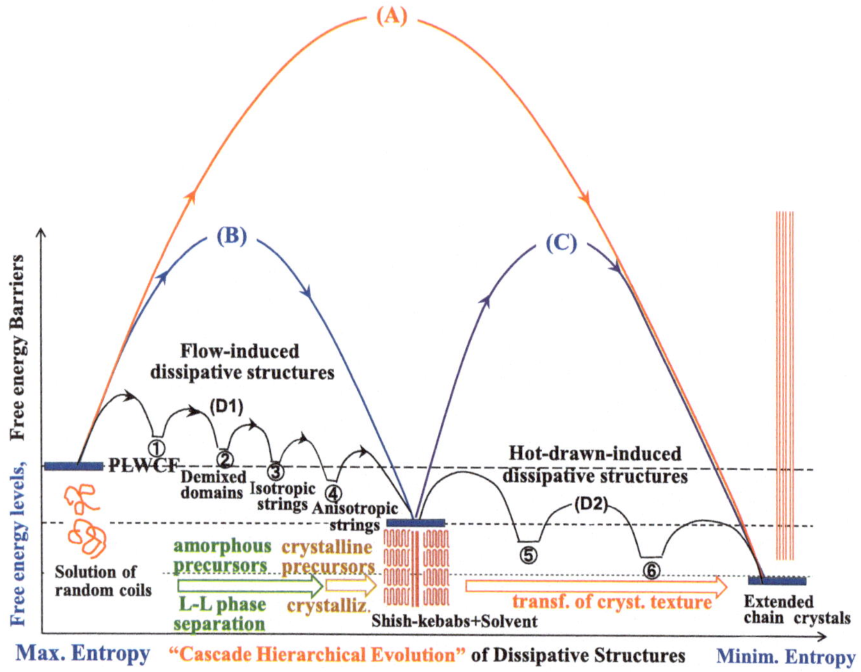

Fig. 6.34 Cascade pattern formation from homogeneous polymer solutions to the shish-kebab crystalline textures (SK) via the flow-induced formation of the dissipative structures ①–④ under the flow-induced the free energy landscape **(D1)**, and cascade pattern formation from the SK to the extended chain-crystal textures ECC) via the hot drawing-induced dissipative structures ⑤ and ⑥ under the hot drawing-induced free energy landscape (**D2**). The dissipative structures ①–⑥ represent as follows, respectively; ①: the PLWCF, ②: the randomly aligned demixed domains, ③: the optically isotropic strings, ④: the optically anisotropic strings, ⑤ and ⑥: the SK having the kebab fraction reduced stepwise and the increased shish length. The barriers (**A**) and (**B**) illustrate the free energy barrier in the absence of the external flow fields from the homogeneous polymer solutions to the ECC or from the homogeneous polymer solutions to the SK, respectively, while the barrier (**C**) illustrates the free energy barrier from the SK to the ECC in the absence of the external flow fields. The barriers (A), (B) and (C) are anticipated to be extremely large, so that the corresponding patterns were never observed in the absence of the external flow fields. From [31]

the lamellar crystals via the sliding diffusion of the chains in the crystals along the chain axis [44], leading to formation of the ECC. The cascade hierarchical evolution of the dissipative structures under the external fields in the stage 1 and the stage 2 leads to the self-organization of the solutions at equilibrium with the maximum entropy into the ECC with the minimum entropy. In this section, we hereafter discuss stage 2.

Stage 1: ***Flow induces***

Phase Separation ⟶ Crystallization ⟶ "Shish-kebab" Crystalline Texture

Stage 2: ***Hot drawing close to T_m induces***

Shish-Kebab ⟶ Phase Separation (Syneresis of Solvent)
& Deformation/Flow of Crystals ⟶ Extended Chain Crystal

Scheme 6.1 Self-organization of the homogeneous polymer solutions into the ECC under the applied flow and deformation

6.7.2 Hot Drawing of Spun Fiber

Figure 6.35 schematically presents the two-step hot drawing process [47]. The first-step drawing with various draw ratios, defined as DR1, involves the drawing of the as-spun fiber (ASF), which is gel-like, containing decalin by 0.3–0.5 in weight fraction, into the drawn fiber 1 (DF1) in the heated air-oven controlled at 120 °C, close to the melting temperature of the lamellar crystals in the ASF (part a). The first-step drawn fibers, designated as DF1, are obtained at the various draw ratios per minute defined as DR1: DR1 = 1.3, 2.0, 3.0 or 4.0. The second-step drawing involves the drawing of the DF1 in the heated air-oven controlled at 145 °C at the various draw ratios per minute, defined as DR2, to obtain the drawn fiber 2, designated as DF2: DR2 = 3.0, 4.0, 5.0 or 6.0 (part b). The hot drawing is conducted as follows. The wound ASF (part a) or DF1 (part b) on the bobbin A is rewound from the bobbin A with a roller S_1 and then fed into the heated air-oven, controlled at 120 °C for the ASF and 145 °C for the DF 1, respectively, and the heated ASF or DF1 is hot-drawn by another roller S_2 and wound on the bobbin B. The roller speeds of S_1 and S_2 controls the draw ratios DR1 and DR2. The total draw ratio (TDR) is defined by

$$\mathrm{TDR} = \mathrm{DR1} \times \mathrm{DR2} \equiv \varepsilon. \tag{6.7.1}$$

The melting temperature of the UHMPE crystals in the ASF and DF1 measured by DSC is 138 and 147.2 °C in the heating process at 10 °C/min, respectively. The drawing temperatures 120 °C for the DR1 and 145 °C for the DR2, close to the respective melting temperatures, enable to achieve the high drawability of the fibers, thereby permitting the long-range molecular rearrangement of the polymer chains within the shish and the lamellar crystals via the sliding diffusion of the chains along the chain axis [44] as will be described later in conjunction with Fig. 6.40. The drawing temperatures for the DR1 and the DR2 is lower than the melting temperature of the ASF and the DF1 to avoid the fiber to break during the drawing process,

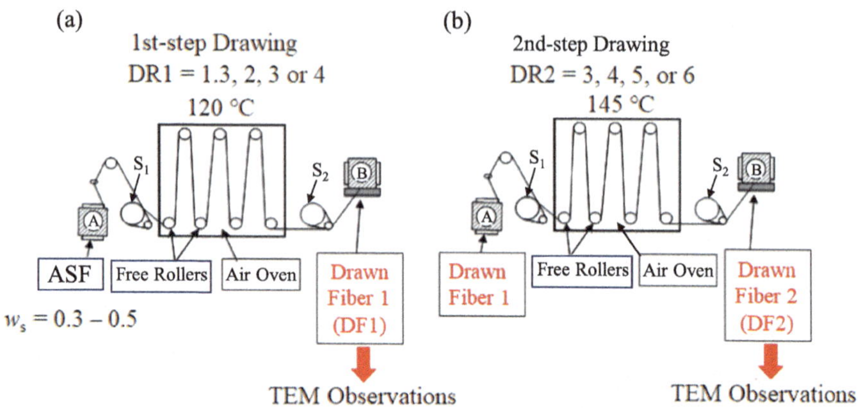

Fig. 6.35 Schematic presentation of the two-step hot-drawing process. The first-step drawing at the various draw ratios per minute, defined as DR1, involves the drawing of the as-spun fiber (ASF), which is gel-like, containing decalin by 0.3–0.5 in weight fraction, into the drawn fiber 1 (DF1) in the heated air-oven controlled at 120 °C. The second-step drawing involves the drawing of the DF1 in the heated air-oven controlled at 145 °C at the various draw ratios per minute, defined as DR2, to obtain the drawn fiber 2, designated as DF2. DR1 = 1.3, 2, 3, or 4 and DR2 = 3, 4, 5 or 6

respectively. The DF1 and DF2 are completely free from the solvent (decalin). These drawn single fibers were subjected to the tensile tests and investigations under TEM.

6.7.3 TEM Observations

The SK structures in the drawn fiber were investigated under the TEM. The ASF to be observed under the TEM contains solvent with its weight fraction $w_s = 0.3$ to 0.5. The solvent should be completely removed before the TEM observations. However, a simple drying of the fiber will cause serious structural changes such as additional crystallization, shrinkage, and formation of cracks. To avoid those structural changes as described above and to conserve the internal structures of the as-spun fibers in the process of the removal of the solvent, we employed the method using the stepwise solvent exchange and the fixation of the internal structures of the fibers, as described in Sect. 6.3.3 (C) and (D) in conjunction with Fig. 6.10, that is necessary to prepare the ultrathin sections stained with RuO_4 for the investigation under TEM. The hot-drawn fibers DF1 and DF2 do not contain the solvent, because the solvent is squeezed out from the fiber during the hot drawing process. The space left out in the single fiber after the complete solvent squeeze from the fiber via syneresis remains as voids.

Figure 6.36 presents the TEM image for the drawn ASF (DF1) with DR1 of 3.0 [47]. Each image in parts (a) and (b) focuses on the single SK texture surrounded by the kebabs of the neighboring two SK textures in the single fiber: parts (a) and (b) represent different areas of the DF1. The lamellae in the neighboring SKs appear to be interlocked. The short and long white bar represents the average diameters of the

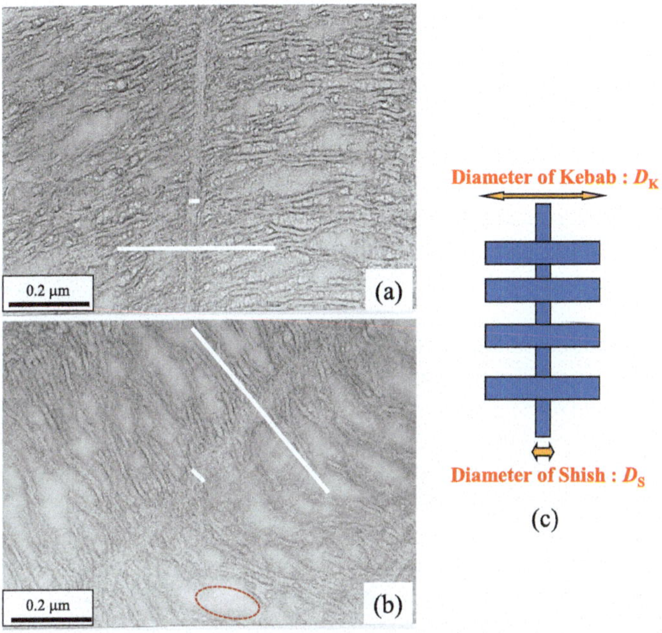

Fig. 6.36 The TEM image for the drawn ASF (DF1) with DR1 of 3.0. Each image in parts **a** and **b** focuses on a given single SK texture surrounded by the kebabs of the neighboring two SK textures: parts **a** and **b** represent different areas of the DF1. The kebab lamellae in the neighboring SKs appear to be interlocked. The short and long white bar represents the average diameters of the shishs and the kebabs in the single SK in each image, respectively, as defined in part **c**

shish and "kebab" in the one SK in each image, respectively, as defined in part (c). Note that the diameter of the "kebab", D_K, is the one which includes the shish.

To quantitatively explore the drawability of the SK texture, we measure the size of shishs and kebabs as a function of the draw ratio. To do this, the diameters of the shish D_S and the kebab D_K are measured based on a simplified model shown in Fig. 6.36c, where the SK is assumed to be composed of the rod (shish) and the disk (kebab), both of which are symmetric around the axis of the shish. By using this simplified model for the SK, we measured the average diameter of both shishs and kebabs (kebab + shish) directly from each TEM image as shown for example by the white bars in part (a) and (b) in Fig. 6.36.

In the TEM image, the crystal phase in both shish and kebabs is not stained by RuO_4. In the case of the kebabs, the amorphous phase located on the surface of the chain-folding loops is stained to appear dark. Therefore, the crystal phase in the kebabs is visualized as gray color against the dark amorphous phase. The shishs also appear to be grayish. The void phase replaced by the epoxy resin can be distinguished against the SK phase; the void phase gives the brighter contrast than the SK phase, because the epoxy resin is not stained by RuO_4. One of the void phases observed in the ultrathin section is illustrated by the red broken ellipsoid in Fig. 6.36b. The bright void phases are observed to exist in many places in Fig. 6.36a and b.

Figure 6.37 presents the TEM image for the DF2 having TDR = 6.0 [47]. The image shows the area is occupied by the volume-filled SK textures. One of the SK textures is illustrated by painting it in red: the parallel yellow lines and the white bar in the sketch painted by red indicate the edges of the shish and the average width of the kebab in the specified SK, respectively, from which D_S and D_K are evaluated, respectively. In other parts of the figure, the parallel red lines and the white bars indicate the edges of the shish and average width of the kabab in the specified SK, respectively.

In most of the areas of the TEM image of the drawn fibers with TDR = 6.0 except for the skin area, the SK texture is found to be the principal morphology even after the hot drawing. In the skin area of this fiber with TDR = 6.0 or the drawn fiber with TDR $\geq$ 9.0, all the parts of the fiber is filled with microfibrillar structures, i.e., the D_K degenerates closely to D_S. Moreover, any given SKs have the low number-density of the kebabs, as presented by the SK painted by the red color in Fig.6.37, i.e., the kebabs in any given SKs have the small volume fraction per SKs and the large average spacing between the kebabs as presented by the red painted SK in Fig 6.37.

The diameters of the SKs (D_K) and those of shishs (D_S) were tried being measured as a function of TDR at various sampling points as many as possible. However, only

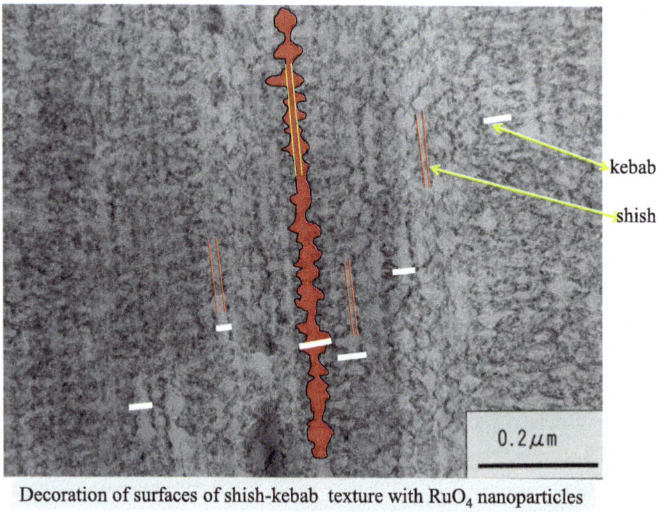

Fig. 6.37 The TEM image for the DF2 having TDR = 6.0. The image shows the area is occupied by the volume-filled SK textures. One of the SK textures is illustrated by painting it in red: the parallel yellow lines and the white bar in the sketch indicate the edges of the shish and the kebab, respectively, from which D_S and D_K are evaluated, respectively. The other SK's, other than the one painted red, they are not painted for the sake of brevity: they are simply indicated by the parallel red lines and the white bar, each of which specifies the edges of the shish and the average width of the kebab in the specified SK, respectively.

49 measurement points in total could be selected for the size evaluation, because there are fluctuations in the clearness of each TEM image which should be highly affected by the alignment and/or the orientation of each SK within the ultrathin section and by the preparation condition of each ultrathin section. The errors of the estimated values of D_S and D_K are roughly $\pm 20\%$.

Figure 6.38 presents evaluated diameters for both the shish and the SK, D_S and D_K, respectively, against TDR ε in the double logarithmic scale [47]. The results clearly reveal that D_K decreases drastically and asymptotically approaches D_S with the increase of TDR, whereas D_S remains almost constant, independent of TDR. The fact that D_K approaches D_S at TDR $\gtrsim 9$ reveals that the drawn fiber at TDR > 9 is essentially composed of the shish crystals only, and hence only the ECC texture, although the texture, i.e., shish crystals, contains some crystal defects.

The results in Fig. 6.38 reveal that D_K decreases with ε, approximately according to the following power law given by,

$$D_K \sim \varepsilon^{-\alpha}, \quad \alpha \simeq 2 \tag{6.7.2}$$

The exponent of the power law, α, is determined from the slope of the straight line in the plot with the double-logarithmic scale presented by Fig. 6.38. When the fiber is drawn according to the affine deformation under the constant volume, the diameter D_F of the single fiber changes with ε, according to the following power law,

$$D_F \sim \varepsilon^{-\beta}, \quad \beta = 0.5. \tag{6.7.3}$$

The power exponent α is much larger than β. Thus, the deformation of the SK in the fiber does not follow at all the simple affine deformation of the fiber, although the hot-drawn fiber as a whole may follow the affine deformation given by eq. 6.7.3 as implied from the mechanical properties of the drawn fiber to be shown in Fig. 6.41 below. The nonaffine deformation of the SK involves such deformation that the kebab crystals are transformed into the shish crystals through the long-range translational

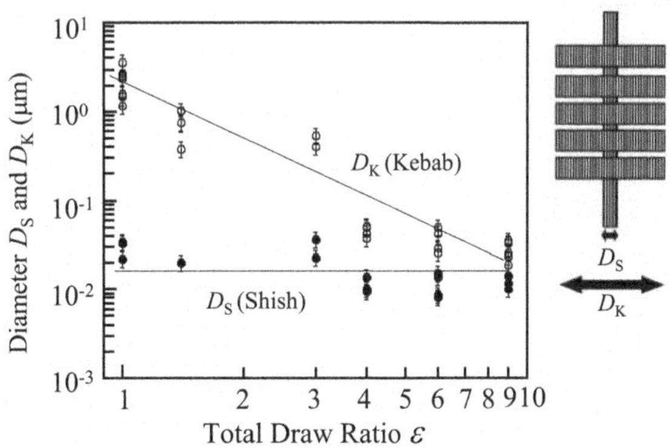

Fig. 6.38 Average diameters of shishs and kebabs measured with TEM as a function of the total draw ratio ε attained by the single-step and/or the two-step hot drawing of the as-spun fiber. From [31] and based on [47]

diffusion of the chain molecules along their chain axes both in the kebab crystals and in the shish crystals, as further detailed in the next section.

6.7.4 Cascade Structural Change from SK to ECC

Figure 6.39 illustrates a plausible mechanism on the cascade self-organization of the SK into the ECC with the increased TDR that can be deduced from the results shown in Fig. 6.38 [31, 47]. Upon the hot drawing of the fiber, each SK in the fiber is predicted to undergo the following changes as shown in Fig. 6.39 [31].

(1) the chain molecules in the kebabs may be incorporated into the shish, which decreases the diameters of the kebabs, D_K, from $D_{K,0}$ to $D_{K,E}$;

(2) the number of the kebab-chains incorporated into the shishs due to the above process (1) may extend the shishs in the drawing direction, keeping its diameter D_S constant as evidenced by the experimental results in Fig. 6.38, which increases the lengths of the shishs, L_S, from $L_{S,0}$ to $L_{S,E}$; the increased length of the shishs is intimately interrelated with the extension of the length of the fiber;

(3) the kebab-lamellae annealed at the given temperature during the hot drawing may cause the lamellar thickness L to increase from L_0 to L_E. The lamellar thickening is evidenced by close observation of TEM shown in Fig. 6.37 with a higher magnification. For example, upon the two-step drawing of the ASF by TDR = 6.0 at the given drawing temperatures, via the drawing of the ASF by TDR1 = 2.0 and TDR2 = 3.0, DR1 = 2.0 DR2 = 3.0 TDR = 6.0, the average lamellar thickness is increased from 9 nm for the ASF to 26 nm for the drawn fiber with TDR = 6.0.

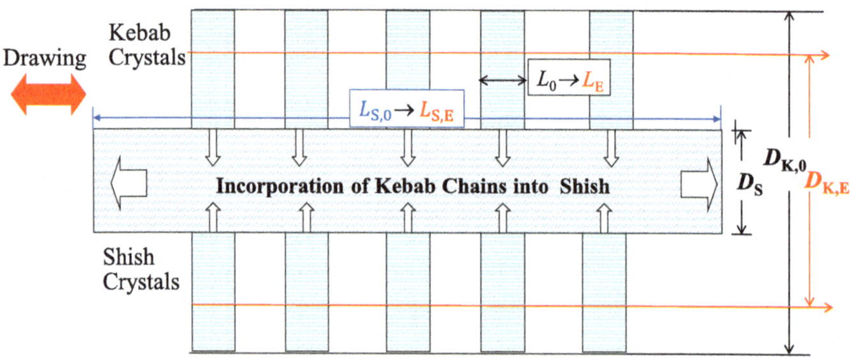

Fig. 6.39 Hot drawing-induced transformation of the kebabs into the shishs via the incorporation of chains in kebabs into shishs that induces the decreased D_K from $D_{K,0}$ to $D_{K,E}$ and the increased length of the shish L from $L_{S,0}$ to $L_{S,E}$ with increased ε under the almost constant D_S. Based on [31]

These changes in D_K and D_S as described above may suggest that the chains can slip one another along their chain axes both in the kebabs and in the shishs at the given drawing temperatures. We predict that the "bridging chains", which extend across the borders between the given shish and its associating kebabs in a given SK, may play a crucial role on the incorporation of chains in the kebabs into the ECC.

A single fiber is composed of an assembly of the SKs with their chain axes oriented along the fiber axis. The stress applied to the single fiber upon the hot drawing is imposed to the SKs. The stress imposed to the given SKs in turn would be transmitted to the kebabs associated with the given SK through the extended chains in the shishs and the bridging chains extended between the shish and kebab. Thus, the stress applied to the single fiber propagates to the given SK, from the given SK to its extended chains in the shish, and further from the extended chains to the folded chains in the kebabs via the bridging chains. The propagation of the stress may drive the sliding diffusion of chains along the chain axes from the kebabs to the shish to result in the extension of the length of the shishs from $L_{S,0}$ to $L_{S,E}$ and the contraction of the diameter of the kebabs from $D_{K,0}$ to $D_{K,E}$ under the constant shishs' diameter D_S.

Figure 6.40 presents the crystal lattice of polyethylene (PE). In the PE crystal at high temperatures higher than the characteristic temperature T_{α_2}, the rotational vibration of the chain around c-axis (the chain axis) (as schematically shown by

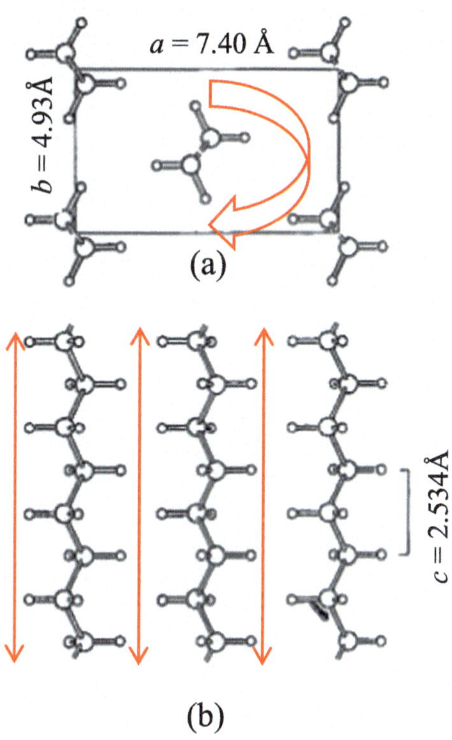

Fig. 6.40 Crystal lattice of polyethylene (PE). Part **a** illustrates the rotational vibration of the chain around the c-axis (the chain axis) (as schematically shown by the red twisted arrow), the anharmonicity of the rotational vibration causes the inter-chain slippage of the chains in the crystal along the c-axis under the mechanical energy imposed to the crystal as shown by the red arrows in part **b**

the red twisted- arrow in part a) becomes anharmonic, and the anharmonic rotational vibrations of the chains in the crystal lattice become incoherent with one another [48–50]. The anharmonicity and incoherence of the rotational lattice vibrations expand the crystal so that the orthorhombic crystal lattice becomes close to the hexagonal crystal lattice. T_{α_2} is the temperature above which the bulk crystalline PE exhibits the mechanical dispersion of crystals in origin, i.e., the so-called $\alpha_2$crystalline dispersion: the characteristic temperature T_{α_2} is around 50–100 °C, depending on the density of PE, the crystallization conditions and/or the heat treatments, and specifically the lamellar thickness. The dynamic mechanical energy imposed to the crystal is dissipated by the inter-chain slippage of the chains in the crystals along the *c*-axis as shown by the red arrows in part (b), mediated by the anharmonic and incoherent rotational vibrations of the chains in the crystal lattice as shown in part (a).

The temperatures at which the hot drawings are applied to the ASF and DF1 are higher than their T_{α_2}'s. Thus, the hot drawing is predicted to drive the sliding diffusion of the chains along the chain axes within both the kebab crystals with the chain-folding loops and the shish crystals. Namely, the folded chains in the kababs flow into the shish crystals, and they cooperatively join to the flow of the chains in the shishs driven by the hot drawing-induced sliding-chain diffusion of the chains. The mechanical energy imposed to the ASF and DF1 in the hot-drawing process is dissipated in cascade for the sliding diffusion of the chains in the SK, which drives the cascade extension of the shish length L_S and the cascade reduction of the corresponding amount of the kebabs and size of the diameter D_K.

The mechanical properties of *a single fiber* such as the tensile strength σ_b and the modulus Ewere σ measured at room temperature for the hot-drawn fibers DF1 and DF2 at different values of TDR ε. Figure 6.41a presents σ_b and E versus TDR ε in a double logarithmic scale [47]. Here, the tensile strength σ_b is the nominal strength

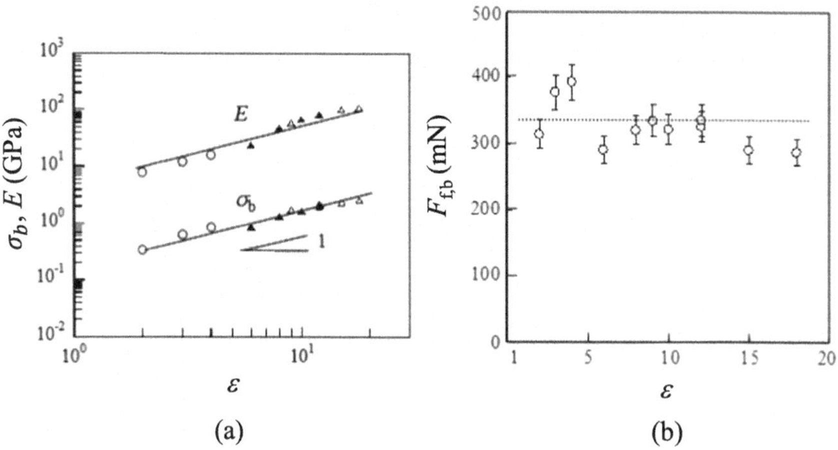

Fig. 6.41 σ_b and E versus ε (part **a**) and $F_{s,b}$ versus ε (part **b**), where σ_b, E, and $F_{s,b}$ are measured for the hot-drawn single fibers at the given drawing-temperatures

defined by

$$\sigma_b \equiv F_{b,f}/D_f^2, \tag{6.7.4}$$

where the force $F_{b,f}$ is the force at the breaking points for the single drawn fibers at the given drawing temperatures, and the diameter D_f is the initial diameter of the drawn fiber before the tensile deformation at room temperature [47]. If the hot drawing of the fibers (DF1 and DF2) occurs according to the affine deformation under the constant volume

$$D_f^2 \sim \varepsilon^{-1},$$

which predicts the tensile strength σ_b in Eq. 6.7.4 is given by

$$\sigma_b \sim F_{b,f}\,\varepsilon. \tag{6.7.5}$$

The experimental result given by

$$\sigma_b \sim \varepsilon, \tag{6.7.6}$$

as shown in Fig. 6.41a, and the prediction given by Eq. 6.7.5 reveal that the force $F_{b,f}$ is constant, independent of ε,

$$F_{b,f} \sim \varepsilon^0. \tag{6.7.7}$$

This prediction is validated by the result presented in Fig. 6.41b [47]. The result of Eq. 6.7.7 predicts that the linear increase of the tensile strength σ_b with the tensile strain ε during the hot-drawing process, as shown in Fig. 6.41a, is merely due to the decrease of the fiber diameter D_f according to the affine deformation under the constant volume, $D_f \sim \varepsilon^{-1/2}$.

If the force at break of the *single drawn fiber* $F_{b,f}$ is given by a product of the number of the shishs in the single fiber n_s and the force at break of the *single shish* $f_{b,s}$, $F_{b,f}$ is given by,

$$F_{b,f} \sim n_s\, f_{b,s}. \tag{6.7.8}$$

Furthermore, the proposed mechanism for the transformation from the SK to the ECC presented in Fig. 6.39 may imply that n_s is also independent of ε. If this is the case, $f_{b,s}$ is predicted to be given from Eqs. 6.7.7 and 6.7.8 by,

$$f_{b,s} \sim \varepsilon^0. \tag{6.7.9}$$

These pieces of evidence that D_s and $F_{b,f}$ are independent of ε and the predictions that n_s and $f_{b,s}$ are independent of ε elucidate that the structures, as revealed by a set of characteristic parameters $\left(D_s, f_{b,s}, n_s\right)$, and the mechanical properties, as revealed by a set of characteristic parameters $\left(F_{\mathrm{b,f}}, \sigma_{\mathrm{b}}\right)$, of the hot-drawn single fibers under the any given drawing conditions are predetermined by the structures and mechanical properties of the ASF.

Figure 6.42 presents a summary on the hot drawing-induced transformation from the SK to the ECC through the cascade evolution of the dissipative structures. As illustrated from part (a) to part (b), (i) the first-step hot drawing at 125 °C $\left(> T_{\alpha_2} \simeq 80\,^{\circ}\mathrm{C}\right)$ by TDR ε_1 reduces the diameter D_{K} of a given SK; (ii) the corresponding reduction of the volume of the kebabs is incorporated into the shishs to increase its length L_{s} keeping the diameter of the shish D_{s} and the number of shishs n_{s} in the given SK unchanged; (iii) the lamellae are thickened, and the inter-locking of the lamellae belonging to the nearest neighbor SKs may progress; (iv) the fiber as a whole, containing a number of the SKs equal to n_{s}, is deformed effectively according to the affine deformation, and the increase of the fiber length gives rise to the increase of the shish length L_{s}. These deformation processes (i) to (iv) are driven by the flow of chain molecules along their chain axes within the kebab crystals and shish crystal mediated by the sliding diffusion of chains within the crystals at $T > T_{\alpha_2}$.

Those deformation processes (i) to (iv) involved in the first-step hot drawing further continue to proceed in the second-step drawing of DF1 at $T = 145\,^{\circ}\mathrm{C} > T_{\alpha_2}$ with the draw ratio ε_2, as illustrated by the change of the fiber from part (b) to (c). When the total draw ratio $\varepsilon = \varepsilon_1\varepsilon_2$ reaches equal to and/or greater than 9, the SK

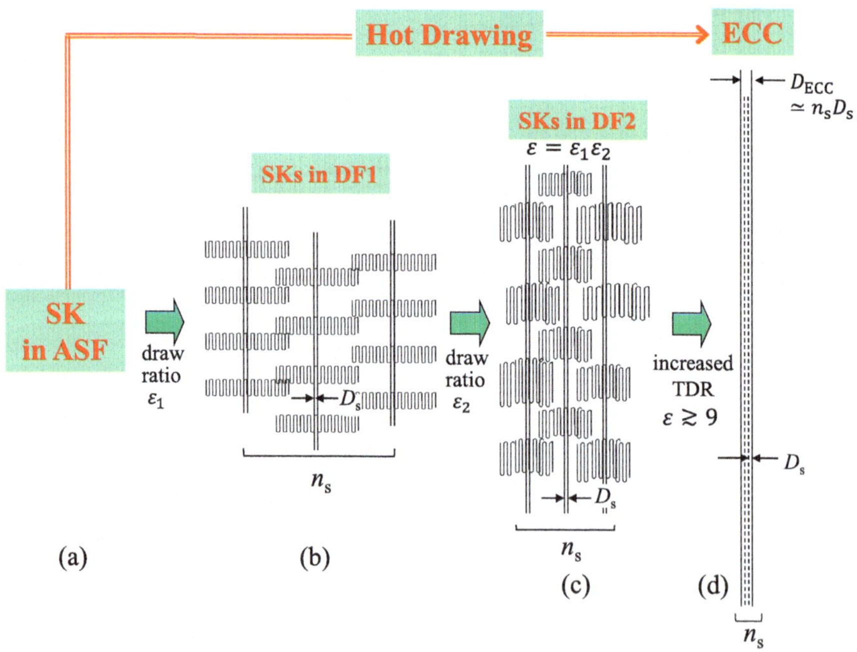

Fig. 6.42 Summary on the hot drawing-induced transformation from the SKs (part **a**) to the ECC (part **d**) through the cascade evolution of the dissipative structures (part **b** and **c**)

is transformed into the ECC "texture", as illustrated by the change from part (c) to part (d). We designate the ECC as ECC "texture" because the ECC after the hot drawing may contain some residual defects. The individual ECC texture has the diameter $D_{\mathrm{ECC}} \cong n_s D_s$ which is predetermined by the SKs developed in the ASF. Moreover, the number of shishs n_s is primarily controlled in principle by the number of strings developed in the spinning line of the spun fiber at P4 (Fig. 6.30), and thereby it depends on the spinning conditions of the solutions.

The extremely high free energy barrier (C), as shown in Fig. 6.34, encountered by the transformation from the SK to the ECC without the applied field is suppressed stepwise under the multistep hot drawing of th ASF via the cascade evolution of the dissipative structures ⑤ and ⑥, as shown by the hypothetical free energy landscape (D2). The free energy landscape in the figure is based on the three-step hot-drawing process to attain the ECC.

Each step of the hot drawing gives the cascade structural change in the SK according to the deformation and flow mechanism, as illustrated in Figs. 6.39 and 6.42, based on the sliding chain diffusion within the crystals as a primary mechanism, which may be responsible for the suppressed free energy barrier. Again, the hypothesis on the free energy landscape D2 is not theoretically verified yet, and the verification deserves future works. Since the melting temperature of the dissipative structure successively increases with the number of the hot-drawing step, the hot-drawing temperature should be raised accordingly to improve the drawability of the fiber to attain the ECC.

6.8 Concluding Remarks

This chapter is devoted to elucidate the self-organization of the SK texture and ECC texture in the nonequilibrium crystallizable polymer solutions subjected to the applied external flow fields. Especially the basic mechanism is aimed to clarify the evolution of the flow-induced ordered structures from the perspective of self-organization of nonequilibrium systems under applied flow field, i.e., the mechanism of ordering from dissipative structures.

It is clarified that the mechanism involves cascade reductions of the exceedingly large free energy barrier for crystallization of the solutions at equilibrium in the absence of the flow fields. The cascade reduction of the energy barrier occurs as a consequence of the cascade evolution of the dissipative structures which is accompanied by the cascade expense (dissipation) of the work done by the flow. The cascade evolution of the dissipative structures eventually result in the evolution of the ordered structures.

Figure 6.43 summarizes the self-organization of entangled, crystallizable polymer solutions (part a) into the SK texture (part b) and the ECC texture (part c) under the external flow fields. The self-organization of the nonequilibrium solutions into the ordered SK texture in pathway (1) is attained by the shear and extensional flow field along the fiber spinning line as the external field (1). The ordered SK texture (part

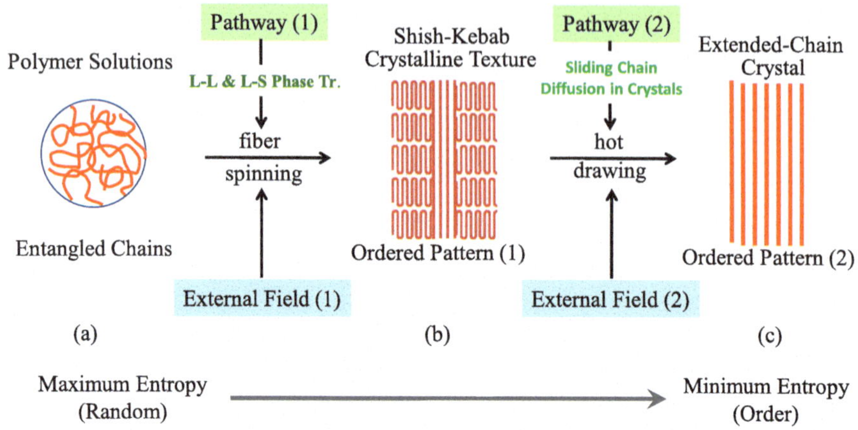

Fig. 6.43 Self-organization of the entangled, crystallizable polymer solutions (part **a**) into the SK texture (part **b**) and the ECC texture (part **c**) under the external flow field. The external flow field (1) first involves the L–L phase transition and subsequently the L–S phase transition in the process from part (a) to (b), and in the SKs already developed (part b), the external flow field (2) involves the the hot-drawing induced sliding-chain diffusion within the shish and kebab crystals along their chain axes, leading to the transformation from the SK texture (part b) to the ECC texture (part c).

b) is attained from the dissipative structures evolved through the flow-induced L–L phase transition (the phase separation and the SD) and subsequent flow-induced L–S phase transition (the crystallization). The crystallization occurs through locally triggered nucleation and growth of the shish crystals in the amorphous template (the strings) developed by the L–L phase transition, followed by the crystallization of the amorphous chains in the templates into the lamellar crystals which are epitaxially grown onto the shish crystals as already grown in the templates.

The self-organization of the nonequilibrium solution with maximum entropy into the ordered ECC texture with minimum entropy involves two-step orderings: the pathway (1) for the ordering into SK texture under the external field (1) applied to the polymer solutions and the pathway (2) for the transformation from the SK texture to the ECC texture under the external field (2) applied to the SK textures by the hot drawing (the extensional deformation and flow) at temperatures higher than the critical temperature T_{α_2}, the temperature above which the $\alpha_2$mechanical dispersion of crystals in origin starts to occur. The ordered ECC texture is attained from the dissipative structures evolved through the cascade reduction of the exceedingly large free energy barrier for the transformation in the absence of the external flow fields. The cascade reduction of the energy barrier is brought by the cascade expense of the work done to the hot-drawn fiber.

References and Note

1. Murase, H., Kume, T., Hashimoto, T., Ohta, Y., and Mizukami, T. (1995) *Macromolecules*, **28**, 7724–7729.
2. Murase, H., Kume, T., Hashimoto, T., and Ohta, Y. (2005) *Macromolecules*, **38**, 6656–6665.
3. Nicolis, G. and Prigogine I. (1997) *Self-Organization in Nonequilibrium Systems*, Wiley, New York.
4. Mitsuhashi, S. (1963) *Bull. Text. Res. Inst. (J).*, **1963**, *66*, 1–9.
5. Pennings, A. J.; Kiel, A. M. (1965) *Kolloid Z. Z. Polym.*, *205*, 160–162.
6. Keller, A., Kolnaar, H. W. H., in *Processing of Polymer*, Meier H.E.H. Eds., VHC: New York, **1997**, vol.18, Chap.4.
7. Seki, M., Thurman, D. W., Oberhauser, J. P., and Kornfield, J. A. (2002) *Macromolecules*, **35**, 2583–2594.
8. Meerveld, J. V., Peters, G. W. M., and Hutter, M., (2004) *Rheologica Acta*, **44**, 119–134.
9. Somani, R. H., Yang, L., Zhu, L., and Hsiao, B. S. (2005) *Polymer*, **46**, 8587–8623.
10. Baert, J., Puyvelde, P. V., and Langouche, F. (2006) *Macromolecules*, **39**, 9215–9222.
11. Ogino,Y., Fukushima, H., Matsuba, G.; Takahashi, N. Nishida, K., and Kanaya, T. (2006) *Polymer*, **47**, 5669–5677.
12. Fernandez-Ballester, F. (2007) *Ph.D. Thesis of California Institute of Technology.*
13. Kimata, S.; Sakurai, T., Nozue, Y., Kasahara, T., Yamaguchi, N., Karino, T., Shibayama, M., and Kornfield, J. A. (2007) *Science*, **316**, 1014–1017.
14. Kanaya, T., Matsuba, G., Ogino, Y., Nishida, K., Shimizu, H.M., Shinohara, T., Oku, T., Suzuki, J., and Otomo, T. (2007) *Macromolecules*, **40**, 3650–3654.
15. Mykhaylyk, O.O., Chambon, P., Graham, R. S., Fairclough, J. P. A.; Olmsted, P. D., and Ryan, A. J. (2008) *Macromolecules*, **41**, 1901–1904.
16. Balzano, L., Rastogi, S., and Peters, W. M. (2009) *Macromolecules*, **42**, 2088–2092.
17. Hayashi, Y., Matsuba, G., Zhao, Y., Nishida, K., and Kanaya, T. (2009) *Polymer*, **50**, 2095–2103.
18. Yan, T., Zhao, B., Cong, Y., Fang, Y., Cheng, S., Li, L., Pan, G., Wang, Z., Li, X., and Bian, F. (2010) *Macromolecules*, **43**, 602–605.
19. Mykhaylyk, O. O., Chambon, P., Impradice, C., Fairclough, J. P. A., Terrill, N. J., and Ryan, A. J. (2010) *Macromolecules*, **43**, 2389–2405.
20. Hashimoto, T. (2005) *Bull. Chem. Soc. Jpn.*, **78**, 1–39.
21. Hashimoto, T. (2008) in *Soft Matter Characterization.*; Borsali, R., Pecora, R. (eds), Springer, Germany, Chap. 8. pp. 378–462.
22. McHugh, A. J. (1995) in Rheo-Physics of Multiphase Polymeric Systems; Lyngaae-Jørgensen, J., Søndergaard, K., Ed., Technomic Publishing, Lancaster, PA, Chap. 6.
23. Hashimoto, T., Murase, H., and Ohta, Y. (2009) *Macromol. Symp.*, **279**, 88–95.
24. Hashimoto, T., Murase, H., and Ohta, Y. (2010) *Macromolecules*, **43**, 6542–6548.
25. Murase, H., Kume, T., Hashimoto, T., and Ohta, Y. (2005) *Macromolecules*, **38**, 8719–8728.
26. Murase, H., Ohta, Y., and Hashimoto, T. (2009) *Polymer*, **50**, 4727–4736.
27. Flory, P.J. (1967). *Principles of Polymer Chemistry*, Cornell University, Ithaca.
28. Keller, A. (1958). *Growth and Perfection of Crystals*, Turnbull ed., J. Willey and Sons, N.Y.
29. Geil, P.H. (1963). *Polymer Single Crystals*, Interscience, N.Y.
30. Pennings, A. J., van der Mark, M. A. A., and Booij, H. C. (1970) *Kolloid Z. Z. Polym.*, **236**, 99–111.
31. Hashimoto, T. and Murase, H. (2015) *IUCrJ*, **2**, 59–71.
32. Matsuzaka, K., and Hashimoto, T. (1999) *Rev. Sci. Instr.*, **70**, 2387–2397.
33. Hashimoto, T. & Noda, I. (2012). In *Polymer Science: A Comprehensive Reference*, Matyjaszewski, K. and Möller, M., eds., Vol. 2, *Chapter 30. Rheo-Optics* pp. 749–792.
34. de Gennes P.G. (1974) *J. Phys Chem.*, **60**, 5030–5042.
35. Dukovski, I., and Muthukumar, M. (2003) *J. Chem. Phys.*, **118,** 6648–6655.
36. Muthukumar M. (2005) *Adv. Polym. Sci.*, **191**, 241–274.
37. Wang, M. (2010) *J. Phys. Chem. B.*, **114**, 3488–3493.

38. Murase, H. Ohta, Y., and Hashimoto T. (2011) *Macromolecules*, **44**, 7335–7350.
39. Murase, H., Jinnai, H., Toriyama, T., and Hashimoto, T. (2019) *Macromolecules*, **52**, 575–591.
40. Lieberwirth, I., Loos, J., Petermann, J., and Keller, A. (2000) *J. Polym. Sci.: Part B: Polym. Phys.*, **38**, 1183–1187.
41. Kume, T., Hashimoto, T., Takahashi, T., and Fuller, G. G. (1997) *Macromolecules*, **30**, 7232–7236.
42. Katayama, K., Amano, T., Nakamura, K. (1968) *Kolloid-Z. Z. Polymere*, **226**, 125–134.
43. Kim, K. H., Cho, H. H., Ito, H., and Kikutani, T. (2008) *J. Polym. Sci.: Part B: Polym. Phys.*, **46**, 847–856.
44. Hikosaka, M. (1987) *Polymer,* **28**, 1257–1264.
45. The rapidly growing shish-crystal chips under the large supercooling (see Fig. 6.32d) brought about by the flow are expected to keep crating the bundle of stretched chains (BOSCs), having size l and w parallel and perpendicular to the chain axes, respectively, in the amorphous-chain phase exiting in front of the growing chips of BOSCs, according to the mechanism proposed by Lieberwirth et al. [40]. In order to keep the longitudinal growth (the growth parallel to the chain axis) of the crystals into the shish inside the BOSCs with a mesomorphic order, the BOSCs may have to satisfy a specific condition of $(w/l) > (w/l)_c$ where $(w/l)_c$ is the critical value of w/l. Otherwise, the lateral growth of the crystals into lamellar crystals may happen to be developed within the mesophase front of the BOSCs [46] to result in the formation of row-nucleated lamellae.
46. Strobl, G. (2000) *Eur. Phys. J. E.*, **3**, 165–180.
47. Ohta, Y., Murase, H., and Hashimoto, T. (2010) *J. Polym. Sci.: Part B: Polym. Phys.*, **48**,1861–1872.
48. Peterlin, A. (1960) *J. Appl. Phys.* **31**, 1934.
49. Tsuge, K., Enjoji, H., Terada, H., Ohzawa, Y., and Wada, Y. (1962) *Japan J. Appl. Phys.* **1**, 270.
50. Okano, K. (1960) *Rep. Progr. Polym. Phys. Jpn,* **3**, 71.

Index

T. Hashimoto, *Polymer Phasing*, https://doi.org/10.1007/978-981-97-9648-9

The manufacturer's authorised representative in the EU is Springer
Nature Customer Service Centre GmbH, Europaplatz 3, 69115 Heidelberg,
Germany. If you have any concerns regarding our products, please
contact ProductSafety@springernature.com

Printed and bound by CPI Group (UK) Ltd, Croydon, CR0 4YY
21/09/2026
02220013-0004